FOOD, NATURE AND SOCIETY

Food, Nature and Society

Rural life in late modernity

Edited by

HILARY TOVEY
Trinity College Dublin, Ireland

MICHEL BLANC
Institut National de la Recherche Agronomique, France

Routledge
Taylor & Francis Group
LONDON AND NEW YORK

First published 2001 by Ashgate Publishing

Reissued 2018 by Routledge
2 Park Square, Milton Park, Abingdon, Oxon OX14 4RN
711 Third Avenue, New York, NY 10017, USA

Routledge is an imprint of the Taylor & Francis Group, an informa business

A Library of Congress record exists under LC control number: 2001091557

ISBN 13: 978-1-138-72913-1 (hbk)
ISBN 13: 978-1-138-72904-9 (pbk)
ISBN 13: 978-1-315-19010-5 (ebk)

Contents

List of Figures

List of Tables

Contributors

Michel Blanc
is Research Director in the Rural Economics and Sociology department of INRA (Institut National de la Recherche Agronomique), France, and editor of *Économie Rurale*. His main research interests are in family farming, rural labour markets, and rural/urban migrations in developed countries.

Yuna Chiffoleau
is based in the National Institute of Agronomic Research in Montpellier, France, in a multidisciplinary action-research team working on innovation processes in co-operative settings.

András Csite (1969)
is a 'János Bolyai' research fellow at the Institute for Political Sciences of the Hungarian Academy of Sciences. His research interests include rural restructuring of post-communist countries and 'Europeanisation' of the Hungarian policy-making processes.

J.C. (Hans) Dagevos
is Senior Researcher in Socio-cultural Dynamics and Consumer Behaviour at the Agricultural Economics Research Institute (LEI) in The Hague, the Netherlands. His research interests include consumer behaviour, consumer culture, consumer concerns, and demand-led food production.

Urs Geiser
is Senior Researcher with the Development Study Group of the Department of Geography, Zurich University, Switzerland. His research focuses on natural resource management and rural development, addressing questions of institutions, organising practices, and actors' 'room for manoeuvre'.

H.J.M. (Harrie) Hansman
works as a Researcher at the Social Research Institute of West-Friesland in the Netherlands, concentrating on consumer research and methodology. He has written several articles about lifestyle, consumer behaviour and food consumption.

Veska Kozhucharova
is Professor in the Institute of Sociology, Bulgarian Academy of Sciences, and President of the Association of the Rural Sociologists in Bulgaria. She has published widely on sociological problems of Bulgarian rural areas, including gender issues.

Andrienetta Kritzinger
is a Senior Lecturer in the Department of Sociology at the University of Stellenbosch, South Africa, working in the areas of the sociology of work, gender studies, feminist theory and youth studies. Her most recent research and publications are mainly in gender studies and rural sociology.

Gwenaël Larmet
obtained his Ph.D. thesis in 1998 on 'food, exchange, and social ties', analysing reciprocity mechanisms in the exchange of food invitations at home. He subsequently worked on food provisioning and is now studying the social context of media consumption.

Lutz Laschewski
is a researcher at the Department of Resource Economics at the Institute for Agricultural Economics and Social Science, Humboldt University, Berlin. His research interests are in agricultural and rural restructuring in East Germany, with a focus on links between sustainability and civil society.

Geoffrey Lawrence
works at the Institute for Sustainable Regional Development in Central Queensland University, Australia, and is co-editor of the *International Journal of the Sociology of Agriculture and Food*. He has published widely on the political economy of agriculture and the countryside.

Kristen Lyons
worked until recently at the Rural Social and Economic Research Centre, Central Queensland, Australia. She has conducted extensive research into organic farming in Australia and New Zealand, with a particular interest in the relations between science, gender and situated knowledges.

Eileen O'Rourke
is a Marie Curie Fellow at the Institut National de la Recherche Agronomique (INRA), Toulouse, France. She has published several articles on nature-culture issues, landscape interpretation and management, and the changing European relations between agriculture, environment, and society.

Giorgio Osti
is Associate Professor of Sociology at the University of Trieste (Italy), and a member of the *Sociologia Ruralis* Editorial Board. Recent publications include studies of the Po Delta Park (One River, Two Policies), and of LEADER partnerships in Italy (*Sociologia Ruralis* 2000).

Bareld van der Ploeg
works on regional studies at the Agricultural Economics Research Institute (LEI) in The Hague, the Netherlands. He is currently preparing a thesis for Wageningen University on future prospects for farming in the Green Heart between Amsterdam and Rotterdam.

Rossitsa Rangelova
is Senior Research Associate at the Institute of Economics, Bulgarian Academy of Sciences, working on macroeconomics and international economic comparisons. Her most recent publication is 'Bulgaria's National Income and Economic Growth, 1913-1945' (*The Review of Income and Wealth*, June 2000).

Rosemarie Siebert
works at ZALF Institute of Socioeconomics in Müncheberg, Germany. She has researched and written on the economic, social and institutional effects of the integration of East Germany into the European Union.

Oddveig Storstad
is a researcher in the Centre for Rural Research in the University of Trondheim, Norway. She has published a number of papers on Norwegian food systems and food cultures.

Hilary Tovey
is a Senior Lecturer in sociology at Trinity College Dublin, Ireland, and currently President of the European Society for Rural Sociology. Her interests include environmental issues, social movements in rural development, alternative forms of food production, and the sociology of animals.

Jan Vorster
is a lecturer in Sociology at the University of Stellenbosch, South Africa. His main research interests include social exclusion and social security.

Introduction

HILARY TOVEY

In the heyday of modernity, it was often taken for granted that the rural symbolised the continuing presence among us of the unmodern. Rurality was treated either as an object for modernisation (industrialisation, penetration by capitalism both as capital and as economic practices, infrastructural development and spatial rationalisation) in which external agents possessed of the correct modern skills and knowledges sought to collaborate with rural people in a process of organised change; or as an object of romantic conservationism, to be kept as far as possible in stasis as a reminder of the values and folkways of a more 'traditional' time. In late modernity, such assumptions start to be opened up for reflection and revision. In August 1999, the European Society for Rural Sociology held its 18th Congress in Lund, Sweden, on the theme of 'How to be rural in late modernity? Process, project and discourse'. The intention was to explore some recent social theorising, which seemed to contain fundamental implications for the rural but which left these largely unelaborated - claims such as, for example, that late modern societies are structured around consumption rather than production interests; that relations to knowledge, rather than to material resources, are what increasingly underpin both social organisation and conflict; or that collective action in the 'new' social movements is oriented to a search for new identities and discourses far transcending the old divisions of 'class' or 'rural-urban'. What do such analytical tales suggest for the meaning and role of rurality in late modernity? To what extent do trends within the rural support or challenge them? In debating such questions it was hoped that the congress would, simultaneously, continue and expand the 'mainstreaming' of rural sociology which has also been a focus of earlier congresses in the 1990s, and also encourage more 'ruralisation' of mainstream sociological theorising.

In a lively and well-attended Congress there were many excellent papers, addressing a range of topics from rural social movements to contested themes in forest management, and from ecologically innovative agriculture to rural household survival practices. Only a small selection can be included in this book, but we hope that nevertheless they give readers a sense of the diversity and range of current research interests among European rural sociologists. In particular, they demonstrate the striking re-orientation of the field of rural sociology in recent years, to incorporate new concerns around consumption and society-nature relations alongside its longstanding concerns with the social and technical organisation of food production, and with social relations and collective action in the countryside. If late modern European rural policy addresses not just agriculture but the shifting relations between agriculture, environment and food policies (Buller 2001), European rural sociologists demonstrate increasing concern with the lifeworld experiences engendered out of such shifts. The book chapters cover a wide geographical spread, from South Africa to Australia and from EU member states to the countries 'in transition' of Central and Eastern Europe. Their focus is on the distinctive ways in which rural social, economic and cultural life is experienced, in relatively developed societies, in late modernity.

'Late modernity' is perhaps an idea which has been more widely established in English-speaking countries (although even here it is still very contested) than elsewhere. In non-English-speaking countries terms such as post-industrialist, post-Fordist or even post-modern appear to have a wider currency. One of the pleasures of cross-cultural and cross-linguistic preparation of the 1999 Congress theme was the experience of trying to develop some shared sense of agreement on what we were talking about. The concept of late modernity as used here includes some elements of all the other concepts mentioned above. 'Late' is used in preference to 'post' largely to indicate that whatever terminology we use to suggest the problematic aspects of the concept of modernity, as a framework for interpreting our contemporary situation it cannot easily be abandoned. Yet there has been an undoubted increase in our sense of detachment from many of its key values and beliefs.

Following Bauman (1991), we might describe late modernity as that era in which modernity has become aware of itself. Within social theory, the idea of modernity has moved from being a background notion (Delanty 1999) - an unexamined frame of reference which helped us to understand, primarily by contrasting them with those of 'traditional society', the specific features of our age and the direction of its change - into an explicit thematic concern. Hitherto dominant conceptions of contemporary society as industrial, capitalist, and/or globalising, are challenged by the claim that behind any of these there lies a larger and more important self-description: modernity as a process of instrumental rationalisation, permeating both its institutional and its cultural dimensions. Theorists of modernity such as Habermas or Castoriadis emphasise the contested and dualistic nature of the modernisation/rationalisation process - captured first by Weber, perhaps, in terms of a contest between formal/instrumental and substantive/value rationalities. Over time, the struggle between rationalities gives rise to dialogue, self-critique, perhaps even moral learning or at least fundamental questioning of the previously taken-for-granted. Central to this is the questioning of what Delanty calls 'the master-theme of cultural modernity', the assumption that change is progress. As the framework of a transition from traditional to modern becomes less and less adequate for understanding our situation, Delanty notes, we also find increasingly problematical the time-consciousness of modernity, as an evolutionary myth of progress understood as the unfolding of a grand narrative: the manifestation of reason, the realisation of subjectivity, the building of institutions to guarantee order and security, the mastery of nature, and the progress of civilisation.

Released from the identification of change with progress, rural sociology is particularly well placed to revisit and re-interpret both society-nature relations, and assumptions about the quintessentially urban location of the civilising process. The rationalising drive of modernity encouraged an assimilation of 'the natural' and 'the traditional', and an assumption that progress resided in the rational transcendence of both. Late modernity however has experienced an undermining of the legitimate authority of science, the dominant form of knowledge through which modern societies relate to nature, as our awareness of its non-neutral status and its contribution to environmental and other forms of risk has grown. 'Traditional' or local forms of knowledge excite interest once again. Nature in late modernity is becoming more and more cultural (Eder 1996), and as such, open to the dualities of cultural meaning. We come to see that nature is increasingly socially created, refashioned by science and technology in their own image, even while in debates around risk we often construct nature as an equal participant, rejecting an ethic of mastery in favour of ethics of communication or of ecological co-survival. But still our management of food remains brutally modern, invoking regulatory sciences for the mundane assurance of food safety/quality, and a response to crisis (mad cow disease, foot and mouth disease) which demands without question the mass slaughter of farm animals.

The project of modernity tended to normalise a particular and limited set of life experiences - industrialised work, transformation of workers into waged or salaried employees, urbanisation of lifestyles, the adoption of time-saving technologies, declining self-provisioning in relation to consumption of food or leisure - which helped to represent the lifestyle of most of those studied by European rural sociology as backward or bizarre, if not substantially disappearing. Yet in late or post-Fordist modernity, the homogenisation and organisation of social life into a narrow set of experiences is tempered by a quest for individualisation and autonomy. Projects for self-realisation, or life-politics (Giddens 1994), may revalue what appear to be communally-based relationships and objects over commoditised ones, the artisanal over the technologically sophisticated, self-sufficiency over a rising pay packet; and such individualised projects link into the public domain through relatively informal environmental and other social movements which provide the space and support for experimentation with new lifestyles. A surprising amount of this experimentation appears to either happen within rural spaces or rest on new interpretations of the meaning and possibilities of rurality. While this reinvention of the rural is largely urban-led, the outcome flows from an interaction between both. If the intermingling between the global and the local which globalisation theory depicts can only be dealt with through the invention of a new term, 'glocalisation', perhaps rural sociology needs a new term - 'rurbanisation'? - to capture the ways in which the rural, as a sign of the non-standard or non-normalised, is being incorporated into lifestyle practices and visions of civil society in late modernity.

These are nonetheless clearly not the only or even the dominant trends within late modernity. There appears to be wide agreement that the key change in the move from middle to late modernity is a shift in societal emphasis from work to play - from productive work to consumptive pleasure as the site in which creativity is expressed and social identity organised. Yet this does not necessarily indicate a change in the dynamic forces which most powerfully shape our life experiences. While the increasing power of food consumers, or at least of consumer food preferences, to constrain the life chances and production possibilities of agricultural producers and rural populations is a topic for a number of authors in this book, there is evidence to suggest that food preferences in their turn are increasingly a construct of food retailers and their growing control over the wider food industry. And while research in the area is less available, there is some evidence to suggest that consumer demands for particular sorts of environment or 'nature' when they visit the countryside are similarly a construction of the material and symbolic practices of the leisure and tourism industries. In both cases the challenge for capitalism is to balance the pressures towards standardisation which result from attention to efficiency, profit and predictability (Ritzer 1996) with the pressures for variety and choice which are central to the discourse and self-understanding of modern consumerism.

Late modernity may be characterised by the rise of 'life politics' but this does not mean the end of emancipatory politics (Giddens 1991) or of the need for them. Even if it is to some extent undermined by globalisation, the modern project to create within a defined territory a legitimate social and political order, and to build the institutions which will sustain and administer this, persists. From its development within early modernity, sociology has tended to operate with a foot in both camps, following both a social engineering role as advisor to the state in controlling and organising civil society, and an emancipatory role as critic of the state on behalf of civil society. Rural sociology, particularly since the development of the modern state project to rationalise and industrialise agriculture, and to construct the appropriate specialised research and advisory institutions, has tended perhaps more than most other subfields within sociology to adopt or to concentrate on the state project. The advent of a late or post-modern scepticism may do little to move its focus away from institutions and policies, but much to encourage emancipatory critique of these. Concerns about the fate of the rural populations in aspiring EU member states in post-socialist Europe help to illuminate both the positive and the problematic aspects of EU rural policy and to make the importance of such critique increasingly clear. To what extent are the extraordinarily complex sets of institutions for rural governance which have been developed within the EU suitable for or relevant to the experiences of the rural populations in these societies? To what extent, indeed, have they provided the sort of legitimate order and basis for social justice within the rural areas of EU states themselves, which would justify their extension over new territories?

The influence of postmodern thinking on rural sociology has been evident in a number of ways over the past decade. Amongst other effects, it has helped to consolidate wide agreement that the future of European rurality will be 'post-productionist'. Farmers have been redescribed as nature-managers for urban populations. Rural sociologists have become familiar with arguments to the effect that, for instance, the countryside is increasingly a location for consumption and leisure rather than for extractive use of natural resources; that the contemporary 'farm family' typically recreates itself as the 'multifunctional rural household', in which incomes from farming occupy a less and less significant position in overall household welfare; that agricultural interests are evidently losing their power to determine how the countryside is used to other interest associations in environmentalism, heritage conservationism, sport and tourism. From a late modern rather than a postmodern perspective, continuity in societal trends persists alongside transformation, and such assumptions can be opened up to further consideration. Two cases can be selected for brief mention which help to illustrate this. A recent issue of *Sociologia Ruralis*, edited by van der Ploeg, Renting and Minderhoud-Jones (October 2000) re-positions smallscale ('peasant') agricultural *production* at the heart of rural development. It calls for a new model for European *agricultural* development which is based in an explicit recognition of the need to transcend the 'modern' assumptions built into current ones. Earlier issues of *Sociologia Ruralis* carried a debate between Goodman (January 1999) and Marsden (January 2000) which revisited modernist accounts of the character of the contemporary food system as one of 'incomplete industrialisation'. For Goodman, in an 'age of ecology' incomplete industrialisation can be recognised *positively*, as a source for the persistence of heterogeneous and diverse practices in farming and food production which in turn provide us with a material base from which to question and critique the arrangements underpinning hegemonic industrial society, rather than *negatively*, as a condition of backwardness. For Marsden the point is to be able to understand how, given this persistence of possibilities for alternatives, the alternatives are themselves constantly marginalised by the continuing dominance of the industrial food model. Food, nature, and food-as-nature, emerging as key issues for a revitalised contemporary rural sociology, provide continuities with earlier concerns of the discipline with farming as social and cultural practice, while opening up new ways of connecting the rural to processes of transformation on a societal and global level.

The book is organised around four themes. However, all of the chapters touch to a greater or lesser extent on all of the themes, so it may not be useful to draw very strict boundaries around each section. The first is that of *agriculture and rural life in late modernity*, and the first chapter a discussion of Bulgarian agriculture in the socialist and post-socialist periods. Such discussions add essential new dimensions to our understanding of the modern European rural experience. They remind us that 'late modernity' may have a very different meaning in Eastern and Central European countries than in the member states of the EU. Kozhucharova and Rangelova emphasise the 'otherness' of the Central and East European experience. Over 70 years, Bulgarian society has passed from agricultural to industrial and back to agricultural, involving extraordinary experimentation with forms of property in land and forms of labour organisation in farming which seem first to be an extreme expression of an attempt to realise the project of modernity, later abandonment of that project as far as rural society is concerned. In the late 1990s, although the market economy has been adopted as the dominant model of development for Bulgarian society, the farm economy has become much more subsistence-oriented and self-sustaining. This might be described as a type of de-modernisation, linked to a programme for land restitution based on a vision of the rurality of the pre-socialist 1930s. Yet the authors suggest that the renewed interest in the rural which appears to characterise Bulgarian society might also be seen as a mark of late modernity, produced from a growing reflexive awareness of the limitations of an urban industrialisation-obsessed modernity, and of the damage which modernity's technology has done to nature and modernity's culture has done to a sense of collective identity.

If agricultural workers in Eastern Europe constitute one sort of 'other' for core European rurality, a second might be constituted by women workers in the agricultural export sectors of developing economies around the world, whose products are an increasingly important element of affluent European diets. Kritzinger and Vorster examine the changing situation of South African women of colour working on export fruit farms, in a highly labour-intensive industry stratified by race and gender. Theses of the 'casualisation' and 'feminisation' of labour in this type of agriculture turn out to be only partially true: globalisation both homogenises and differentiates production localities, depending on how the local state responds to it in the context of its own local history and culture. For the women workers in this case study, current outcomes seem highly contradictory. Political democratisation in South Africa has increased their access to permanent employment status, but on the other hand, 'modernisation' of agricultural labour relations has deprived them of employment benefits and facilities that were available to permanent workers in the past under more paternalistic labour arrangements.

A third 'other' might be constituted by alternative agricultures, such as organic farming. Does organic production offer a way for small farmers to survive in late modernity? Or should we rather ask (see Marsden, above) how alternative forms of food production are kept marginal - how the unsustainable is sustained? One way to marginalise alternative food production may be to institutionalise it (Tovey 1999). The chapter by Lyons and Lawrence examines how organic production is becoming integrated into the conventional food system in Australia and New Zealand, and how this is creating a new rift among organic growers themselves. There are several ironic twists in their story. Until recently organics was given little serious state or corporate consideration as it was seen as a primitive, underproductive, unscientific technology with no hope of ever meeting world food needs; but the 'greening' of Western society (also sometimes called 'environment-proofing', or the routine incorporation into decision-making by state agencies and economic institutions of an assessment of the environmental costs or benefits of any proposed course of action) has now made organic production thinkable as mainstream. This attracts the attention of large commercial interests, so also does the routinisation of codification and certification of organic production which opens up opportunities to penetrate new export markets. Thus state and corporations re-position organics as a modern, formalised, commercial industry in contradistinction to the older, more 'orthodox' informal and localised practices. This draws in new growers who see organics as a way of expanding market access and returns from farming, often through arrangements with food-processing companies who offer premium prices to organic growers, while the more 'orthodox' growers start to resist what they see as a move to reduce organic production standards to the lowest acceptable levels.

How does the concept of late modernity help to make sense of the situation of these 'others' in agriculture across the world? Should we abandon the modernist assumption of a global convergence towards the same forms of organisation of agricultural work, the same work goals, experiences and conditions? Is diversity the key to understanding rural life in late modernity? These questions also permeate Part II, in which chapters try to bring together *food producers and food consumers* as interwoven concerns of rural sociology. First here is Larmet's discussion of how, even in an increasingly consumer-driven society, diversity persists in French food consumption practices, much of it due to the persistence of small farmers. Larmet straddles the boundary between Parts I and II, and reminds us that the consumer-producer relationship is not identical with an urban-rural relationship since rural people are also food consumers. He looks at rural food provisioning practices as a way of studying the impact of modernisation (rationalisation, impersonalisation) on the rural world. Over the past half century, food provision practices in rural areas have shifted from auto-provisioning, to buying in small local shops, to using supermarkets and even hypermarkets, so rural practices have converged with those of urban food consumers. At the same time, some distinctively rural practices remain. His study reveals an increasing heterogeneity within the French rural population itself, as evidenced by their consumption practices. Farmers remain particularly distinctive, even if less so than before: they still provide food for themselves outside the market nexus far more than others do, although they are abandoning methods of provision which use more time (e.g. animal breeding), and much depends on whether or not their spouse is also a farmer. Other rural groups, especially those who have moved to rural from urban settings or who commute to work in urban professions, appear to be becoming largely indistinguishable in their food buying habits from urban patterns. The survival of a distinctively rural consumption culture seems to depend, paradoxically, on the survival within the rural locality of agricultural producers.

Whether or not small French farmers trust food which they have not provided from their own resources or those of their neighbours is not directly discussed by Larmet. Storstad's chapter looks at the nature and effect of consumer trust in food, or more specifically at how high levels of Norwegian trust in Norwegian food creates something of a cultural barrier to trade in imported food. 'Cultural' and 'mental' barriers to trade in food increase in significance, she notes, as technical and political barriers decline. 'Trust' can be understood either as what makes it possible to live with a knowledge of risk, or as what makes it possible to live with insecurity or lack of knowledge. In both cases we are talking about trust in abstracted or impersonal expert systems, which become more routine elements in our daily lives as the world gets more 'stretched in space and time' (Giddens). Trust in a food product means trusting in a range of unknown or invisible actors. Norwegian consumers appear to have remarkably high levels of trust in Norwegian food producers, processors, inspectors and food safety controllers. But this may be related more to their own national culture and history, and the imagined community of fellow Norwegians which these have created for them to inhabit, than to objective evidence about high standards maintained by these actors.

The issue of global convergence versus diversity in food consumption patterns emerges again in Dagevos and Hansman's discussion of consumer behaviour in the Netherlands. They argue strongly for diversity. In a post-industrial era, when demand and consumer desires substitute for supply and producer constraints as the main driving forces of economies, variety in individual preferences, tastes and inclinations would be expected to translate into a diversification of consumption patterns. But beyond the development of the consumer society, we are witnessing the advent of the 'emotion economy' based on the satisfaction of non-material needs: consumer choices are dictated not solely by functional reasons, but also, and increasingly, by the emotions provided by purchase and consumption themselves (pleasure, identity affirmation, distinction, animal welfare concerns, etc.). Dagevos and Hansman construct four distinct ideal-types of consumer behaviour on this basis: the 'calculating consumers' who have a 'fuel' approach to food; 'traditional consumers' also mainly preoccupied by the material qualities of food, but influenced by the habits and conventions prevailing in their close social environment; 'unique consumers' whose consumption pattern is directed at novelty, variety or exclusivity, the feelings of fun, self-esteem and distinction they derive from consuming foods with certain characteristics; and 'responsible consumers' who are concerned about the impact their consumption choices have on the society at large, future generations, or the physical environment. Unique and responsible consumers are presented as epitomising post-modernity (or the post-materialistic era), while calculating and traditional consumers are figures of modernity or pre-modernity, although the chapter does not provide evidence of change over time in consumer attitudes. Whether, except in very specific and rare cases (famine, cosmonauts' diet), foods have ever been regarded as mere 'fuel' by consumers might be questioned; what is changing over time may not be so much the relative significance of consumers' non-material needs as the kind of non-materialistic attributes consumers are concerned about. Among responsible consumers we might wish to distinguish between those whose behaviour is shaped by social norms and those who explicitly take into account the impact of their own decisions on the well-being of others, whether humans, animals, or natural objects - that is, to distinguish among them a subset of 'reflexive consumers' characteristic of late modernity.

The third theme of the book is called *living with nature* but in some respects it is more about how farmers find themselves having to live with the expectations of other social actors, such as the 'reflexive consumers' above, about how one should live with nature. O'Rourke studies relations between farmers, conservationists and tourists in the Cevennes National Park in France, a park with the distinctive (for those used to more Anglo-American understandings of what national parks are about) founding mandate of combining protection of nature with sustainable development. Farmers in the park area continue to produce food, and even to integrate it into a 'culture economy' (Ray 1998) by using meanings attached to the park to add value to their local products, and by selling local foods to tourists who come to visit nature. But O'Rourke suggests that, increasingly, what is produced and consumed in the park is images and symbols, used to interpret the relations of present to past and to possible futures. And different groups interpret and use the symbols in different ways. Local farmers see themselves as modern professionals but this is not how conservationists or tourists want to see them. Tourists experience local nature and local society as timeless and unchanging; conservationist experts recognise the role of farmers in creating the local heritage they try to conserve but do not trust local farmers to appreciate its full value. Distinctions between representation and reality become blurred and Cevennes rurality is constructed more around the 'lifestyles' of the visitors than around the 'way of life' of the farmers.

Bareld van der Ploeg's chapter on 'the city's countryside' emerging in the Netherlands in recent years also identifies relations between farmers, conservationists and tourists as a critical nexus for the future of rurality. Between the 1950s and the 1980s, Dutch agriculture took an extremely 'modern' orientation towards the countryside, emphasising rational site reconstruction and deliberately altering physical features, such as water table levels, to make them more favourable to agricultural production. That has been replaced by a growing recognition that agricultural output and its continuous intensification is not the only goal society might hold for rural areas. Netherlands farmers find themselves a minority in a society that is interested in the countryside as a location for amenities much more than for agriculture. But there is a widespread assumption that such rural amenities (wildlife, flowers, unusual ecosystems, manmade features like landscapes or historic sites) are resources which can be harnessed to contribute to local economic development, and should be preserved by farmers in their own self-interest. Van der Ploeg argues that the contributions are not easy to establish. The relationship between agricultural development and countryside amenity conservation is as likely to be negative as positive. He looks for some positive way to combine the interests of both, in which amenities could be internalised into farm economies as benefits rather than costs and without jeopardising farmers' necessary participation in a world market which demands cheap food. His study, located in the peat-meadow district in the western Netherlands known as 'the city's green heart' (and a powerful symbol in Dutch politics and national identity) suggests that very few farmers are able to turn themselves in a commercially successful way into the sort of farmers who match urban ideals (a combination of organic farming, artisanal local production sold in a small farm shop, and educational experiences for tourists). Many farmers and their wives would prefer on both financial and social grounds to combine farming with off-farm jobs than with nature management, and the demands on farmers in terms of occupational conversion, education and organisational development in these types of development programme are very great. However where farmers' groups are themselves trying to come to terms with and take control over the urban demands made on them, there seem to be better chances of a positive solution.

Van der Ploeg thus raises implicitly the issue which is the explicit focus for the final chapter in this section - getting those dependent on nature for a livelihood to 'participate' in the management of their own practices for some 'larger good'. Nationally and internationally, 'participation' has become a primary ingredient in programmes to achieve sustainable development. Geiser's discussion of forest management practices in S. India, N. Pakistan, and Switzerland shows how, by the end of the 1990s, it had become part of mainstream thinking. However Geiser makes an important distinction between a 'modern' and a 'postmodern' understanding of 'participation in forest management'. A modern perspective sees participation as a more effective way of achieving management goals defined by forest experts: it is a tool to alter state-population relations, to overcome distrust of authorities, get better access to local knowledges, and get the public to treat forests more responsibly. A postmodern perspective gives equality to the claims of the multiple actors involved in using any given forest; it relativises the claims of scientific forestry to privileged expertise on how to manage the forest, seeing all users as interest groups acting in pursuit of their own 'development programmes'. The postmodern perspective allows us to hear a wider range of voices - but gives us no guidelines on how to construct what will necessarily be in some way hierarchical or exclusionary principles and structures for managing forests. If once we admit that human beings do not have a single 'species interest' in the preservation of forests, does 'development' become impossible?

The last part of the book takes up a theme which has been central to much general sociological debate on late modern society but has perhaps been rather underdiscussed in rural sociology, that of *rurality and civil society*. Questions about the relations between civil society, politics, and rural or agricultural development are posed particularly by the experiences of the post-socialist societies on Eastern and central Europe, and two chapters here address that issue: Siebert and Laschewski discuss the reorganisation of local government in the former East Germany, and in the course of that, illuminate how a collapse of civil society can hold back rural development; and Csite analyses the activities of state actors and civic intellectuals in developing new images of Hungarian rurality which in turn shape rural policy development.

Post-socialist societies need to look at the condition of rural civil society and not just at agricultural or rural economy issues. Siebert and Laschewski say that much less attention has been paid to reconstituting rural society than to agricultural reform, despite socialist thinking's emphasis on the unity of the economic and the social. The agricultural firms established in socialist East Germany were much more than just agricultural production units: they contributed significantly to the provision of social security, social integration, education, infrastructural services, and local power relations. The appropriation of much of public socialising for political purposes encouraged withdrawal into private social lives, but that in turn gave rise to informal social networks and exchange relationships, constructed in particular around household production plots. Unification destroyed this socialistic rural order, but the transfer of new institutions from the west has not led to social re-integration and the authors suggest that models which are rooted in the particularities of East German culture and history might be more successful than the importation of institutional and developmental practices which in some cases are themselves already under pressure from critics in the West.

Csite's chapter is also concerned with the westernisation or Europeanisation of post-socialist rural policy - this time in Hungary, a country which is not yet a member of the EU, yet where EU models are nevertheless driving change in its domestic policy structures, particularly in regional policy legislation and in the rural development domain. Europeanisation is effected through transformation of the perceptions and expectations of policy actors in Hungary, not through direct instructions from the EU or legislative incorporation into EU policy projects. Hungarian policy actors are discursively influenced by EU bureaucracies and respond by introducing discursive changes in rural policy (introducing new representations of the countryside) and institutional changes in policy implementation arrangements. The main beneficiaries so far appear to be the professionals and civil servants, for whose services Europeanisation creates a need; whether other rural actors do or will benefit also is less clear. There are parallels here with the activities of modernising knowledge elites who imported American models for farm development into Western Europe in the 1950s and 1960s. The impact of intellectuals in mobilising and directing rural social change is a topic which has not been discussed as much by rural sociology as it needs to be.

Osti's chapter returns us to the question of the civil society underpinnings of sustainable rural development, investigating the thesis that rural localities which achieve development possess a form of 'sociality' or 'social capital' (capacity for co-operative or collective action) which is built up in the community through the experience of belonging to and participating in local associations. His Italian case study suggests that association membership may build up a disinterested civil consciousness in individuals, particularly if the associations concerned are democratic and inclusive, but that people need to belong to more than one or they may simply remain narrowly focussed on their own group interest. However Osti also remarks that the small associations he studied are about producing sociality (interaction, entertainment, mutual aid) but that this does not appear to have anything to do, in an explicit sense, with development. Most did not want to be part of a larger development process or even to co-ordinate among themselves to deal collectively with public institutions or companies. It would seem that social capital is not necessarily development capital; yet the values which are created and carried within civil society are important resources to legitimate, or to challenge, the development projects of more political actors.

Chiffoleau's richly detailed discussion of transformations in the Languedoc-Rousillon wine co-operatives over the past decade addresses a similar issue: how can a producer co-operative which is, or has become, a largely economic organisation mobilise the 'sociality' or solidarity among its members which it needs if it is to carry through a 'collective action for innovation'? Is it possible, in late modern conditions of consumerist individualism, for co-operatives to abandon their modernist preoccupation with production efficiency and reinvent themselves in an earlier, multifunctional co-operative form? The question is posed for these co-operatives by their decision to shift Midi wine from a 'mass' to a 'quality' product. The technical requirements of this change seem to have weakened the loyalty of members, many of whom withdrew into a purely economic inter-pretation of their relationships which in turn impaired the capacity of the co-operatives to continue to innovate technically and economically in response to changing world market conditions. By the mid 1990s, many members were not bothering to attend AGMs, some were deliberately corrupting the technical procedures they had been advised to follow, some were resisting the production model their co-operative was imposing on them. As part of the switch to 'quality' production the wine is to be marketed on the basis of locality, so the co-op had to develop ways of controlling the production of a 'local identity' which would fit the targeted market niches, including getting members not just to plant new varieties and adopt new viticulture techniques but also to care for the local environment and to take part in promotional activities such as local festivals. Ironically, members most likely to accept these roles seem to be not the core full-time wine growers of the locality but newcomers from outside, often part-time wine producers, who are motivated as much by a search for 'community' as for profits. But a change in the core type of membership has implications for local solidarity and for the co-op's capacity to think of itself as an actor in the local development process as well as an economic actor in a global wine supply chain.

The last chapter in this section on rurality and civil society concerns the emergence and development of the co-operative movement in Ireland in the end of the nineteenth and early twentieth centuries. Readers may legitimately ask what on earth can this have to do with late modernity? In terms of most of the definitions of the 'new social movements' which supposedly characterise late or post-industrial modernity, the early Irish co-operative movement was in fact a 'new social movement'. In an odd sort of way that supports Chiffoleau's suggestion that conditions of rural late modernity may precipitate a return by local co-operatives to the early stages of their development. Alternatively, we might find ourselves wanting to question evolutionary theses of history which posit reflexivity and civil society as later outgrowths from an earlier self-confident, state-led, unreflective modernity. If late modernity is modernity becoming aware of its own presuppositions, and in the process, less confident of their obvious truth and rightness, that may be a situation which can come and go in the life of a group in whatever historical period they find themselves located. Perhaps some historical conjunctures, for some groups, are more likely to engender a reflexive care about the development of civil society than others. A useful programme might be to start to try to identify and categorise these.

If the book suggests any conclusion, it is probably that there is no single way in which 'to be rural in late modernity'. There has been a lot of interest in recent rural sociology in debating the question whether there is just one European rurality, or many (see e.g. Jollivet 1997, Gray 2000, Buller and Hoggart 2001, Granberg *et al* 2001, forthcoming). The 'Europeanisation' of food and nature should probably more accurately be described as the 'EU-isation' of them. On one side, EU legislation, policy documents and development programmes constantly seek to reinvent a single sense of European rurality. On the other, this in turn provokes resistance and the re-introduction and re-statement of 'non-normative' visions of the rural (Mediterranean *versus* Northern European, for example), while European rural space becomes increasingly segregated between locations for high intensity agriculture, rural desertification, ex-urbanite invasion, mass tourism and, perhaps, revived small-scale farming of one or another 'alternative' kind. Debates about the likelihood or effects of Europeanisation parallel and closely resemble, although they are not identical with, debates about globalisation. The recent decline in social theory of linear thinking and of grand narratives about the normal offers rural sociology an opportunity to live more comfortably with difference and disorder. We may not be able to see clearly where we are going, but at least, as Geiser tells us, we can hear more voices and understand a greater range of points of view than was possible when we were simply 'modern'.

Part I

Agriculture and rural life in late modernity

1 Rurality and late modernity in transition countries: the case of Bulgaria

VESKA KOZHUCHAROVA and ROSSITSA RANGELOVA

Introduction

Agriculture is historically important in the Bulgarian economy. The country is well endowed with natural resources for agriculture and has some of the most potentially productive lands in Central and Eastern Europe. In the 1990s Bulgaria's agriculture underwent its second major transformation in fifty years, and in the context of a severe recession and radical restructuring of the national economy. Thus the country is an interesting example of rapid and contradictory changes in its rurality.

This chapter is devoted to the formation of rurality under the conditions of late modernity in transition countries such as Bulgaria. First of all our understanding of the concept of late modernity, and the specificity of Central and Eastern European countries (CEECs) related to it, is described.

The historical destiny of Bulgaria and its geopolitical location were the factors which led to the specific results and paradoxes of rural and agricultural development. As agriculture is a basic characteristic of the village as a type of settlement, we focus our attention on metamorphoses in agriculture.

The chapter presents the basic features of changing rurality in Bulgaria, within three sub-periods:

- the pre-war period and during World War II;
- under the conditions of a centrally planned economy up to 1989;
- during the transition to a market economy in the 1990s - in particular the land reform - and the emergence of new forms of ownership and rights to its use.

This historical excursion serves as a background determining changes in present and future Bulgarian rurality, and its specificity in the context of 'late modernity'. We compare the modernity of the Bulgarian village with that of the 1930s rather than that of contemporary developed countries.

On the basis of the experiences described, the chapter tries to define the probable pattern of rurality after 'late modernity' in such a transition country. In particular the following aspects are analysed:

- what kind of rurality could correspond to 'late modernity';
- Bulgaria's belated modernity and postponed late modernity;
- paradoxes of rural Bulgaria and to what extent these processes could be related to the 'otherness' of Central and Eastern European experience. Finally we give some conclusions on how emerging Bulgarian rurality could adjust to the nature of 'late modernity'.

The authors' contributions to the paper are comparatively self-contained. The problems are considered by Kozhucharova from a sociological point of view, and the socio-economic aspects are presented by Rangelova.

Late modernity and the 'otherness' of Central and Eastern Europe

Modernity changed the societies referred to by Toffler as societies of the first wave of social development. The latter was connected with acceleration of development in all spheres of public life, but its main characteristics were industrialisation and urbanisation, speeded-up social time and expanded social space. Most people gained goods, making them more self-confident and more independent. As a result humanity ventured to think itself a master of nature.

Rurality was considered a sign of traditional societies. It could be said that agriculture, which in its essence is in the utmost degree a successor of the first wave societies, found its own reincarnations through mechanisation, usage of chemicals, creation of new species and races, hugely increased production. Toffler's general conclusion about modern societies is that they were created using non-recoverable energy sources. For the first time, civilisation began to impair 'the basic capital' of nature instead of living from its 'interest'. Industrialisation threatened civilisation.

The notion of *late modernity* seems vague but we take it to mean a period when the common models of modernity have exhausted their positive, creative potential and when *new zones of transition* to something significantly different emerge. Of course, this 'something' has a pessimistic and an optimistic sense. We assume a positive variant of possible balanced development. In this connection one of the marks of late modernity is the increased interest in rurality. The very advent of modernity and industrialism stimulates urbanism and displaces rurality.

Interest in rurality, in the authentic natural environment, in folklore, in the paternalistic world, may even be taken as signs that a new stage has begun in the development of modernity. It is a form of reflecting on the fact that rurality, rural culture and local distinctions *dissolve* into the uniform appearances of modernity; planetary resources visibly melt away and irreversible processes of permanent damage to nature are triggered by the already uncontrollable new technologies. The new value attitude towards ruralism is a reaction against the general loss of identity due to the pervasive incursion of a new type of culture, to the all-engulfing ocean of information.

That late modernity is assessed as a time when modernity has exhausted its potential does not make late modernity equivalent to de-modernisation and destruction of modernity; rather, it signifies a new kind of global development. This implies a search for new forms of *harmonising* highly modern technologies with the ecological culture and technologies of traditional societies, by which means natural resources and specific local cultural models will be preserved. Within its globalisation processes, late modernity ought to include the rich variety of local identities, and this means harmonising multiple 'otherness', not obliterating differences.

The nature of late modernity imposed a certain model of development for all CEECs, but it was balanced by the historical and geopolitical particularity of each country. It is true that the necessary changes were not given a chance to develop naturally. The great changes of the late 1980s led to transformations which practically obliterated the 'otherness' of Central and Eastern Europe. Although all of these countries had previously been developing within the common economic and ideological framework of socialism, each of them had also been following an individual course of development that both corresponded to the remainder and differed from them.

Evolution of Bulgarian rurality

Rurality before the end of World War II

Before World War II Bulgaria had one of the lowest incomes *per capita* in Europe. This was a backward agrarian country, suffering from overpopulation and over-employed land. Most farms were smallholdings with scattered plots and enormous hidden unemployment.[1]

Data for the development of this sector are taken from the Census of Agriculture held first in 1934-1935, second in 1944-1945. The decade between these two dates was very important because this was a period of comparatively stable social and economic development in Bulgaria, and from the viewpoint of world events this was the time of the Great Depression.

According to the 1934 census, nearly 84 per cent of the total economically active population were engaged in agriculture. In the 1930s the country had virtually exhausted the possibilities of extensive growth in agriculture. For the period 1934-1945, i.e. the decade between the two surveys, we note the following fundamental changes (Kemilev 1946):

- A negligible increase in farm land, by only 0.17 per cent, whereas the number of farms increased by 15.8 per cent. As a result the average area per farm decreased from 5.04 hectares in 1934 to 4.33 hectares in 1944.
- The groups of smallest farms have shown the biggest increase, especially holdings of 2 to 3 hectares in size, whereas the biggest farm holdings, between 40 and 50 hectares, showed a decrease.

Evidently, a distinct process of fragmentation and consequently differentiation was going on in Bulgarian agriculture. The biggest fragmentation was among the smallest farms (under 1 hectare) and the biggest enlargement was among the largest farms. By the end of World War II there were 1.2 million small private farms, and the largest farms possessed only 2 per cent of total cultivated land in the country.[2] The pattern described gave rise to a culture of very small landowners, with the specific features of working in small-scale plots, mainly for self-sufficiency. Together with the limited funds available to them, this meant that the rural population could not think much about modern farming.

Another important feature was that the industrial sector in Bulgaria was underdeveloped, and so not able to absorb a redundant labour force. At that time, the co-operative form of property-holding was already known and practiced in Bulgaria, not only in the agricultural sector but in other branches too, operating as an effective form of economic activity organisation. The situation provided fertile ground for the development of a rural co-operative movement. In fact, attempts to form agricultural co-operatives in Bulgaria dated from the turn of the 19th century and built on the Bulgarian tradition of collective agricultural labour. However, many of these early co-operatives lasted only a few years.

In the beginning of the 20th century politicians and state policy advisors moved to support credit co-operatives as a means of modernising agriculture. Although production co-operatives were becoming more popular, in particular in the 1930s, the vast majority remained credit co-operatives. By 1939 there were 3,502 co-operatives in the country. Rural intelligentsia, including teachers, priests and clerks, were significantly represented among their members, sometimes working as book-keepers or agronomists in the co-operatives.

These features of Bulgarian agriculture shaped the country's rurality at that time. They explain the slightly different nature of the land reform during the central planning period which followed, in comparison with the land reforms in other CEECs which predetermined rurality through aspects of collective farming.

Rurality under the conditions of a centrally planned economy

At the end of World War II, Bulgarian agriculture was amongst the least developed in Europe. Modernisation of agriculture and the use of an agricultural surplus to promote industrialisation were considered important government priorities.

Agricultural modernisation, nevertheless, was not the only goal driving collectivisation under socialist rule. In accordance with socialist ideology, private ownership in Bulgaria was eliminated as a 'birthmark of capitalism'. The Bulgarian model of collectivisation however was different from the Soviet model, being closer to the real co-operative form. For example some new principles of organisation were initiated: members would retain ownership of their land, including the right to sell it and to receive rent on pooled land from the co-operative, although the rent was limited to 20-40 per cent of distributed income. The co-operative members could keep small plots of 0.2-0.5 hectares and a few animals for so-called personal use. But the common feature of the two models was that both foresaw early support for collective forms of agriculture as a means of modernising agriculture.

During the following four decades or so, rurality in Bulgaria took shape under the influence of different stages of agricultural development:

- collectivisation of individual (private) farms, including 'socialisation' of the land through its expropriation from big landowners and redistribution to landless and poor villagers (1944-1958);
- consolidation of small-scale co-operatives (1959-1967);
- establishment of agro-industrial complexes (APKs) and promotion of individual (private) farms (1968-1989).

Collectivisation of Bulgarian agriculture: 1944-1958

1944-1949: Beginnings. The government promoted the formation of agricultural production co-operatives but also expanded private property in land through the Law on Work-Based Land Ownership (1945). This meant the land reform law was based on the principle of 'the land to the tiller'. This law limited privately owned land to 30 hectares in the Dobrudja plains and 20 hectares in the rest of the country. People who did not work the land were limited to 5 hectares in the plains, 3 hectares in other areas. Land above these limits was nationalised, and the owners reimbursed according to rates established by the state. Nationalised land was redistributed among land-poor peasants or given to co-operatives free of charge. This law enjoyed broad popular support, in part because of the very small number of large landholders in Bulgaria. In Bulgaria, nationalised land comprised only 7 per cent of total arable land, while it was 38 per cent in Czechoslovakia, 35 per cent in Hungary, 25 per cent in Poland, and 14 per cent in Romania.

Benefits to newly established co-operatives (the so-called TKZS) included complete tax reliefs for these units and their members for three years, extensive credit for mechanisation, and a 50 per cent reduction in the price of land purchased from the state. Relative to other sectors of the economy, however, state support was quite limited. Only 6 per cent of investment planned for the period 1946-48 was targeted on agriculture, which employed nearly 70 per cent of the labour force. State-controlled agricultural prices also remained low, preventing accumulation within agriculture itself.[3]

1950-1954: Increasing state pressure on the private sector. In 1950 there was a marked renewal of the collectivisation campaign. Facing falling output, discriminatory rationing policies were undertaken which weighed heavily on private producers. Peasant resistance often took the age-old form of reducing production and slaughtering livestock which had become uneconomical to keep. At the same time, positive incentives for co-operative work were offered in response to falling production and peasant hostility. Delivery quotas for collective farmers were reduced and member households were permitted to keep more livestock for household use. State-run Machine Tractor Stations (MTS) were developed to provide mechanised services to the new co-operatives, as the number of stations increased from 81 in 1950 to 149 in 1954. State support also promoted an increase in chemical inputs.

One can judge the speed of collectivisation by the fact that in only a few months after September 1944, nearly 110 co-operative farms were established. During 1946-1950 their number reached 2,501, and their land exceeded 51 per cent of the total arable land in the country (Table 1.1). The number of TKSZs also rose rapidly, from 1,601 in 1949 (14 per cent of arable land) to 2,739 in 1951 (56 per cent of arable land) and 2,723 in 1954 (61 per cent of arable land). Landholdings per member recovered significantly: from 2.4 hectares in 1948 to 4.5 hectares in 1952-1954. The full triumph the of co-operative system in agriculture was declared official in 1958.

The process of collectivisation took place at a different pace in different regions. By 1954, collectivisation levels in grain districts ranged from 61 per cent to 80 per cent of arable land. In the mountain regions, however, TKZSs controlled only 16 per cent to 40 per cent of arable land. On average in 1954, mountain zones had only 30 per cent of arable land in TKZSs, about half of the national level. Collectivisation of livestock production, which was concentrated in mountainous areas, was also difficult. About 22 per cent of cattle and 28 per cent of sheep were collectivised in 1953.

Table 1.1: Process of collectivisation in Bulgaria, 1945-1959

year	*number of cooperatives*	*percentage of arable land*	*collectivised households per cooperative*
1945	382	3.1	34
1948	1,100	7.2	124
1949	1,601	13.6	156
1950	2,501	51.1	502
1951	2,739	56.4	582
1952	2,747	60.5	553
1953	2,744	61.1	569
1954	2,723	61.3	569
1955	2,735	62.5	591
1956	3,100	77.4	911
1957	3,202	86.5	1,017
1958	3,290	93.2	1,244
1959	972	98.0	1,290

Source: Meurs *et al*, 1999: 100

1955-1958: The final push. Over the period under review the collectivisation process was completed. To achieve this, both incentives and pressure were used. Incentive programmes targeted the lagging mountainous regions in particular, and gave additional credits and tax breaks. In 1955-1956, the prices of many agricultural products were increased, and the prices of chemicals, veterinary assistance, and services from the MTSs, were reduced. New tax breaks were also given. As a result of mechanisation, peasant working conditions were substantially improved. The earnings of TKZS members began to rise relative to industrial wages. At the same time, however, local control was further reduced, and the remaining rent payments to members were eliminated.

Initial results of collectivisation. In fact, *the collectivisation*, which meant the pooling of private farms into big collective farms, was performed hastily, without conviction or deliberation, i.e. without observance of democratic principles of free choice. At times unlawful methods were applied against rich farmers and the upper stratum of middle-ranking peasants, aimed at bringing about their economic destruction. Mass co-operation was carried out without first creating the necessary basis for machine cultivation of land.[4] After this period machine and tractor stations were set up. Thus in collective farms different types of machinery, equipment and technologies were introduced, and gradually a large number of production processes in crops and livestock was mechanised.

Despite the massive dislocation of this period, the agricultural sector recovered quite rapidly. The consolidation of landholdings and intensification of production paid off with rapid productivity increases. This resulted in an increase in net output of 2.5 per cent per year from 1948 to 1958, so that net output reached 128 per cent of the 1939 level (Meurs *et al*, 1999: 105). The productivity increases in agriculture also led to improvements in the standard of living for the general population.

Consolidation of TKZSs: 1959-1967

In 1959 a consolidation of the TKZSs was undertaken. In this process, many TKZSs were unified into one unit, the so-called OTKZS (unified TKZS), with their centre in one of the villages of the municipality. In principle, farms were to consolidate on a voluntary basis, but the process was pushed along quickly by local party functionaries. In terms of forms of ownership, while co-operatives were seen as a form of 'socialism in a village setting', state ownership (ownership by the 'whole people') continued to be regarded as a higher stage of socialism than ownership by a group of villagers. Thus the move from group to social ownership was a natural evolution. Consolidation into OTKZS was a first step in this direction, bringing ownership and control from the village to the municipal level.

Intensification of production. Investment in agriculture continued to increase with consolidation, peaking at 27 per cent of national investment for the period 1958-1960, then falling back slightly to 21 per cent for the period 1961-1965. The number of available tractors and combines rapidly increased, and chemical intensification also continued to rise. But even so, the level of mechanisation remained twice as low in Bulgaria as in the GDR (East Germany) and Czechoslovakia, and lower than in the less mechanised USSR, Poland, Hungary and Romania (Berov 1974: 266).

Control and incentives. On the new farms, the village-level TKZSs were brought under a common plan and leadership, although they continued to control their own assets and liabilities. Without a means to ensure just treatment by the 'outsiders' however, many villagers felt themselves victims of free riding, and reduced their work effort in response. To address the growing incentive problem, the state supported the organisation of the 2,000 worker farms into permanent production brigades. These sub-farm units were designated to recreate the smaller work groups of the TKZS, in which individuals might identify with production results and social pressures could be used to promote quick work. In the brigades, wage payment was to be linked to the brigades' final production.

The brigades proved a poor substitute for village organisation, however. First of all the measurement of production of a single agricultural brigade proved difficult. In addition, if production did not even cover loans (which was not unusual given the state policy of low agricultural prices), state contributions covered the difference. Thus no links were achieved between work effort and rewards.

Performance of the OTKZS. In the period 1957-1960, the number of co-operative farms decreased more than threefold. As a result the average area per co-operative increased from nearly 1,000 to 4000 hectares.

In the early 1960s, agricultural growth (5 per cent yearly) was accompanied by continued improvements in the economic conditions of TKZS members. By the late 1960s however, agricultural growth began to slow.

While investment rates remained high throughout the late 1960s, it appears that slowing growth rates resulted from the exhaustion of easy gains from technological 'catching up'. New growth would have to come through improved use of existing technology or from innovation. In any event, this stage of agricultural development was viewed by the villagers as the most successful for the co-operative farms in Bulgaria. There was a belief that the rational size of the farms had been attained. Total agricultural output increased sharply. Changes came in villagers' consciousness and psychology in favour of the co-operative system because of greater social fairness and the possibility of joint settlement of problems that could arise.

Establishment of agri-industrial complexes (APK) and promotion of individual (personal) farms (1968-1989)

From collectives to state farms: further reorganisation, growing distance. In response to slowing agricultural growth at the end of the 1960s, the state moved to further 'industrialise' and 'socialise' collective agriculture. The collective farms were integrated into state-run Agro-Industrial Complexes (APKs). Thus collective agriculture was replaced by state agriculture.

In principle, under the conditions of a centrally planned economy in Bulgaria as in other Eastern European countries, state policy attempted to hustle private and individual farms out of the agrarian sector. In the 1970s it was acting in the same way in respect to the co-operative sector, transforming the co-operative farms and machine and tractor stations into *agro-industrial complexes* or *industrial-agrarian complexes* (depending on the prevailing sector). The latter were large-scale firms, designated to combine agricultural development with complementary industrial activities. This changed the forms of property and labour organisation. Former owners working in the farms became divorced from ownership and began to treat themselves and act as hired workers. Along with the low ratio of capital investment, stagnating or even declining production in the agricultural sector, non-conducive state policy on farm redemption prices, underdevelopment of industry connected with agriculture, and other factors - all contributed towards a gradual deterioration in farm productivity.

In the 1970s, nationalisation of private and co-operative property was marked.[5] As a result, the economic performance of the agricultural sector took a turn for the worse. The incentives and work motivation of active farmers decreased, while social apathy increased. As farm productivity fell off, problems in feeding the population emerged.

Agricultural performance under the APK. As could be expected, further centralisation of agricultural production and management failed to restore growth rates. In a study of Bulgarian agricultural growth, Boyd (1991) found that the rapid technical change that had driven output and productivity improvements through the 1960s had ceased by the 1970s. The lack of technical change could not explain all the growth slowdown, however. It is evident that organisational and motivational problems associated with the large-scale farms were also significant factors in declining productivity and growth.

Promotion of individual private farms. The rather limited capacity of the public farms to meet the rapidly growing needs of the country for some products was one of the reasons for promotion of *individual private farms*. Some other reasons were as follows: the impossibility of supplying the entire village population with farm production; the necessity for the population to become self-sufficient; and the evident potentialities of the villagers to cultivate extra production.

The emerging individual farms in Bulgaria could be divided into two groups: individual farms of co-operative farmers, and individual farms of workers and other employees. The latter group appeared later and consisted mainly of people, most of whom had previously been skilled workers in towns. The number of individual farms gradually increased as they not only created relations with the co-operative farms, but developed themselves as an individual branch. A considerable contribution was made by the advancement of agro-industrial contacts covering also the individual farms. Under the new circumstances, the individual farms stopped producing merely to meet the needs of the single household. The re-orientation of production in individual farms towards more labour-consuming, market gardening crops and towards livestock breeding increased their share of public production and of national income.

Generalising the collectivisation. During the 1970s and 1980s, due to continuing industrialisation accompanied by large transfers of labour from agriculture to non-agricultural sectors, the proportion of employment in agriculture was falling. Thus, by the mid-1980s in all former centrally planned economies the proportion of agricultural labour in the total national labour force declined to less than one-third, except in Czechoslovakia and in East Germany where the decline was 12.5 and 10.2 per cent respectively.

Under conditions of intensive industrialisation and rapid urbanisation, the village was most vulnerable in its demographic potential: it was constantly shrinking. Some basic figures show the trend of change in this respect. In 1946 nearly 75 per cent of the total Bulgarian population was rural, while in 1992 the rural population amounted to 32 per cent. Up to the end of the 1940s over 80 per cent of all workers in Bulgaria were engaged in agriculture. Within only one decade, the percentage of those employed in agriculture had decreased by one third, and in 1960 this category had declined to 55 per cent of total employment.[6] In 1989 the share of employment in agriculture was about 18 per cent (increasing to over 24 per cent in 1996-1999). Compared with advanced Western industrial countries, the relative importance of agriculture in CEECs measured in the terms of percentage of total employment in 1989 was still five to eight times larger, which clearly indicates its lower productivity.

Migration from village to town was significant at that time and was a result of collectivisation of agriculture. Industrial centres attracted village populations, which on one hand led to a decrease in hidden unemployment in the village, but on the other hand, led to de-population of villages and imbalances in the rural age structure.

In the economic history of Bulgaria the early co-operatives will be remembered as a favourable form of production organisation. This is partly because they coincided with an 'easy' stage of economic growth, resulting from the mobilisation of resources under-utilised in Bulgarian pre-war small-scale agriculture, and from the provision of the credit needed for agriculture to 'catch up' technologically with industrialised countries. But the co-operatives also offered real advantages to producers compared with the APK, or even the OTKZS. They facilitated mechanisation and risk sharing, and greatly improved the working and living conditions of rural households.

All these circumstances contributed to the need for transition from an administrative-command to a market-type economy. The latter should gradually have created legal, organisational and economic bases for intensifying farm production and overcoming the existing agrarian crisis. Compared with Western Europe, however, Bulgaria like other Central and Eastern European countries did not and could not adjust to the structural and technological changes of the 1970s. Central and Eastern European economies remained 'modern', based on the post-war certainties of industrialisation and urbanisation (although as it turned out, they were not pursued very effectively). 'Late modernity', with its uncertainties concerning employment and traditional sources of identity, did not develop in the socialist period. Western European rurality needed over two decades to cope with the process of 'late modernity'. In Eastern Europe, including Bulgaria, 'late modernity' was the 'shock' of the 1990s.

Agricultural restructuring to a market economy: what kind of rurality emerged in the 1990s?

Initial conditions

The change of ownership, in particular land restitution, was a very significant element of ongoing economic reform in Bulgaria. The new landowners had the following options: to sell the land, lease it, form a co-operative with other owners, or to develop private farming. But at the beginning there were a number of very complicated problems:

Restitution of what? As well as land for purposes of production, farm machinery, outhouses, etc. were also necessary. Who owned these means of production? They were produced neither by the co-operators, nor by their heirs.

Restitution to whom? It turns out that only about one-third of those working on farms were to inherit land. Nearly 55 per cent of the new landowners lived in town and many were either unable or unwilling to return to farming.

Restitution, but how? After so many years (about 50), technical aspects of this act may have proved of crucial importance. In many cases the necessary documents could no longer be found, or were not valid.

Restitution for what purpose? If one puts the 'historic fairness' issue aside, the economic grounds of the land restitution seem questionable. If the heirs got the land, they would decide what to do with it. Probably part of the land would be sold, another part would be used for self-sufficiency and for building villas, but part would not be cultivated at least as long as there was no developed land market. Thus the area of cultivated land would reduce.

Psychological attitudes were of great importance at the beginning of the process of agricultural privatisation. There are research findings indicative of the reaction of the population in some agrarian and industrial settlements. A useful survey was carried out in June 1990 by Club 'Alternatives' ('Bulgarian Business' newspaper, 1991). About 1,600 residents living in three communities were included in the survey: one with *highly developed agriculture* (Dolna Mitropolia), one with *moderately developed agriculture* (Kaloyanovo) and one with *less developed agriculture* (Krivodol). The survey findings differentiated these people into two groups. The first group was larger in number than the second and could be called *conservative*. This group maintained that the co-operative form of property had to be kept dominant. In their opinion, land reform should concern only (or mainly) the organisation of co-operative farms. Generally people in this group had lower education and income levels and were older. Most preferred to keep their basic job and get land for private use, preferably up to 1 hectare in size. But it would be wrong to accuse this social group of lack of initiative. If we take into account that until 1989 the private (or so-called individual) farms were regarded as more effective in comparison with the already pseudo co-operative farms, this shows their skills and enterprise.

Much of the rural population of Bulgaria is advanced in years, and has already withdrawn from social activity. Given the obscure outlines of the future market economy, most of them do not see any other alternative except that of working in a co-operative farm.

The second group of people was *pro-reforms*, and was rather smaller in size than the first group. These people supported the idea of an agriculture based on private property. Most were young, with secondary and university education, did not appear to be people who just wanted to grow rich. They were convinced of the advantages of private farming as the best future for Bulgaria. But we could not explain the fact that a large part of this group did not wish to take land for private use. Only about 10 per cent of them declared that they would acquire land and actively engage in agriculture.

It should be added that at that time many people were waiting for the process of transition to develop further before they would take their own final decision, depending on future prospects. In any case we suggest that the results of the survey reflect to a great extent the dominant attitude of the Bulgarian population in general to land restitution in 1990, i.e. at the beginning of the agrarian reform (see Table 1.2).

Other subsequent surveys which were representative of the whole country, and carried out later, provided similar data on the preparations and the expectations of the rural population for the future; for example 'Collectivization and its Alternatives' (1992) and 'Bulgarian and Agricultural Labour' (1994). According to them, the best variant for future agriculture was a combination of family farming and modern co-operatives.

Land reform and rural restructuring

Land reform started in 1991 without any clear objective. The only point on which there was agreement was the necessity for land restitution. The Land Law aimed to restore land to those who had owned it before 1946 when the communist regime forced collectivisation on farmers. There was a lack of understanding about the basic features of modern farming, as well as some exaggeration of the advantages of old small-scale farms and under-estimation of collective or large-scale farming. One of the most important weaknesses, however, was that people were not prepared for the changes and the difficulties which they would have to face during the economic transformation, in particular during rural restructuring.

The Land Law required that reinstatement of property be done *within the real old boundaries of land where they still existed.* Where land boundaries no longer existed, reinstatement in ownership was to be done within the boundaries of farmland of equivalent area and quality in compliance with a plan of land division. This specificity in comparison with other Eastern and

Table 1.2: Questions and replies in the Inquiry '90

Questions	Answers (%) Workers – total number	Agricultural workers
1 If there are favourable market conditions, as well as reasonable taxation and bank credit, would you like to start a private farm?		
I do not want a private farm	58.1	63.3
I hesitate	19.9	16.2
Yes	20.2	18.7
No reply	1.8	1.8
2 What kind of farms will be most efficient?		
State	13.6	8.4
Co-operative	47.9	63.1
Private	26.5	18.3
I cannot judge	9.9	7.8
No reply	2.1	2.4
3 Is it correct for former owners to get back their land?		
Yes	19.2	15.7
No	64.1	71.6
I cannot judge	13.8	10.1
No reply	2.9	2.6

Source: *Bulgarski Biznes* (Bulgarian Business), weekly newspaper for economic information, 3-4 June and 10-16 June 1991

Central European countries made the process of land restitution in Bulgaria more difficult.

The process of land restitution was very complicated and included many stages. The Land Law has been amended over 20 times so far, and is actually the imposition of a politically inspired model for which most of those involved in agriculture were unprepared. Despite this, the Government was much more successful in destroying the old system than in providing the institutional support required by the new one. The delay of agrarian reform, the specific approach for its implementation in Bulgaria, and the other parallel socio-economic processes such as privatisation in other sectors, restructuring of production and markets, and deep economic crisis, have produced a very fragile rurality which should be treated only as temporary (transitional).

Concerning 'late modernity' in the 1990s, we argue that there was a double shock: on one hand, the collapse of socialism and the introduction of 'shock therapy' measures to promote economic restructuring, including changes in the social structure; and on the other, rapid technological progress and globalisation in the world economy, as well as integration of the former socialist economies into world structures, in particular the European economic space, all of which impose adjustment to a modern market-type economy.

Basic features of the present day rurality

As a result of land restitution there is a large number of new owners. Land restoration involves 5.5 million hectares of agricultural land and when it is complete, there will be approximately 2.4 million owners of 10.5 million plots. After land has been sub-divided among heirs, the number of owners could increase to 4 million. Consequently, agricultural land fragmentation should also increase. An important feature of private farms in Bulgaria is polarisation in farm size and structural dualism: small subsistence farms, large commercial farms and co-ops. According to official data, most of the farms are less than 1 hectare in size.

After the liquidation of the old co-ops under the new Land Law, two types of new co-ops appeared (Swain 1997), some of which are quite similar to the former co-operatives. They include a lot of members, not living in the village or even in the same region, who cannot cultivate their land by themselves, and a restricted number of members able to work on farm. These structures are undermined by the incentive to make more profit for members, because most are either unable or unwilling to control how the co-operative is managed. Some of these co-operatives are actually not legal, because Co-operative Law stipulates that a members' assembly is legitimate only if more than half of the membership is present.

The other type of co-operative is mostly established by people interested in cultivating the land. They may be business-oriented, but sometimes their main goal may be to deliver employment to members. Normally such co-ops have a much smaller membership than the older co-op types; their members participate with both land and labour and the main conflicts existing in the old type of co-ops do not exist here.

Since 1991, with the progress in land restitution, the private sector's share of agricultural production has gradually become prevalent. In 1990, less than a third of total agricultural production was produced in the private sector; however, since 1996 over three-quarters has been produced in the private sector (Table 1.3), including since 1997 about 97 per cent of grain, 100 per cent of industrial crops, 96 per cent of vegetables, 98 per cent of fruit, 97 per cent of meat and 99 per cent of milk.

The household plot is an important income source; however, the small contribution which household production makes to cash income shows that this production is used mainly for self-sufficiency and is instrumental to the survival of the rural population. In other words, most private farming has been made up of a small number of self-sufficient farms, which do not produce for the market (Rangelova 1998).

Table 1.3: Percentage share of private sector in agricultural output, 1990-1997

1990	1991	1992	1993	1994	1995	1996	1997
30.9	37.3	48.7	59.7	73.8	74.1	75.4	76.8

Source: Statistical Yearbook of Republic of Bulgaria, 1998, National Statistical Institute

The delay in land reform, in particular the lack of official titles, posed severe obstacles to private investment and development in Bulgarian agriculture. The unresolved situation concerning land ownership caused a considerable proportion of uncultivated arable land (from 10 per cent to 30 per cent in various years) to become available. Regarding investment in agriculture, the official data show an accelerated process of de-capitalisation of agriculture. The sector needs significant modernisation and initial financial support, including development of an agricultural credit system for farmers. The newly established co-operatives have neither enough modern machinery nor sufficient capital. Co-operative farmers are motivated to mechanise agricultural production, but since they were not the owners of the land in the 1990s, they could not exercise efficient control over its administration. At present one of every four people employed in Bulgaria is engaged in agriculture, which is a high rate by modern international standards. (The share of agricultural output in gross value added changed from 12.3 per cent in 1994 to 26.2 per cent in 1997 and 20.7 per cent in 1998).

Table 1.4: Percentage share of agricultural employment in total employment, 1990-1998

1990	1991	1992	1993	1994	1995	1996	1997	1998
18.5	19.6	20.6	21.7	22.7	23.4	23.4	24.3	24.7

Source: Statistical Reference Book of Republic of Bulgaria, National Statistical Institute, various issues

About 40 per cent of the population lives in villages, and 20 per cent in small towns or so-called towns of rural type, which means that nearly 60 per cent of the total population lives in the countryside, the majority of whom are relatively aged. The population of less than working age (up to 15 years) is only about 17 per cent of the total rural population, while that beyond working age is over 40 per cent. While the average age of the urban population in 1998 was 37.3 years, that of the rural population was 43.8 years. We also observe a gender imbalance among the active rural population. At present the ratio of women to men is 45:55. There are many more lonely old women in the villages than in the big towns. Their pensions are the lowest in the country. The most well-to-do-families are those where three generations live together. There are also regional differences in the living standards of villagers. The lowest is found in the rural population of the Northwest and Southeast, including the mountain regions. Villages located in the central part of the country are wealthier.

An important feature of transition economies is the high level of education inherited from the socialist era. People are far better educated than in other countries at similar levels of income *per capita*. This reflects some specific features of rurality in Bulgaria. Among the present rural population, 7 per cent have less than basic education, 28 per cent have basic education, 55 per cent have secondary education, and 10 per cent have a college or university education.

The preferred pattern of rural lifestyle is to work in the nearby town but to possess and cultivate plants on a houseplot. Part-time farmers are mainly middle-aged, predominantly skilled people who work in towns but live in villages. In some cases when a worker is made redundant, agricultural activity becomes the major source of income, but more often than not they regard this as only a temporary situation. Small villages or towns offer a number of advantages: easier opportunities to buy a one-family house, the availability of fresh air and clean nature, and the stronger morality of the local society.

Bulgaria's belated modernity and postponed late modernity

Modernity and individualism were late in coming to Bulgaria, and modernisation processes have followed an uneven and forcibly accelerated course over the last 50 years. The village pursued its mode of development under conditions of rapid industrialisation and in a rather stable connection with paternalism. In the four decades following World War II the Bulgarian village and Bulgarian agriculture have managed to achieve the general dimensions of modernity, having undergone all the perturbations of social acceleration, the tensions resulting from rapid restructuring of a primitive mode of agriculture on small, parcelled plots of land into large-scale, modernised agriculture, in addition to industrialisation and mass migration from village to town. The acceleration of social time in rural Bulgaria, the extension of social space and the sudden substitution of cultural models led to a basic paradox: *urbanised ruralism never lost its connection with traditional paternalism* and this has created an ambiguous position for the village, susceptible to both modernity and to tradition.

Urbanised rurality was considered a mark of progress, whereas its traditionalism was judged to be a sign of backwardness. The village became urbanised, yet to a very great extent the town included a concealed rural space. For its part, the village maintained its status and prosperity through what Creed has called 'the domestication of industry' (Creed 1997). Bulgarian society matured and was gradually overcoming the anti-rural paradigm that pictured the future world as a place without villages and rurality; the role and contribution of the village to economic modernisation, as well as to national identity, were beginning to be appreciated. Bulgarian society was seeking new paradigms of development and adaptation to the globalisation processes, assessing its own values and achievements, but also rejecting whatever separated it from the mainstream of world development. Passionate rejection and distrust of past achievements were a necessary form of catharsis before the new stage of development could commence.

The problems of the late 1980s were met with hopes for forthcoming improvements on what had already been achieved, with expectations of only restructuring the inefficient variants. In looking forward to changes, no-one foresaw the crisis and breakdown of social-economic structures. The actual turn of events was far from what had been expected. The new political platforms proclaimed several formulae: democracy and political pluralism, market economy and privatisation, encouragement of individual initiative versus collectivist attitudes and statism, the construction of a new social stratification and a new middle class, etc. In public, these transformation formulae were presented without alternatives, but also without a clear agenda for their implementation. These slogan-formulae did not define objectives, but posed the means - i.e. political change, market economy, privatisation, etc. - as an ultimate goal. Euphemistic formulae were offered in place of an objective, for the unattractive image of certain social notions had to be concealed from the public, both in Eastern Europe and elsewhere.

The lack of a legitimate programme and an agenda for achieving it led to excessive politicisation of the social environment and of the economy, chaos in the economic situation, and what we provisionally call 'antisocial phenomena': breakdown of structures, unemployment, criminalisation, corruption, poverty. This was not the only risky outcome; there was also the hazard of an unknown, unidentified change that cut short the natural course of evolution of Bulgarian modernisation and industrialisation in its socialist variant. These were not just repercussions from correcting the speed of social time; these were the tremors of destruction, of indifference to unique local development, to local traditions and values.

We will not go into the debate as to whose interests this model of change has been serving, and why this of all models was imposed. We will conclude that *what shocked Bulgaria was not the advent of late modernity, but the fact that the coming of late modernity was postponed indefinitely in the country.*

Listing the social paradoxes of rural Bulgaria

Some characteristics of Bulgarian agriculture have been strikingly constant during the whole period under review (from the 1930s up to the present) and show some typical features. For example, sociological surveys suggest that about 96 per cent of the rural population works its own plots whether currently employed in agriculture or not. Second, as in the 1930s, at present only about 35 per cent of agricultural products are directed to the market; the other two thirds are for self-sufficiency. Third, about 65-69 per cent of the rural population supported collectivisation in Bulgaria in the late 1940s and early 1950s - nearly the same as the percentage (63 per cent) supporting modern co-operatives (see Table 1.2). Fourth, as in the 1930s, only a very small proportion (about 5 per cent) of farmers report that they are making a good profit and doing well.

This social situation is accompanied by a number of myths introduced in the 1990s for the purpose of 'remaking' rural space (Kozhucharova 1999):

- the model of a future 'golden age' of the village was retrospectively oriented to the past (to the 1930s) defined as the most prosperous time for the Bulgarian countryside;
- co-operative farming was condemned without right of appeal, and the private farm run by an individual producer was presented as the ideal for Bulgarian agriculture;
- the solution to the demographic problem of an aging population was to be the re-orientation to villages of the unemployed urban population, the homeless, and the poor; their migration to villages and adoption of farming once the structural reform was completed.

At the close of the century we can identify two groups of paradoxes:

- those referring to the *village*;
- those referring to *agriculture*.

The first group includes several important paradoxes regarding *the village as a specific social organism*. Late modernity is organically related to emerging globalisation processes and the general consolidation of a new type of information society. But already evident in rural Bulgaria is the closing of the village to external society, its narrowing down in terms of social space - the smaller the village, the more defined its *isolation*. The village is increasingly becoming a niche of economic degradation and poverty. Second, there is a decline in the rural institutions which were agents of modernity. Either their activity has become minimal, or they have completely disappeared from the circle of activities provided by the village community, e.g. schools and other educational institutions, health institutions, and especially the co-operatives and cultural centres. Third, resources are being drained out of the village - people and material goods - just as they were at the start of modernisation. After a lull in the early 1990s, migration from village to town (and even abroad) has recommenced, this time young and educated people who are unable to find employment suiting their qualifications and interests, and who cannot give their children a desirable quality of education are seeking alternatives. Fourth, all the revenues from the village (communal land, forests, enterprises, taxes and charges from village enterprises) are going to the local urban municipal centre. Village resources are being extracted without profit to the village. Farmers sell their agricultural production several times cheaper than the corresponding price on the town market.

Paradoxes related to *the development of agriculture* reveal the fact that rural de-modernisation and the inability of rural producers to cope with the market situation, for which they have proved unprepared. In the context of large-scale modern farming, farmers were used to working small parallel farms which existed in partial symbiosis with large-scale agrarian production and were mutually complementary with the latter in respect to technology. These family farms absorbed unskilled and non-mechanised labour; such para-economic farms made a definite contribution to the national economy. However, the changes that occurred in the early 1990s have destroyed this symbiosis before any new integral production structures could replace it. Thus, de-collectivisation in the Bulgarian village has become equivalent to de-modernisation, and the restoration of private property rights over land was accompanied by actual destruction of the enormous amount of co-operative assets: buildings, machinery, irrigation equipment, as well as livestock and other property.

The *market economy* has become the dominant model of development for society. But what is actually emerging is a form of self-sustaining farm economy. Two thirds of farms work merely to supply their own needs, and for almost 5 per cent, agriculture is the only source of income. Only 2-3 per cent of producers connected in some way with agriculture gain a good income from it. The gravest problem for rural producers is in realising their produce; there is a lack of markets and whatever they manage to sell is sold at very low prices. Even on the domestic market, Bulgarian farmers have to compete with imported subsidised goods. Many of the co-operatives function as simple co-operatives, without reproduction, dividing whatever they produce among their members.

Agriculture has not become an attractive occupation for the urban unemployed. On the contrary, even the rural unemployed look for work in spheres other than farming. Those who are occupied on family farms define themselves as unemployed. Small rural industries are more attractive than agricultural work. The amount of arable land is relatively small (less than 0.5

hectares *per capita*), yet the proportion of land which is uncultivated has been increasing over the last few years. Agricultural producers connected with the market are little inclined to innovate and prefer to cope through whatever means and skills they already possess. Level of education and age make little difference in this respect. There are many unemployed agricultural specialists with a secondary or higher level of education, but they do not seek work in agriculture. In both private co-operative farming, the predominant survival strategies are passive ones.

Nevertheless, some outward signs of modern times can be seen in Bulgarian villages: satellite antennae, cable TV, computers, use of Internet. Despite the demographic crisis, the proportion of people with a secondary or higher level of education is growing. This includes active farm producers working for the market.

What kind of rurality could correspond to 'late modernity'?

The changes in Bulgarian rurality during the transition to a market economy pose more questions than they provide answers concerning further changes in the context of 'late modernity':

- A basic question concerns what kind of rurality could be created where the average size of the newly returned plots is nearly 1.6 hectares, i.e. too small for modern farming. In any case, the 'new' rurality presents itself in different forms: family farms, co-operatives, and large-scale farms.
- Taking into account Bulgarian rurality in the pre-war period and ongoing land restitution, we should expect some reconstitution of 'old' rurality which in the case of Bulgaria means 're-peasantisation' in a context of very different internal and external conditions. The previous very high degree of industrialisation, in particular heavy industry, and the effort to overcome this disproportion by implementing the current structural reform, has uncovered the important role of agriculture and the service sector for the country under the new economic conditions. In this context we could hardly speak about future 'de-agriculturalisation' of rural areas.
- Further rural development will inevitably involve a combination of farm and non-farm activities at the level of the rural household in response to policy or market pressures. In this context, capital movements and the growth of agri-food companies will create a new environment for modern rural households. They will have the chance to explore different household strategies - in other words to gradually determine their own strategies aiming at self-sufficiency or market relations.

- The combination of farm and non-farm activities is a typical feature of modern rurality. Family farms can progress only where they have a choice of combining farming with other kinds of activities, providing additional income for family members, even on farms with 'full-time' engagement in agriculture. That requires a new state policy for the recovery of Bulgarian agriculture. For the real progress of the family farms, manufacturing and craft services have to be developed in the rural regions with a view to organising a modern and civilised existence.

Bulgaria has a natural advantage in tourism development. Agri-tourism, also called 'holidays on a farm', is the second developed activity of farmers in Germany, Austria, Switzerland, The Netherlands. In some mountain regions like the Alps, it is the most important source of villagers' income. In Austria about 9 per cent of national income comes from tourism. According to forecasts up to 2020, tourist flows will increase threefold, and preferences for 'holidays on a farm' will also increase. Bulgaria has very favourable natural conditions for the development of both summer and winter tourism (respectively at the Black Sea and the mountains respectively). As experience to date had shown, hospitable rural people contribute a very good environment for recreation and entertainment. In the context of European integration, the only way of adjusting Bulgarian agriculture to international standards is to orient it competitively to ecologically and biologically pure farming.

- Modern family farming in Bulgaria needs machinery and funds for effective production activity. Family farm-owners already need, and in the near future will increasingly need more advice and information about modern agriculture, new technologies, machinery, marketing studies and other services, i.e. everything which is typical of modern knowledge-based technological progress.
- Adapting Bulgaria's economy, in particular its agriculture, to the more developed economies in Europe is the country's main policy objective. This will influence national rurality in 'late modernity' and will provoke a need to adapt to the CAP in rural household strategies, state measures and market conditions.

Conclusions

Socio-economic implications

During the second half of this century we have observed very radical (even polar) changes in rural consciousness in only two or three generations: from a predominantly small-holder rurality up to the end of World War II, to co-operative and state-worker rurality in the socialist period, and again to small-holder rurality in the era of transition to a market economy. While at the beginning of this period young people had dominated the population, at present the elderly do so. The socio-economic and technological environments of the three types of rurality are quite different.

Land reform and agricultural restructuring in the 1990s which have resulted in a new rurality have proved much more difficult than at first envisaged. Land restitution has been slow and costly. The model accepted tries to reproduce the old structure of 50 years ago, which does not adequately promote modern farming. Whereas the process of land restitution is almost finished, other important and difficult activities follow. These include support for vital farm units, infrastructural development (schools, roads, medical services, cultural activities) necessary to make the countryside an attractive place in which to live and work, and improving farmers' qualifications to produce agricultural products in accordance with modern standards. Continuation and consolidation of reform in the whole economy will better sustain the future development of agriculture in different rural areas. Agricultural policy recommendations which influence changing rurality should relate more closely to demographic and social, industrial and environmental aspects.

On the specific state of rural development policy, we can say that economic transition should be more concerned with non-agricultural activities. The modern development of individual rural regions should be based on the development of industry and ecological agriculture. Additional sources of income are needed for agriculture via agri-tourism, small-scale food industry, and the service sector; these could revitalise the existing structure of villages. The European Commissioner for Agriculture recommends that Bulgaria develops ecological agriculture. Also called organic, biological or alternative, this is regarded as the most radical form of transformation of the old rural models, and could be successfully developed in the country, particularly since there is insufficient funding for fertilisers and other chemical inputs. Some studies have suggested that conditions in Bulgaria are very favourable for this activity. The Bulgarian and Switzerland Foundation for Biological Agriculture reports some achievements in this sphere in the central part of the Balkan Mountains.

Policy reform should be based not only on improving competitiveness and increasing participation in the world market, but also on promotion of integrated multi-national rural policy. Many initiatives for the smooth harmonisation and convergence of Bulgarian agriculture to the European agriculture sector, anticipating accession to the EU, have been undertaken. The development policy of the reformed CAP in relation to the transition countries should take into account the diversity characterising the agricultural sector and rural areas across the EU and the associated CEECs. With moderate help from the West and an adequate national policy, in the medium term we could expect Bulgaria to show its advantages in agriculture, find its own place among the other EU member-countries and create a 'proper' rurality for 'late modernity'.

Social aspects

The Bulgarian village does not yet correspond in our view to the parameters of late modernity, since all elements in its present structure are those of a pre-modern era, and the territories of modernity seem to have been destroyed. But this is only an approximate judgement, reflecting the change in values brought about by the sharp break in the natural processes of development and by the rapid changes of perspective. In the first place the village has to return to modernity by means of a well-constructed strategy of development. So far, the state cannot be relied upon to provide such a strategy.

The ill-chosen model of development by the state has had an impact on rural development as well. It would be utopian to expect that any solution could be found outside the overall stabilisation of the country and of the entire Balkan region. But given their specific features as a type of community, villages command considerable techniques of self-preservation at the local level, and these techniques ought to be availed of. Local initiative could contribute by stimulating the development of village organisations. The mechanisms of non-governmental organisations (NGOs) should probably be used with their firm traditions and outstanding adaptability. Although co-operatives and cultural centres in villages are latent in their activity, they still have a wide potential to adapt to the challenges of social development. But a national strategy will have to give general direction and decisive support to villages. The problem is in deciding government strategy and what this rural model strategy implies for the countryside. So far, villages remain unprotected.

A question remains unanswered: can separate countries like Bulgaria and regions like Eastern Europe retain their 'otherness' in the context of late modernity? We believe that, in general, the elimination of 'otherness' would not be healthy for the planet. And this includes rural 'otherness', which must, therefore, be part of the urbanisation paradigm of late modernity.

Notes

[1] The overpopulated agricultural sector at that time was the main reason for economic migration of Bulgarian farm workers to other countries, like Hungary, Czechoslovakia and Austria, where Bulgarians had shown their experience and skills in horticulture.

[2] For the sake of comparison the average area of land owned by a farmer in Bulgaria was 2-3 times smaller than those in other Eastern European countries like Rumania, Hungary and Czechoslovakia.

[3] Characterising the collectivisation in the period 1944-1989 some ideas and data are borrowed from M. Meurs, V. Kozhucharova and R. Stoyanova (1999).

[4] Compared with the other former centrally planned economies in Central and Eastern Europe, Bulgaria was the first to begin mass co-operation of private farms.

[5] The process of property nationalisation, including urban property, was initially realised (the crucial phase) at the end of the 1940s and the beginning of the 1950s; and for the second time at the end of the 1960s and the beginning of the 1970s. Bulgaria was one of the CEECs with the highest proportion of state property, producing over 90 per cent from the national income and nearly 95 per cent from the total output during the period 1960-1990.

[6] Nevertheless, until the middle of the 1960s, agriculture - as measured in terms of its share in total employment - was the largest economic sector not only in Bulgaria but also in several of the CEECs. In 1965 Bulgaria still had close to half of its labour force in agriculture, followed by Poland with nearly 40 per cent, Hungary with more than a quarter; Czechoslovakia and East Germany each having less than one-fifth of the labour force in agriculture.

2 Food industry responses to global integration: fruit producers, state legislators and women farm workers in South Africa[1]

ANDRIENETTA KRITZINGER and JAN VORSTER

Introduction

The study of economic restructuring and the global food system has become a major theme in current international debates within the field of Rural Sociology (Arce and Marsden 1993; Marsden 1992; Ward and Almås 1997; McMichael 1997; Goodman and Watts 1994). Much of the literature on the global integration of Third World/Developing countries' food systems has emphasised the 'casualisation' of labour in agri-food export sectors as a response to global competition. Associated with this is the so-called 'feminisation thesis', that is, the increasing employment of women as temporary and seasonal workers within these sectors. In a recent publication, Barrientos *et al* (1997; 1999a: 25) provide an overview of case study research indicative of this trend. Women, for example, account for half of the rural labour force involved in fruit and vegetable production in Mexico and 70 per cent of the grape labour force in Brazil.[2] Within the Chilean fruit export sector women approximate half of the total labour force. Given the seasonal nature of these agri-industries, women constitute a reserve army of labour as they are only employed for a few months per year and enjoy little employment security. Gender relations have been shown to underlie the employment status of women, their specific skilling and the division of labour between men and women.

It has been argued that the integration of the fruit sector into world export markets and the subsequent integration of women into paid employment in the Chilean case, for example, has had contradictory consequences for women (Barrientos et al 1999a; Bee and Vogel 1997). Prior to their integration into the fruit export sector and as wives of subsistence farmers, women's farm and domestic labour were highly invisible and, for the most part, went unrecognised. While women's entrance into paid employment has empowered them, their temporary status has left them insecure and extremely vulnerable to exploitative practices. Research (see Bee and Vogel 1997; Matear 1997) has shown how the temporary nature of these women's work excludes them from the legal right to organise, affects their pension and social security payments, contracts and job security guarantees while giving employers total freedom to set wages and dismiss workers. Women also have to carry a double burden as paid employment has not resulted in these women relinquishing their traditional domestic and household responsibilities. As Wilson (1985: 1020) comments, studies have often assumed that women's entry into paid agricultural labour would 'undermine traditional links of personal dependency and subordination based on gender' and would 'at least produce the formal freedom that men are supposed to enjoy in class society'.[3]

Contrary to the Chilean experience and due to historical developments, women working on South African export fruit farms have not only been integrated into agricultural wage labour since the inception of the industry, but the majority of female (and male) wage labourers have always resided on farm property. However, these women have much in common with their counterparts in Latin America. Although the situation appears to be changing[4] research undertaken in 1994/5 (Kritzinger and Vorster 1995, 1996) found that, while women account for just over half of the total wage labour employed on fruit farms, they only represent 26 per cent of permanently employed workers. On the other hand, of the temporarily and seasonally employed labour they represent 75 and 65 per cent respectively.[5] Like women in other producer countries, those occupying temporary and seasonal positions are exempted from certain rights and employment benefits enjoyed by permanent workers. Women farm workers on South African farms, like their counterparts elsewhere, continue to carry the double burden of paid work and unpaid domestic labour. Gender relations also continue to structure a number of aspects of their paid work on farms.

However, while similarities exist between the situation of women wage labourers within the South African fruit industry and that of women agricultural labourers in many other parts of the world, global integration has also generated ways in which their situation is distinctly different. This relates to an important theme in the debate on globalisation, that is, the growing awareness of heterogeneity and differential responses to global integration at the local level, the need to theorise cultural and historical diversity and to explain how 'restructuring processes are being experienced in localities'. The 'production of difference' in agri-food systems has consequently become a major focus within the new political economy of food (Ward and Almås 1997). Recent theoretical developments emphasise the need to not only analyse agricultural production within the context of the global food system, but also to examine how globalisation takes on a fragmented and heterogeneous character within different localities (Arce and Marsden 1993). In a recent publication on New Zealand's apple industry, for example, McKenna *et al* (1998: 394) suggest that food regimes research in agricultural political economy has opened up key debates about 'how subnational and regional farming practices, social organisation and "rural lifestyles" are being restructured and realigned throughout the world'. They also emphasise the need to closely examine the links between agri-food restructuring at the global level with a 'new geography of agricultural employment'.

Within the context of this wider debate and a greater awareness of localisation, the paper examines the linkages between South African fruit producers' responses to global integration and the 'new geography of employment' on fruit farms - particularly as it pertains to the employment and position of women wage labourers. As Arce and Marsden (1993: 300) suggest, we need to examine the responses of producers as significant actors in the global food system from an 'active rather than a passive' perspective. More to the point and, following Rickson's (1997) suggestion, the paper explores how the interplay of global trends and the role of the state coupled with local, cultural and historical forces mediate labour arrangements and the employment position of women labourers on fruit farms. In doing this, the paper traces the ways in which fruit farming in South Africa is 'locally specific'. The focus on women is prompted by not only the 'feminisation' thesis, but also by an effort to keep women of colour on Rural Sociology's research agenda.

In the first section of the paper we attempt to contextualise the South African fruit industry and the changing 'geography of employment' on fruit farms. We do this by examining the pressures experienced by fruit export producers following the global integration of the fruit sector and the extent to which the South African state has retreated or intervened within the broad context of global integration. It also examines the local cultural context of paternalism within which producers' responses and existing labour arrangements are embedded. Based on research undertaken in 1994/5 and again in 1998/9, the main section of the paper attempts to trace fruit producers' responses to these pressures and cultural specificities. We specifically attempt to identify trends/shifts in the employment arrangements and position of women farm labourers over a period of five years.

Contextualising the South African export fruit industry

Race and gender in historical perspective. Within the Western Cape Province of South Africa, the deciduous fruit industry makes the largest single contribution to the value of agricultural output estimated to account for 25 per cent. While the initial development of the fruit export industry in South Africa dates back to the mid-17th century, exports to the UK and other European countries took off at the end of the 19th century (Birch 1993; Stander 1983). Following the lifting of sanctions in the 1980s, exports of fruit have grown - Europe being the main export destination - earning R2.5 billion in foreign exchange in 1996 (Capefruit 1998). In 1996 52,650,000 cartons of deciduous fruit were exported (Eurofruit Oct 1997). The industry is highly labour intensive with total employment estimated at 283,000 in 1994 and also the biggest employer of black people (mainly coloured) in the Western Cape (De Klerk 1992, 1994).[6]

The composition of the labour force employed on fruit farms is predominantly structured along racial and gender lines and based on the distinction between 'permanent' and seasonal labour. From the very beginning of agricultural enterprise in the Cape Colony, labour was provided by slaves and, to a lesser extent, the KhoiKhoi. With the opening of the British market to Cape wines, slaves were sold to farmers at high prices. Also, slave women and children were used more extensively alongside men in field labour (Armstrong and Worden 1990). During the 19th century, fruit farmers increasingly came to depend on a permanent work force as well as a large labour force during picking season. Given the fluctuating supply of labour during this period, wealthier farmers established a system of tied rent in order to secure the supply of required labour (Scully 1987; 1989).[7]

Intervention by the state in the supply of labour has also had a profound impact on the racial composition of the labour force on fruit and wine farms. The Coloured Labour Preference Policy introduced by the Nationalist Party in the 1960s, for example, stipulated that Africans could only be employed in the Western Cape when no coloured workers were available. The movement of Africans was further curtailed by influx control measures which determined that they could not seek employment in the Western Cape. The abolition of these measures did not significantly alter the composition of the labour force on farms. The core of permanently and temporarily employed farm workers has remained - at least until now - coloured, while Africans and coloureds are employed as seasonal workers. Women represent a substantial proportion of seasonal labour on farms.

Global competition, labour costs and productivity levels. The South African fruit export industry has experienced radical changes since the early 1990s. The most important emanate from political democratisation in 1994, the deregulation of the industry in 1997 and the sector's re-entry into global markets. Global integration has resulted in competition becoming much more intense. South Africa is currently occupying second place as a Southern Hemisphere exporter of deciduous fruit after Chile (Barrientos *et al* 1999b). Prior to deregulation the industry exported through Unifruco (now Capespan) - a single channel export marketing system - and a pool system was in operation whereby all growers' produce which met with minimum standards was accepted. Within a deregulated environment some producers have opted to operate as independent exporters in an effort to obtain better terms for their fruit at higher quality levels or have joined other marketing organisations.[8] Deregulation means that Capespan now negotiates individual contracts with each grower/producer to market and sell produce within a particular market sector. The ability to earn high prices will depend on the ability of the grower to supply high quality produce at the required volume that a particular market demands. As Barrientos *et al* (1999b) correctly point out, growers operate at the most labour intensive point of the supply chain. In the case of the South African fruit export sector labour represents 43 per cent of gross costs at farm level. This means that labour productivity is a crucial consideration if producers want to be efficient in global terms.

An important theme highlighted by the debate on the globalisation of the agri-food system concerns the implications of the increasing dominance of transnational corporations and the impact that globalisation is having on forms of agricultural employment. This is often seen as being similar to other sectors of export industry (Marsden 1992; Barrientos *et al* 1999a). The internationalisation of food often results in efforts by growers/producers connected to agribusiness to reduce labour costs in order to remain competitive within global markets. This not only leaves labour extremely vulnerable and open to exploitation by growers/producers, but low-paid, insecure and flexible employment is generated through the increased employment of temporary and seasonal labour. Research indicates that women often constitute a significant proportion of such temporary and seasonal labour.

Control along the supply chain: supermarkets and codes of conduct. At the retail end of the global supply chain within many developed countries to which fresh produce is exported, there is an increasing tendency by large supermarkets to increase control along the supply chain (Barrientos *et al* 1999a). As noted by Arce and Marsden (1993) supermarkets are reinforcing the importance of consumerism and choice in modern societies. This tendency has facilitated a situation 'where consumers' choices are not just directed to save money but generally to allow them to superimpose on the market their high and low standards of quality'. Thus food purchases are increasingly associated with 'quality of life, the appearance of food, and the actual environmental context in which food is purchased and consumed' (1993: 301-302).

Prompted by demands of consumers and environmentalists, companies are not only making strict demands on farmers to supply food at the right quality, appearance, in correct volume and price, but also concerned that export countries will cut labour costs by lowering their employment standards in an effort to compete globally. South African fruit producers have experienced the introduction of codes of conduct especially by UK supermarkets - Sainsbury, Tesco, Safeways, Somerfield, Waitrose - following the formation of the Ethical Trading Initiative (ETI) in 1997 (Barrientos *et al* 1999b).[9] Should farmers fail to raise productivity levels of their labour force the introduction of codes of conduct is expected to raise labour costs.

The role of state policies. The role of state policies has been shown to mediate the context for action in rural areas (Marsden 1992; Arce and Marsden 1993). In Chile, for example, state economic adjustment policies have resulted in the expulsion of peasants from the land followed by permanent workers who previously lived on the land. This resulted in the creation of a pool of proletarian labourers in search of employment in the early 1970s (Arce and Marsden 1993; Bee and Vogel 1997). Under the pressure to maximise profits and revenue, the neo-liberal state is also seen to retreat and 'turn a blind eye' to the conditions under which labour participates in the global agri-food system. Within neo-liberal economies the state does not always act in the interest of agricultural workers in securing good employment conditions and sustainable livelihoods. Commenting on the Chilean fruit export sector's experience, Barrientos *et al* (1999a: 188) suggest that adherence to the neo-liberal export model means that 'the ability of the state to regulate at the national level is inherently limited if the model itself is to be successful in a competitive global environment'. They also suggest that when proposed efforts by the state to improve labour conditions conflict with the interest of producers and exporters, the needs of the latter tend to take precedence over the interests of (temporary/casual) workers.

Contrary to developments elsewhere in producer countries, South African fruit producers have not been able to arbitrarily shed labour or follow a 'hire and fire' strategy in order to reduce labour costs. Contrary to the 'retreat' tactics of the neo-liberal state elsewhere, the fruit and other agricultural sectors have since 1993 been exposed to a radically changed legislative environment aimed at providing farm workers with greater protection and security. Of particular significance are The Basic Conditions of Employment Acts (BCEA) 1993 & 1998 which stipulate basic conditions of employment and determine length of work week, maternity leave, sick leave and overtime pay. In 1993, following negotiations between the South African Agricultural Union and the Congress of South African Trade Unions, the Act was extended to the agricultural sector. Until then agricultural labourers were excluded from labour legislation.[10]

While farm labourers were excluded from taking strike action in the past, the Labour Relations Act (LRA) 1995 extended the right to strike action and protection against unfair dismissals to farm workers. However, unions have organisational rights only if they are 'sufficiently representative' and participation in decision-making through workplace forums is limited to farms with more than 100 members. Of special significance for producers is The Extension of Security of Tenure Act (ESTA) 1997. Until 1997 farm workers had no legal protection against arbitrary evictions from farms. Introduced by the Department of Land Affairs, the Act now provides farm workers with some measure of tenure security by preventing unlawful evictions. Farmers can obtain an eviction order only under very strict conditions and even then, workers can contest the order in the labour court.[11] The Employment Equity Bill (EEB) 1998 provides for the implementation of affirmative action programmes for particular categories of employers, for example, women and the disabled.

It is tempting to ascribe these changes in legislation solely to efforts by the state to secure good employment conditions for agricultural workers in the wake of global integration and consumer pressure. However, these legislative measures must also be interpreted in terms of a human rights culture establishing itself during and after political transformation in 1994. This has taken place within a historical context characterised by exploitative practices on the part of many white farmers in the treatment of their black wage labour force. However, contrary to the apartheid state which regulated and protected the agricultural sector, the post-apartheid state has been 'in retreat' in at least one sense. Government agricultural policy assumes that the path to the most effective utilisation of production factors on commercial farms is to allow the exposure of commercial farming to increased levels of competition. Exposure to international competition could easily encourage cost-cutting efforts and employment-shedding strategies. While legislation (and trade unions) can make particular processes of retrenchment or dismissals less arbitrary in the short term, the destruction of jobs cannot be prevented.

Pressure to increase wages and improve employment conditions has also been put on producers by unions. Even prior to legislative changes and political democratisation, unions started to organise farm workers in the Western Cape.[12] However, union growth within the agricultural sector has been relatively slow up to now. It was estimated that by 1995 only between 2 and 8 per cent of farm workers were active members of unions (Murphy 1995).[13] Producers are highly suspicious of unions which leaves union organisers little opportunity to access workers. Workers on the other hand, distrust unions and are also exceptionally uninformed about unions (Kritzinger and Vorster 1995). However, there are indications that, with the support of NGOs and bolstered by state legislation, unions are having more success in organising farm labourers.

Paternalism and farm worker communities. Whilst producers have had to calculate their response within a context of tighter integration into the supply chain, state legislation, union pressure and high interest rates, their responses cannot be disembedded from the local cultural context of paternalism. Paternalism has governed relations between growers/producers and wage labour since the 19th century.[14] The close-knit character of South African farm communities finds expression in the ideological construct of the 'farm as family', the origin of which can partly be traced to slavery - more specifically the management of slaves - and the Masters and Servants Act of 1856 following the abolition of slavery in 1836.[15] Shell (1994: 213) argues that an important factor determining the style of slave management/control was the size of the farm holding. Given that most farm holdings were small, 'family management was a much more subtle and more appropriate means of control than physical punishment would have been'. For both the Dutch and British, the concept 'family' (*familie*) included slaves and servants in the immediate household. For slaves who have been uprooted, this household provided their only 'home' while the slave owner insisted that the slave was part of the 'family'. The later incorporation of the free KhoiKhoi into family management was a natural extension of the slave family model (Shell 1994: 215-216 and 221; Kritzinger & Vorster 1997). Traditional paternalism is characterised by its 'organic' view of labour relations and the absolute power of the farmer.

On most fruit farms at least traditional paternalism has since the beginning of the 1980s been increasingly transformed into so-called neo-paternalism.[16]

This has not only reduced the unilateral power and authority of the farmer, but has motivated workers to see themselves as 'partners in the farm business enterprise rather than "children" dependent on the master's goodwill' (Du Toit 1992: I). This is manifested in more formalised wage and employment relationships with the emphasis on human resource management through productivity measurement, training, procedures and work rates, performance appraisal, motivation and greater worker participation through farm committee structures (Mayson 1990). Global and legislative pressures have provided greater momentum to this process and an ideology of reciprocal obligations characteristic of more paternalistic arrangements between farmers and workers has increasingly been substituted by one of reciprocal rights. However, it is also true that vestiges of paternalistic practices remain intact on fruit farms. This is partly due to the inherently diverse nature of 'new managerialism' itself, for the paradigms that inform management in the deciduous fruit industry are often contradictory. 'New managerialism' borrows from Taylorism with a focus on cost-reduction efforts and piece-rate systems as well as from a human resources tradition with an emphasis on participation and motivation.[17]

Although there exists little disagreement amongst South African scholars and researchers as to the origin and features of traditional paternalistic wage labour arrangements, there seems to be less agreement on the extent to which neo-paternalism or a 'new managerialism' has come to manifest in different agricultural sectors. This can be ascribed to the internally heterogenous nature of this approach already noted, the erratic way it has been applied between, as well as within, particular sectors, the extent to which researchers apply the industrial development model to agribusiness development, and ultimately, the ideological position of the researcher. Within this context the development of clearly specified criteria to be applied when assessing transformation is highly problematic.

Nevertheless, due to limited opportunities to mechanise[18] and the pressure to provide high quality produce complying with specifications of freshness, colour, size and shape to niche markets and retailers in time, fruit producers appear to have been more willing to invest in their labour force. Our previous study (Kritzinger and Vorster 1996) suggests that deciduous fruit producers (especially on larger farming enterprises) have made substantial financial and social investments in the training and 'social development' of their wage labour force over the last fifteen years. Of significance is that many of these community projects, activities and programmes originally developed and implemented by the Rural Foundation[19] were directed at and involved women in particular. Projects included crèches, after school care, literacy classes, women's clubs, old age clubs, health programmes, bible study classes, choirs, health and life skills programmes. Previous research, for example, also indicated that on 73 per cent of fruit farms some of the above projects were being implemented.

Research method

In a study on the labour situation on South African deciduous fruit export farms in 1994/5 a survey of 105 producers in nine fruit growing regions in the Western Cape Province and one region in Northern Cape Province was included in a probability sample. Due to time and budget constraints it was decided in the 1998/9 follow-up study to interview only fifty producers (50 per cent) from the previous sample – all farming in the Western Cape. This study forms part of a larger research project initiated by the Centre for Rural Legal Studies. The 1994/5 sample consisted only of members of Unifruco (the sole exporter of deciduous fruit at that time). However, as pointed out above, some producers have since left Unifruco (now Capespan) and are presently either exporting fruit themselves or have joined new marketing organisations. Capespan sponsored this part of the study and requested that producers who are still members of the organisation be considered for inclusion in the 1998/9 sample.[20] The exclusion of producers who no longer market through this organisation can be considered a limitation as one does not know whether this group significantly differs from the Capespan exporters, in terms of labour practices for example. Future research will hopefully shed some light on these producers' responses to global markets, consumer pressure and state legislation as well as on their employment arrangements. Face-to-face interviews were conducted with the fifty producers from November 1998 to January 1999 and one experienced fieldworker conducted the majority of these interviews.[21]

Fruit producers and state legislators: implications for 'on-farm' women workers?

The majority of women farm labourers on fruit farms are coloured, between the ages of twenty and forty, married (or cohabiting), Afrikaans-speaking and have been in employment for five years and longer. While 47 per cent of women have seven years of schooling, 21 per cent have four years and 10 per cent no schooling at all. A fifth of women claim illiteracy. Family-related and financial considerations account for these women dropping out of school or having had no schooling at all (Kritzinger and Vorster 1996: 343). Women on average enter farm employment at 18 years[22] and, within the fruit industry at least, exhibit relatively low rates of mobility. They tend to remain for long periods on one farm.

An analysis of the position of women living on South African fruit farms (see Kritzinger and Vorster 1996; 1998) showed that, similar to women labourers in Chile, gender relations structure women's paid farm work. This happens not only through recruitment practices and employment status, but also for reasons of division of labour, wages, opportunities for training and promotion as well as the occupation of supervisory and leadership positions.[23] Women farm workers have had little independent claim as workers to either employment or access to housing on farms. A woman's employment and access to housing is largely conditioned either by being a member of a farm worker family or by her relationship to a male worker - as either a spouse, partner or child.[24] As a general rule, a woman has an obligation to work on the farm, is usually not allowed to take up employment elsewhere - even on a temporary basis - and is expected to vacate her house should her husband's/partner's/father's employment be terminated.[25]

On South African fruit farms the division of labour means that, while women and men do comparable work for the most part, physically demanding tasks remain the exclusive domain of men. Women are especially allocated to tasks which require 'nimble fingers' or manual dexterity in order to 'minimise damage to fruit' (Bee and Vogel 1997: 86). They are also allocated to tasks requiring the ability to 'work at a fast pace' like picking, selecting and packing etc. and which they are able to sustain for longer periods than men. The secondary status of women is reflected in the income disparity between men and women. On the whole women earn less than men do[26] and men dominate the higher paying jobs of irrigation, tractor driving and supervision. Wage differentials between men and women also result from training practices in that men are much more likely to receive formal training than women are. Women's work is also less likely to be submitted to appraisal and formal evaluation than men's work and women are less likely to be members of worker committees. In addition, while women are utilised as team leaders, they are used less often than men as supervisors.

In the following section we attempt to trace the most significant shifts in the employment arrangements and position of women labourers on fruit farms over a period of five years. The discussion is structured according to the following themes: structuring of the labour market; employment contracts and women's off-farm employment; job-segregation and training; conditions of employment; and fringe benefits and facilities.

Labour market: race, gender and employment status. In the 1994/5 study it was found that although numbers vary in terms of size of farms, on average 60 per cent of workers live on farms on a permanent basis. The remaining workers are predominantly seasonal workers who live in nearby towns. Of those workers living on farms, 61 per cent enjoy permanent status while 39 per cent have temporary status. As noted earlier, of the total number of workers per farming enterprise in 1994/5, more than half are women employed on a temporary and seasonal basis. Men, on the other hand, account for approximately three-quarters of permanently employed farm workers (Kritzinger and Vorster 1995: 6). In our more recent survey we found that the average percentage of permanent workers of the total labour force has increased by almost ten per cent over the past five years (51 compared with 42 per cent).[27] This increase, however, was not experienced on the majority of farms.[28]

The increase in the percentage of permanent workers is also reflected in a decrease in the percentage of seasonal workers employed. During 1994/5 the median percentage for the latter group was 46 per cent of the total labour force compared with 36 per cent in 1998/9. The average percentage of seasonal and temporary workers in the total workforce also decreased slightly.[29] What is significant is that in the majority of cases the increase of the percentage permanent workers reflects a change in the employment status of women workers, that is, from temporary to permanent status. Although there has been no statistically significant change regarding women workers as a percentage of the total labour force between the two surveys, the average percentage of women occupying permanent positions has increased by 17 per cent.[30] Therefore, the increase in the percentage of permanent workers does not necessarily represent an increase in the size of the labour force living on the farm. Producers indicate that, should the need for more labour arise in future, they will recruit off-farm seasonal and/or contract labour. Our recent study also indicates an increase of farms employing African workers over the past five years. This can be ascribed to an increase in the percentage of African workers as part of the seasonal and/or contract labour component.[31]

The vast majority of producers (almost 70 per cent) foresee no change in the employment of permanent workers as a component of the total labour force, while only 12 per cent envisage an increase and a fifth foresee a decrease. Given that a third of producers indicated an increase in 1994/5, it would appear that on the majority of farms the component of permanent workers has reached saturation point. This is partly related to the permanent status that women have acquired over recent years. Most of the producers who envisage an increase currently employ a relatively small proportion of women holding permanent status. Thus, an increase in the component permanent workers will most likely involve women with temporary status already living on farms.

Employment of seasonal labour is increasingly becoming an option for producers as more than 40 per cent report that their seasonal component will increase over the next five years. While only 4 per cent of producers will not enlarge their seasonal component, the corresponding percentage was 33 per cent in 1994/5. This trend can partly be ascribed to two considerations. Firstly, in the event of an expansion of production and the corresponding need for more labour, producers are reluctant to build more houses on their farms. Providing new houses means escalating labour costs. In accordance with trends elsewhere, producers view the employment of seasonal and/or contract labour as a means to cut labour costs. Housing, for example, is not provided for these categories of workers on a permanent basis. As a rule they are also exempted from certain conditions of employment and benefits. Secondly, new legislation on land tenure and property rights (ESTA) poses a threat to producers' property rights. Some producers are even unwilling to fill existing vacant houses.

Producers are reluctant to voice preferences regarding the gender composition of their future seasonal and/or contract labour component. If past as well as existing employment arrangements are anything to go by, it will be predominantly women who will be employed as seasonal workers. On the other hand, the high unemployment levels currently existing in South Africa - estimated at 37 per cent - could result in another scenario manifesting itself in future. Men farm workers, by virtue of being the 'breadwinner' and 'head' of the family household, could lay claim to seasonal jobs in future. Thus existing gender relations could define some women out of the seasonal labour market.

Employment contracts and women's off-farm employment. There has been a significant increase in the percentage of producers who have signed written employment contracts with the various categories of workers over the last five years. As can be seen from Table 2.1, the increase is especially significant in the case of permanently employed workers. However, for all categories of employment, a lower percentage of women than men have secured written contracts with farmers.

Although the percentages have dropped, a substantial percentage of producers still include women in their husband's/partner's employment contracts and fail to grant women interviews when appointing them. Women are seen as being 'included' in the interviews conducted with their husbands/partners. Despite the fact that married women often enjoy non-permanent status and have no employment contract, the majority of producers only allow women to take on other work during those times of the year when the particular farming enterprise is not in need of her labour.[32] Similar to our findings of 1994/5 half of producers will appoint a woman applicant only if she is married. The above findings suggest that, although women are gaining in employment status and in being recognised as 'independent' workers, a substantial percentage of producers still perceive women as somehow 'tied' to a male worker.

Table 2.1: Type of employment contract

	Verbal %	Written 1998/9 %	Written 1994/5 %
Type of contract with:			
Permanent men	22.4	77.6	49.0
Temporary/casual men	62.5	37.5	25.0
Seasonal men	67.7	32.3	22.0
Independent contract men	50.0	50.0	X
Permanent women	35.7	64.3	48.0
Temporary/casual women	75.0	25.0	17.0
Seasonal women	72.7	27.3	19.0
Independent contract women	100.0	0.0	X
	N=50		N=101

Job-segregation, training and programmes. In our previous study we found that producers had started to appoint women in jobs traditionally demarcated for men, for example, tractor drivers, orchard monitors and supervisors. In anticipation of the possible extension of labour legislation to agricultural workers at that stage, a substantial number of producers (two-thirds) envisaged greater utilisation of on-farm women in future - especially in jobs like pruning, general orchard work and driving. This has been confirmed by our most recent research results. Half of the producers indicate that women have in fact been performing tasks previously performed only by men over the past five years. This trend is also expected to continue over the next five years as a sizeable percentage (more than a third) of producers indicate that in future women will be utilised for traditionally male defined jobs. The greater utilisation of women will in all likelihood contribute to an even greater proportion of women being employed on a permanent basis. However, given the significant drop in percentage of producers envisaging greater utilisation of women for previously male defined tasks, it would appear that a saturation point has been reached here as well.[33]

Table 2.2: Conditions of employment[a]

	Yes 1998/9	No 1998/9	No 1994/5
Women workers receive:	%	%	%
Paid holidays	80.0	*20.0*	*48.3*
Paid sick leave	74.0	*26.0*	*48.3*
Paid maternity leave	20.4	*79.6*	*92.2*
Paid occasional leave	50.0	50.0	49.4
Work pension scheme	52.0	*48.0*	*75.9*
Overtime pay	78.0	8.0	X[b]
Paid public holidays	80.0	20.0	X
UIF	73.5	26.5	X
Medical aid	8.2	91.8	X
Long service bonus	12.2	87.8	78.1
Production bonus	26.0	*74.0*	*28.7*
Finance for major purchases	38.8	*61.2*	*35.6*
Loans	84.0	*16.0*	*31.0*
Burial scheme	50.0	*50.0*	*67.8*
Work clothes	92.0	*8.0*	*21.7*
Study bursaries	14.3	85.7	85.1
	N=50		N=87

[a] Not all conditions stipulated by legislation.

[b] Items marked with X were not covered in the 1994/5 survey.

As far as training of workers is concerned, there has been a major shift. Compared with five years ago when almost all producers provided industry-based training courses, approximately two-thirds provide training at this point. Most of the in-service training is now done by the producer/manager himself rather than by outside agents. There has also been a significant drop in producers offering leadership training. On some farms formal training courses have not been presented in the last two to five years. Producers attribute reduction in training to cost-cutting efforts and, as one producer explains, 'In an economic crisis courses are the first to go' and 'we (the producers) have to show them (workers) how the job needs to be done'.

Conditions of employment, fringe benefits and facilities. In an attempt to establish whether women have made any gains over the past five years regarding formal employment rights, producers were asked to indicate which of the conditions listed in Table 2.2 they were complying with. It would appear that there has been a major improvement in the position of women workers regarding basic conditions of employment since 1994/5.

As can be seen from Table 2.2, more farmers are making provision for paid holiday leave, paid sick leave, paid maternity leave and a work pension scheme. These improvements are primarily linked to the changing employment status of women in that more women have gained permanent status over the last five years. As 'permanent' workers, women are now entitled to benefits stipulated by the Basic Conditions of Employment Act. There has also been an increase in producers willing to secure financial loans to women, contribute towards a burial scheme and provide women with work clothes. Despite the fact that women workers have made some gains and have secured formal employment rights, it is noteworthy that a substantial percentage of fruit producers are still not complying with conditions stipulated by legislation. Paid maternity leave and contributions to a work pension scheme are indicative of this neglect. In the case of the former 80 per cent of producers are unwilling to comply with the law.

Our findings also demonstrate that the position of women has deteriorated in other respects. There has been a substantial decrease in the percentage of producers who offer women workers production bonuses, bonuses for length of service and financial schemes for major household purchases. These benefits are not covered by legislation. They are based on more personal and informal employment arrangements between producers and farm labourers and which have been institutionalised over time. Financial schemes for household purposes, for example, enable women farm labourers to make more expensive purchases (usually household), for example, stoves, furniture etc. While farmers buy goods on a cash basis, the scheme allows women to make monthly payments to farmers without being charged interest. While almost two-thirds of producers provided such a scheme five years ago, the percentage has now dropped to approximately 40 per cent. One can safely assume that the withdrawal of these more traditionally institutionalised benefits relates to producers' efforts to scale down their labour costs.

Findings pertaining to the availability of fringe benefits and other facilities on fruit farms are presented in Table 2.3 in terms of categories of employment. Provision was made for the following categories: all women irrespective of employment status,[34] permanently and temporarily employed women (of whom the vast majority live on farms); seasonally employed women only (of whom the vast majority live elsewhere) and women with permanent/temporary as well as seasonal status. Percentages presented in the last two categories provide insight into changes over the past five years.

Table 2.3: Fringe benefits and facilities available to women workers

To whom is . . . available?	All Women %	Only Seasonal Women %	Perm /Temp Women %	Perm / Temp & Seasonal Women %	Other %	Not available 1998/9 %	Not available 1994/5 %
Farm appointed daycare	30.0	2	24.0	18.0	0.0	*26.0*	*42.6*
Transport to church	16.0	2	26.0	10.0	2.0	*44.0*	*24.5*
Subsidy on medical fees	16.0	2	20.0	10.0	0.0	*52.0*	*28.7*
Cooked lunch in season	0.0	4	2.0	0.0	0.0	94.0	89.4
Farm stores	24.0	0	10.0	14.0	2.0	50.0	40.4
Old age home	2.0	0	0.0	0.0	0.0	98.0	97.0
Women's club	4.0	0	6.0	2.0	0.0	88.0	77.2
Literacy classes	4.0	0	2.0	2.0	0.0	*92.0*	*76.0*
Library	2.0	0	2.0	4.0	0.0	92.0	84.9
Training in labour law	6.0	0	6.0	10.0	0.0	*78.0*	*67.4*
						N=50	N=98

Our findings indicate a major improvement in the availability of day care for children on farms. Whereas 57 per cent of producers provided for day care five years ago, this percentage has now increased to 74 per cent. Given that women are increasingly employed in a permanent capacity and utilised for tasks previously allocated to men, the provision of day care facilities by producers relates to the efficient utilisation of women as paid workers. Our previous research, in fact, highlighted women's concerns regarding the safety of their children when not adequately attended to during a normal working day (Kritzinger and Vorster 1996).

Improved day care facilities appear to be somewhat of an exception, however, as our findings indicate a reduction in the availability of all the remaining fringe benefits and facilities listed in the table. The trend is especially notable with regard to subsidised transport of women workers to church, subsidy on medical expenses of workers, the availability of literacy classes and training of women in labour legislation. Previous research established the central position occupied by religion and the church in the lives of women farm workers (Kritzinger and Vorster 1995). Approximately 95 per cent of women on fruit farms view themselves as being religious and belong to a church - with more than 40 per cent belonging to the Uniting Reformed Church.[35] The reduction in subsidised medical expenditures relates to the fact the state now provides free medical care for mothers and their children under the age of six years. In light of a reduction in industry-based training and implementation of programmes previously referred to, producers' unwillingness to facilitate workers' knowledge and understanding of labour legislation comes as no surprise.

Although less dramatic, there has also been a decrease in the provision of the farm stores,[36] library facilities and the availability of women's clubs on farms. Although no longer a key player, women's clubs were initiated in 1980s by the Rural Foundation – a major agent for implementing life-skills programmes as well as other community development projects as part of the new managerial approach associated with the liberalisation of a paternalistic labour regime on farms.[37]

Are on-farm women gaining and/or losing?

In view of the findings presented in the previous section, any attempt to provide a straightforward answer to the question as to whether the position of women farm labourers has 'improved' or 'weakened' over the last five years is highly problematic. It would appear that the implications of fruit producers' responses for on-farm women are indeed contradictory. On-farm women have made important gains in that women who have had temporary status in the past are increasingly gaining permanent employment status. These gains are closely linked to a changed legislative environment that is embedded in the new South African Constitution as well as international treaties.

Women have been increasingly utilised for tasks previously monopolised by men. Apart from the more efficient utilisation of women, more producers are providing on-farm day care facilities for the children of these women. Because women's formal employment situation has substantially improved, producers also appear to be more willing to secure financial loans to women. Legislation has also improved women's position in the short-term as far as tenure rights are concerned. Producers are no longer allowed to arbitrarily retrench workers and evict them from farms. Considerations of improved rationalisation and efficiency appear to underlie these efforts and relate to producers' reluctance to enlarge the permanent component of their total labour force in future. As noted, some farmers are even envisaging a reduction of this labour component.

Although women have gained, and are in the process of gaining, formal employment rights - even if at a lower rate compared with men - there has been virtually no improvement in their position as 'independent' workers allowing them independent claims to employment and housing. Tied employment and tied housing remain institutionalised. A substantial percentage of producers only appoint women if they are married, do not grant interviews when appointing women, include women in their husbands'/partners' written contract of employment and do not allow women do undertake off-farm employment at certain times of the year. Thus, for many producers the family and not the individual remains the unit of employment. This is also evident given reluctance on the part of many producers to employ single women – especially in light of new tenure legislation.

On-farm women are clearly losing out on certain employment benefits and facilities that were available to them by virtue of more paternalistic labour arrangements. End of the year bonuses based on production and farm net income are the most crucial of these benefits. More producers are also doing away with bonuses for length of service. While the reduction in work-related training and programmes affects all workers, many of the programmes and facilities that were targeted at women specifically have been terminated. Thus, some of the most dramatic shifts which seemingly are taking place are those which have special significance for women as *women*. These include financial assistance for major household purchases, transport to attend church services, women's clubs, literacy classes, library facilities, on-farm stores and subsidised medical expenditures.

There is, of course, some irony in the fact that producers are withdrawing benefits and facilities from on-farm women labourers which were associated with 'paternalistic' labour arrangements and practices – either in its more traditional or in its 'modernised' version. As noted earlier, producers have in the past viewed most of these benefits and facilities as part of their 'social responsibility' to 'develop' their on-farm labour force. More and more farmers are voicing an unwillingness to take responsibility for the personal and social 'problems' of their labour force.[38] Linked to this shift and following tenure legislation is the present tendency of more producers to consider off-farm housing for their permanently employed labour force and to make a financial contribution to such housing.

While the position of women living on farms has been the main focus of our research, our findings suggest that growing numbers of producers are considering off-farm housing for their permanently employed workers, reducing their permanent labour component and enlarging their seasonal component. While the shedding of labour has become more difficult, it would appear that in some cases the increased level of formal protection has hastened the process by which particular producers have reduced the numbers of permanently employed workers.

As noted earlier, women are expected to represent a significant proportion if not the majority of seasonal workers. Experiences in other fruit producing countries as well as employment patterns on South African fruit farms, all point to this scenario. To the extent that this scenario plays itself out, women will be in an extremely vulnerable position and open to exploitation. Although existing legislation (e.g. BCEA) covers seasonal workers, it is already evident that it will be difficult to enforce and to monitor compliance with legislation. That some producers are still not complying with legislation in their employment arrangements with permanently employed labourers is indicative of this point. As noted elsewhere (Kritzinger and Vorster 1998), women living in rural areas and only able to work for three months per year as seasonal labour are not always reached by organisations which aim to empower rural women.[39] Within this context it will be crucial to develop the necessary resources to safeguard these women against exploitative employment practices and other forms of marginalisation. However, while women are expected to comprise the majority of seasonal labour in future, high unemployment levels could result in men rather than women claiming seasonal work.

Conclusion

The discussion attempted to address an important theme in the debate on food industry responses to global integration, that is, the awareness of heterogeneity and differential responses at the local level and how agri-food restructuring links with a 'new geography of agricultural employment'. The main purpose of the paper was to explore how the interplay between global trends and the role of state legislators coupled with local, cultural and historical forces mediates labour arrangements on South African fruit farms – specifically the employment position of women labouring and living on farms. It has attempted to trace fruit producers' responses to a range of social, political and economic forces, and to identify what these have come to mean for women. This was no easy task as the implications were shown to be highly contradictory. In fact, whether or not the changes and apparent shifts in the position of women farm workers that we have identified are indicative of long-term trends, remains to be seen. In the short-term political democratisation and the non-standard role of the state within a context of an increasingly liberalised economic order have undeniably secured important formal rights for women on fruit farms. However, in the long term this role could be eroded by the dictates of global market forces and competition.

Notes

[1] We would like to thank Stephanie Barrientos and Andries du Toit for valuable comments on an earlier draft of the paper.

[2] In Ecuador the participation rate of women is estimated as approximately 70 per cent while research points to a 50 per cent rate in Guatemala and Costa Rica.

[3] In a recent publication, Matear (1997: 103-104) discusses attempts by the state to provide *temporeras* with childcare facilities to enable them to enter the labour force. Given that this provision has not been converted into national policy thereby reacting to women's gender-specific needs, she argues that the initiative's primary aim is to assist private enterprise to maximise profit.

[4] Our most recent study conducted in 1998/9 indicates a substantial increase in the percentage of producers who foresee changes in their housing arrangements for workers over the next five years. In 1994/5 major plans involved the upgrading of already existing structures and building of new houses. At this stage the majority of producers have no plans to build new houses. In fact, their response to rising labour costs and greater tenure security for farm workers guaranteed by legislation has resulted in producers seriously considering demolishing existing structures in the event of houses becoming vacant.

[5] Permanent workers are in the service of the farming enterprise on a full-time basis. Temporary workers do not work for the farm for the whole year, but work only seasonally (mainly women workers who live on the farm). Workers are sometimes incorrectly treated by producers as temporary in terms of employment benefits as determined by the Basic Conditions of Employment Act. Seasonal workers are workers who are used only for seasonal tasks - mainly teams from nearby towns and, in a few cases, women who live on the farm. Contract labour is usually utilised for a specific task and most contract workers are housed on the farm for the duration of the completion of the task.

[6] The apartheid governments of South Africa classified the population into four so-called racial groups: white, African, coloured and Indian. In a study like this it is necessary to use this classification as apartheid policies affected these groups in different ways.

[7] The house/plot was let at a nominal sum in return for the assurance that the lessee would work during the picking season.

[8] While some growers supply sole exporters others choose to supply a number of exporters on less preferential terms, or receive lower payments if they fail to supply highest quality.

[9] NGOs and consumer pressure have been key reasons why companies initially adopted ethical trade. The initiative is now examining ways to monitor and verify compliance with codes of conduct in different localities.

[10] Under the BCEA seasonal workers employed for 4 months or less were not entitled to sick leave or annual leave. The BCEA of 1998 provides greater protection to seasonal and temporary workers. The Act also reduces working hours to 45 hours per week and increases sick leave from 30 days over 3 years to 6 weeks. Holiday leave has also increased to 3 weeks per year.

[11] Workers can also obtain lifelong tenure. After working and living on the farm for more than 10 years and/or reaching retirement age, the farm labourer is allowed to remain in his/her house. Potentially, this legislation has far-reaching consequences for farmers' housing and labour market policies.

[12] There are several unions operating in the agricultural sector of which the Food and Allied Workers Union (FAWU) and the South African Plantation and Allied Workers Union (SAPAAWU) are particularly active.

[13] In 1994/95, for example, we found that unions were active on only 2 per cent of deciduous fruit farms (Kritzinger and Vorster 1996), while our recent survey indicates the figure to be 14 per cent. Despite union activity and a greater percentage of women farm workers having knowledge of and positive sentiments towards unionism, membership remains low. Membership of women has increased from 0.8 per cent in 1994/95 to 7 per cent in 1998/99. There has also been an increase from 66 to 83 per cent of producers indicating that they would accept it if unions were to organise on their farms, albeit under certain specified conditions.

[14] The historical origin, characteristics and transformation thereof have been discussed in various publications (see Shell 1994; Van Onselen 1992; Du Toit 1992, 1993; Mayson 1990; Kritzinger and Vorster 1997; Ewert and Hamman 1999).

[15] This Act rendered every command from an employer lawful and punished disobedience with forced labour, fines and forfeiture of wages (Rycroft and Jordaan 1992, p. 28). Thus the abolition of slavery in 1836 did not substantially improve the material position of agricultural workers.

[16] This transformation has been ascribed to a range of considerations which include the following: sanctions instituted by South African export markets prior to democratisation and the need to improve the image of the industry, producers' anticipation of the extension of labour legislation to agriculture, the role of community development organisations - especially the Rural Foundation - as well as efforts to improve productivity levels of farm workers.

[17] Du Toit (1993), for example, has argued that farmers (and managers) themselves have exposed the contradictory nature of their 'new' approach. While farmers were prepared to reform labour and employment practices, their efforts had important unintended consequences in that workers' expectations were raised. However, farmers (and managers) were not prepared to relinquish power and authority over workers.

[18] Mechanisation is limited to the mechanical processing of fruit according to colour and size. However, labour is still required, for example, to discard fruit with blemishes and russet by hand. There is also a limited degree of chemical thinning and mechanical pruning possible.

[19] Established in 1982 and with backing from government, overseas donors, community organisations and farmers, the Rural Foundation initiated the 'development' of farm workers and farm worker communities. Its primary focus was on issues of health and by the early 1990s, the Foundation realised many of its original objectives; for example, a reduction in teenage pregnancies, infant mortality as well as fertility rates on farms. Other important objectives included convincing farmers to establish consultative mechanisms (committee structures) to facilitate workers' involvement in decision-making processes as well as performance appraisal systems (Groenewald 1993).

[20] Because of refusals, 8 producers (16 per cent) who did not participate in the first study were randomly selected per production region to be included in this sample. Given that this

percentage is low and given the homogeneous composition of the population, this is not expected to have a significant effect on the representiveness of the sample.

[21] In collaboration with the CRLS a questionnaire was constructed which included a range of questions used in the 1994/5 study. Some items were modified to accommodate legislative changes since 1994. These modifications as well as the fact that certain original items were excluded complicate comparisons and in some instances even make comparisons impossible. There is, however, sufficient information available to identify certain trends.

[22] While 16 per cent of women started farm work at 16 years of age and younger, only 0.6 per cent were 16 years of age in 1994/5.

[23] Very little research has been done on women farm workers who perform seasonal work only. In terms of our definition of seasonal workers, most of these women do not reside on farms but live elsewhere which makes it difficult to locate/access them. Due to the fact that research is usually conducted during off-season periods women doing seasonal work only are not accessible via farm owners and managers.

[24] Under the *hacienda* system which persisted in Chile until the mid-1960s peasants were tied to large estates as *inquilinos*, permanent resident workers. In exchange for this residence women and family members were obliged to work for the estate, mainly on a seasonal basis. Thus, women's role as workers was seen as dependent on their husbands' or fathers' position as *inquilino*. This situation is quite similar to the system of tied housing operating on South African farms in that the majority of women living on farms are obliged to work (although paid) for the farm in exchange for right of tenure.

[25] On the issue of tied employment and housing also see Hill-Lanz 1994; Hofmeyr 1995; Hill-Lanz and Mitchell 1997.

[26] The wage differential between women and men relates to the fact that men usually work more hours per day and more weeks per year. Due to women's unpaid domestic labour women normally start work half an hour later and finish half an hour earlier than men.

[27] Sixty per cent of producers had 44 per cent and fewer permanent workers during 1994/5 compared with nearly 60 per cent permanent workers during 1998/9.

[28] Twenty per cent of producers reported that there was in fact a reduction in permanent workers as a component of the labour force, while in 60 per cent of cases the component of permanent workers remained constant. Thus only 18 per cent of producers reported an increase.

[29] This decrease does not necessarily pertain to the majority of farms, as only 6 per cent of farms reported a decrease.

[30] Thirty-four compared with seventeen per cent in 1994/5.

[31] The increase has been from 28 to 40 per cent.

[32] More than a third (36 per cent) of producers do not allow women to take on other employment irrespective of any other considerations.

[33] The percentage has dropped from 66 to 36.

[34] The first category (all women) also includes women who are employed on a contract basis. They are utilised for specific tasks for shorter periods of time.

[35] Attendance at church services in most cases involves the whole family.

[36] The farm store is run by the owner family (usually the farm wife or the farmer's mother/mother-in-law) and stocks basic food items that are sold to farm workers.

[37] The Cape Women's Forum (CWF) and its 'upliftment programme' in now the key player. This organisation is an industry-based initiative and part of the sector's social investment drive. It offers a Life and Business Skills Programme, Health Extension Programme as well as a Home and Family Management course. The latter two are exclusively aimed at women (Kritzinger and Vorster 1998). Membership of the CWF is not restricted to fruit producers as farms in other agricultural sectors are increasingly buying into existing programmes.

[38] There is also evidence which points to a substantial proportion of women farm workers preferring to live in a town or village rather than on the farm. Home ownership has always been a major consideration for women - some have identified home ownership to be their ultimate dream. State subsidy for acquiring land and/or housing means home ownership for women has now become a realistic option.

[39] Both the Cape Women's Forum and the Women on Farms Project (NGO) are involved in the empowerment of women workers living on farms.

3 Institutionalisation and resistance: organic agriculture in Australia and New Zealand

KRISTEN LYONS and GEOFFREY LAWRENCE

Introduction

The expansion of the organics industry in Australia and New Zealand is premised upon the continued institutionalisation of what was once considered to be a marginal 'unscientific' approach to farming. What is emerging is the integration of organic practices within conventional food systems - something that many commentators would not have predicted, but which appears to fit well with theories of 'greening'. With growth in the number and influence of organic certification (and other regulatory) bodies, and the evolution of profitable international markets, commercial growers as well as transnational companies have entered the industry. As a consequence of seemingly entrenched philosophical and other differences there has been a growing rift in Australia and New Zealand between the 'new' commercially-focused organic producers, and the more orthodox organic producers whose practices continue to be based upon a rejection of scientific agriculture. Many within the latter group have devised sites of resistance which form the basis for ongoing contestation between their own attitudes and practices, and those of the 'new' growers.

This chapter reports findings from the first large-scale comparative study of the organics industry in Australia and New Zealand. It examines the ways organics is becoming increasingly institutionalised, and explores the motivations of producers in 'going organic'. It concludes by suggesting that the future shape of the industry is likely to remain bifurcated as the two main groups struggle to assert their own definitions of organic.

'Greening' and the place of organics

The developing interest by sociologists in organic farming parallels what has been termed the 'greening' of agriculture and the food industry. 'Greening' describes the process of change in the ideologies and practices of (largely) western social systems as they move toward the incorporation of ecological discourses, and of practices which seek to address environmental concerns (see Burch *et al* 1999). For writers such as Thomashow (1995: 139) a growing environmental consciousness is linked firmly to the development of the 'ecological citizen' as someone who

> takes responsibility for the place where he or she lives, understands the importance of making collective decisions regarding the commons, seeks to contribute to the common good, identifies with bioregions and ecosystems rather than obsolete nation-states or transnational corporations, considers the wider impact of his or her actions, is committed to mutual and collaborative community building...and acts according to his or her convictions (Thomashow, 1995: 139).

For Eder (1996) the western world is viewed as having entered the era of modern environmentalism in the late 1960s. Since that time a three-phase development has unfolded. The first phase was characterised by the simple realisation that environmental problems arose from the incompatibility between existing production regimes and ecological systems. The second phase was that of regulation; in order to protect the environment the state must act to prevent certain practices and outcomes from occurring (regulating for sustainable production is seen to fit here). The third phase - that emerging in the late 1990s - has been one based on knowledge, and acceptance of the underlying principles of eco-systemic processes and typified by the 'cultural normalisation' of environmental concerns and their acceptance within established patterns of thinking (see Eder, 1996: 163). It might be posited that there is a new opportunity for farmers to remain rural in late modernity. This might entail producers eschewing artificial chemicals and returning to a more sustainable system of production - embraced by 'organics'.

Despite such speculation, we know little of the link between the process of greening, and the development of organics. Greening is, as suggested above, generally interpreted as a growing concern for the environment - something manifested in a changing consciousness, in new (green) political movements, and in altered production regimes. In Sarah Harper's (1993) *The Greening of Rural Policy* sociologist Fred Buttel sought to distinguish between 'greening' and what he termed 'environmentalisation'.

> By greening, I mean the process by which modern environmentally-related symbols have become increasingly prominent in social discourse. By environmentalization, I mean the greening of institutions and institutional practices - or, in other words, the trend towards environmental considerations being brought to bear in political and economic decisions, in educational and scientific research institutions, in geopolitics, and so on (Buttel, 1993: 12).

For Buttel, there were some important connections between greening and social structures (particularly the rural social structures that are the focus of his work). One was that, in the context of the rise of neo-conservatism, social justice discourses were being progressively replaced by green discourses (where concerns for sustainability replaced those of 'saving' the family farm, for example). Another was that greening has contributed to the 'scientisation' of social movements, and therefore helped to legitimise calculation and measurement, justifying technocratic approaches to the understanding of environmental problems.

For Buttel (1993), greening is a phenomenon which has emerged in the rhetoric of the new social movements, and can be seen as a voice of reason as social democratic institutions move toward what he describes as 'neo-conservative post industrialism'. However, the ultimate importance of greening will be judged by how well it can lead to positive outcomes relating to 'the quest of the majority of the world's population to earn an adequate living, have economic security, and live in dignity' (Buttel, 1993: 25). It would appear from his writings that Buttel is suspicious about the ability of such a malleable term as 'greening' - or indeed its practical manifestation in 'environmentalisation' - to produce such outcomes.

Buttel's (1993) view has been reflected in the work of Johnson (1998), who sought to establish if there was any relationship between green business operations, and green consumerism. Johnson (1998) concluded that green consumerism did not translate into better long-term environmental strategies by businesses. Where marketing taps into green consumer sentiments, the products are superficially more desirable, but they do not lead to changes in production practices which result in more positive environmental outcomes. Yet, in a marketplace where the purchase of products has been transformed from a concern with social reproduction into 'an act of creative expression' (Sachs *et al* 1998: 119), consumers are demanding products which improve individual and family health, and are produced in ways which do not harm the environment.

The aims of the green consumer are, therefore, to seek improvement to personal, family, community and environmental health via the strategic choice of products which are more 'natural', are biodegradable, can be re-cycled, are simply packaged, and are from less polluting, and energy-draining, production systems.

Organic production has been an important, albeit small, part of these changes. Because of the hegemonic position of productivist agriculture, organics has struggled to be recognised as a legitimate and alternative approach to food and fibre production in agricultural economies dominated by productivity and efficiency concerns. While 'sustainability' has had increased saliency in agri-food discourse and its practical manifestations have included new environmentally-friendly production options such as minimum tillage, it has tended to endorse the modification of current, 'high tech' strategies, rather than contributing to their abandonment and replacement. It has allowed producers to feel they are farming sustainably by embracing industry best practices, while leaving the basic elements of a polluting and ecologically-damaging agriculture in place (Lawrence, 1996; Lawrence, 1999; Lockie, 1999). Who needs to consider 'going organic' when one can farm in a sustainable way by tinkering at the edges of conventional production?

Organics has also faced the wrath of those who want to ensure it retains its marginalised position, so that there is no doubt in the minds of governments or consumers about the need for continuation with high production chemical farming. Avery (1995) in his provocatively titled, but ideologically crass, *Saving the Planet with Pesticides and Plastic* argues that organics, *inter alia*, produces lower and erratic yields, is incapable of sustaining soil fertility, increases soil erosion (through tillage), cannot result in any 'natural' balance between production and pests, and has no ability whatsoever to provide a foundation for feeding the world's burgeoning population. Along with the views that organic production is a quaint, self-sufficient, option for 'greenies', and that export-producing nations could never expect the volumes of production from organics that were achieved with high tech agriculture, Avery's sentiments are basically those that have underpinned scientific agriculture in Australia and New Zealand (see Malcolm *et al*, 1996). More pointedly, such views have been in accordance with what Youngberg *et al* (1993: 298) have summarised as the US government position - that organic agriculture was a 'primitive, backward, nonproductive, unscientific technology suitable only for the nostalgic and disaffected back-to-the-landers of the 1970s'. Until recent times, these sentiments have largely prevented serious consideration of an 'organic future' for rural production in countries such as the US, Australia and New Zealand.

However, as Thomashow (1995) and Eder (1996) have suggested, the growth of a 'green' environmental consciousness - present in Australia and New Zealand as much as anywhere else in the world - has encouraged the development of what are perceived to be less polluting forms of farming. In this sense, western 'greening' has allowed organics to emerge from its previously marginalised position. Two related factors - health/'food security' issues and the rise of 'green protectionism' (see Campbell and Coombes 1999a; Lawrence *et al* forthcoming) are viewed as important elements in the move to organics. Premium prices for organic foods in countries like Japan (where the market is estimated to be some A$40 billion) has demonstrated to Australian and New Zealand producers, and particularly New Zealand transnationals, that the export of organics is an important 'keyhole' market channel (see Monk, 1999; Penfold and Miyan, 1999). Within this context, the Australian and New Zealand organic industries have undergone recent and rapid growth. Estimates indicate the Australian organic industry has increased in value from A$28 million in 1990 toA $200 million by 1999, while in New Zealand the industry has grown from NZ$1.1 million in 1988 to NZ$33.5 million by 1996 (Kinnear, 1999; RIRDC, 1996; Saunders *et al*, 1997).

As we will discuss below, the move toward what might be termed the 'institutionalisation' of organics – referring to the integration of organics within conventional food systems – highlights both a positive response to the export opportunities for organics, and a growing tension between the principles and practices of what we have termed 'orthodox' organic growers and those of the more recent entrants to organic farming. Processes of institutionalisation are evidenced in the entry of corporate actors, scientific advisory services as well as government regulatory organisations to the organic industry. Importantly, these processes are evidenced more prominently in New Zealand than in Australia. Within this context of formalisation, there appears to be something of a bifurcation amongst growers engaged in the industry. Without wishing to oversimplify the diversity of values and beliefs amongst those engaged in the organic movement, there appears to be some challenge and contestation between those orthodox farmers attracted by the possibilities of organics to develop a more sustainable agriculture, and the newcomers, who have been significantly swayed by profit-making opportunities that have driven the current (and unsustainable) system of agricultural production. Examining the historical formation of the organic industry highlights the significance of the symbolic meanings associated with food and food production in initiating this movement, and provides the background to contextualise current contestation amongst growers as an outcome of recent transformation of the industry, marking a substantial shift from its historical roots.

Institutionalising organics

The organic agriculture/food movement that began to expand early this century represents, today, both a coherent philosophical challenge to reductionist thinking in agriculture (see Lyons and Lawrence, 1999) and a practical alternative to those intensive farming methods – ones reliant upon the continued application of agri-chemicals and mechanical innovations - that emerged during the Scientific and Industrial Revolutions. It is now widely acknowledged that productivist (high tech) agricultural practices have led to an array of adverse ecological, health and social impacts throughout both the developed and developing world (Allen 1993; Redclift and Woodgate 1997). Despite the growing costs associated with these impacts, intensive agricultural practices have been rigorously promoted by both Australian and New Zealand governments since the late nineteenth century as an attempt to maintain and improve production efficiency (Barr and Cary 1992; Campbell 1996; Lawrence 1987; Tennant 1978). Additionally, chemical-intensive farming methods have been heralded by government and other research institutions as the key to economic prosperity of agricultural and rural communities. While there remain challenges in counteracting the environmental and other 'costs' associated with the adoption of, and continued reliance upon, European farming methods throughout Australia and New Zealand, there is little acceptance of the need fundamentally to abandon such practices. Continued acceptance and encouragement of these strategies has confirmed their pre-eminence and 'normalisation', while organic systems - often positioned in agri-food discourse as the antithesis of conventional agriculture – have been ridiculed and discredited, and consequently constrained from expanding. As an outcome of these constraints, the initial genesis of the organic agriculture industry throughout the 1920s and 1940s has been located at the periphery of broader changes occurring within agriculture. Concomitantly, and in an 'oppositional' sense, membership within the organic movement has represented the rejection of dominant conventional approaches to agriculture. It has, instead, embraced those practices involving acceptance of experimentation and 'local knowledge' which reside on the 'margins' of acceptability, and has grown in spite of, rather than because of, state actions in supporting agriculture (Belasco 1993; James 1993).

The genesis of the organic agriculture movement (and other low input systems) throughout this century has been driven initially by some concerned producers, and later health-conscious consumers. Points of departure from conventional agriculture have been in relation to the impacts of chemical-intensive practices upon the soil, as well as the questioning of the nutritional benefits of industrial food production (Ritchie and Campbell 1996). As a consequence of the culmination of these driving forces, alternative food distribution networks - many with prospects for long-term viability - were established throughout Australia and New Zealand (and indeed worldwide - see Lyons forthcoming a) by the 1980s. While this growth-initiated expansion resulted in a wider variety of organic produce being available in the marketplace, organic agriculture and food production still remained peripheral to dominant systems of food provision (Monk 1998; 1999).

Although remaining marginal in terms of contribution to the proportion of total agricultural production, organic food production systems have - especially during the last two decades - been significantly transformed. Without suggesting notions of 'convergence', they have come to resemble their conventional counterparts in many ways. This transformation has largely occurred as actors previously involved in conventional food production have found new and innovative ways to enter the organic food system – referred to here as institutionalisation. For example, the organic industry has undergone significant change as an outcome of the entry of corporate capital at various sites in the organic food system (Buck *et a.* 1997; Burch *et al* 1999; Guthman 1998; Lyons, 1999). In New Zealand, Heinz Wattie Ltd, McCains and Only Organic have contracted growers to supply organic inputs for the processing of frozen vegetables and baby foods while, in Australia, various companies including Uncle Tobys (until 1997), Berrivale and Sandhurst have undertaken similar activities to produce a range of breakfast cereals and fruit beverages. Supermarkets, worldwide, have also begun to stock increasing quantities of organic food lines (Burch *et al* 1999, Lockie, *et al* forthcoming; Michelsen 1996). The recent entry of these actors has occurred alongside the formulation of national and international standards for organic produce. Guthman (1998), for example, argues that it is the codification of the organic industry through regulation and certification that has attracted corporate interest and has resulted in the 'dramatic' increase in organic production and consumption. Others (Burch *et al* 1999; Campbell and Coombes 1999b) have pointed to green consumerism (which incorporates food security issues) as at least partly responsible for market growth within the organic industry segment.

This 'codification' of the organic industry in Australia and New Zealand occurred throughout the early 1980s, resulting in the formation of organic certification bodies. These bodies attempted to regulate producers, processors and retailers involved in this expanding industry. To date, seven nationally-recognised organic certification bodies exist in Australia, alongside a number of other, less formal, groups. These recognised bodies include; the Bio-Dynamic Research Institute of Australia, the Biological Farmers of Australia, the National Association for Sustainable Agriculture, Organic Food Chain, Organic Herb Growers of Australia, Organic Vignerons of Australia, and Tasmanian Organic-Dynamic Producers (Lovisolo 1997). Across the Tasman Sea, Bio Gro New Zealand, and the Biodynamic Farming and Gardening Association, are the two leading certifiers for organic and biodynamic produce.

Commencing in the early 1990s - and paralleling the 'self regulation' approach described above - both the Australian and New Zealand governments sought to play a more active role in the regulation of this industry. This occurred in response to growing recognition of the potential revenues that might be obtained from organic exports. As an outcome, in 1990 the Australian Government - under the auspices of the Australian Quarantine Inspection Service (AQIS) - established the Organic Produce Export Committee (OPEC) to administer a national export standard for organic produce (Lyon 1994; Lovisolo 1997). The purpose of OPEC is to approve an export-accreditation system for organic produce, with the aim of enhancing the export trade in organic products (Lovisolo 1997). As part of this mission, OPEC provides and oversees an internationally-recognised and government-approved set of standards for organic produce which are equivalent to the guidelines stipulated in Codex and EU Regulation (Lovisolo 1994; 1997). OPEC has also established accreditation procedures for the production, labelling, processing, and handling of organic produce (Clarke 1991). Government support in the form of a national standard is based, more or less exclusively, on the intention to expand export earnings. That organics might be the vehicle appears to be of much less importance. The establishment of this regulatory structure does, nevertheless, represent an historical break from the pattern of government ridicule of organics. It also provides at least partial acknowledgment of organics as a 'legitimate' farming system within the context of capitalist agriculture.

A similar pattern of 'codification' has also occurred in New Zealand, where the government has involved industry in plans to expand organic export markets. State interest became evident in 1995 with the establishment of the Organic Products Exporters Group (OPEG), which aims to facilitate trade of New Zealand's organic produce. OPEG is a joint action group (JAG) established by the New Zealand Trade Development Board (Tradenz) to provide services to producers, retailers and distributors in order to increase exports of organic produce. The OPEG is comprised of a network of businesses, scientific institutions, government agencies and two independent organic certifying bodies. Many people involved in the New Zealand organic industry suggest that increased involvement by large companies, including Heinz Wattie and Zespri International, was the catalyst for government support - with government accepting that an 'organic' flavour might be a key to the generation of increasing levels of export revenue. Although OPEG was established to expand exports of organic produce, there has been little other government support for the industry's expansion. In particular, the New Zealand Ministry of Agriculture and Fisheries (MAF) has remained distant from growers and their concerns since the neo-liberalist restructuring of the New Zealand economy in 1984. As a consequence - and in stark contrast to what is occurring in Europe - MAF has played only a small, indirect, role in the expansion of organic farming systems. Although many within the organic movement would like to see MAF establish a national organic standard (similar to that administered by AQIS in Australia), until very recently MAF has shown no interest in this idea (see Fairweather 1999).

The recent changes which have occurred within the Australian and New Zealand organic industries highlight the various strategies employed by both corporate interests and the state in an attempt to expand export earnings from organic products. These changes have, in turn, led to a significant shift from the historical genesis of the organic movement. The changes highlight the increasing integration of actors previously unassociated with organics, and the subsequent 'resistance' to such integration by many of those wedded to what we have termed 'orthodox' systems of organic food provision. It is presently no longer accurate to conceptualise *all* organic food production as marginal to the broader changes occurring throughout agriculture and food production. Indeed, these changes represent the repositioning of organics as a formalised industry, that is one in which a 'modern' form of organics is becoming institutionalised alongside both high tech agriculture, and older - more orthodox - forms of organic production. The irony here should not be lost. What we appear to be witnessing is the entry of a cohort of growers who would once have been hesitant to undertake organic practices, but who, in economic rationalist terms, see market advantage in moving to something which can be conceived as 'organic agriculture'. While many of these producers also share similar concerns with their 'orthodox' counterparts, they have been largely driven by economic imperatives. Many of the original - and once 'radical' and marginalised - organic producers have expressed concern with the entry of these growers. They have resisted incorporation into what they believe is a corrupted and/or inferior system of modern organic production. Despite their approach to agriculture now receiving general state endorsement, and being driven by the enthusiasm of elements of the corporate agri-food sector, this cohort of producers opposes the 'new' organics as much as it does high tech agriculture!

Corporate connections – towards a new organic industry

These recent processes of institutionalisation – characterised by agribusiness involvement and increasing state and industry regulation - have led to significant changes within the organic industry. Such changes are reflected most starkly by the increasing diversity of growers undertaking organic production. As described above, many 'new' growers have been encouraged to enter the industry in both Australia and New Zealand as a direct outcome of corporate involvement in the organic industry. For many, the sourcing of organic inputs by the food companies ensures a reliable market for growers - as well as premium prices. In order to gain an understanding of the diversity of growers involved in the organic industry throughout each of these locations, in-depth face to face interviews were undertaken with 72 women and men organic farmers – 31 located in Australia and 41 in New Zealand. Farmers included in this study were accessed via lists supplied by two certification organisations in Australia (BFA and NASAA), and by Bio Gro in New Zealand. In order to expand the diversity of growers included in this study, growers were also accessed from the WWOOF[1] list, and a snowball sampling technique was also utilised to enlist growers who might otherwise have been excluded from the study. Throughout these interviews, participants were asked a number of questions related to their decision to undertake organic methods, the markets for their produce, attitudes towards certification, government and corporate involvement in the industry, gender relations within the farm household and general concerns they believed the industry faced. Interviews were semi-structured, and there was scope throughout interviews for participants to raise issues not included within the interview schedule.[2] Prior to examining the responses of growers generated throughout these interviews however, it is important to account for the recent shift many food corporations in Australia and New Zealand (and indeed worldwide) have made towards sourcing and processing organic food lines.

Throughout both Australia and New Zealand the shift towards corporate involvement in organic agriculture has been understood in competing ways. It has been posited that export-dependent food corporations have begun to alter their production practices as an outcome of the imposition of green protectionist trade barriers (Campbell and Coombes 1999a). In particular, companies such as Zespri International (kiwifruit) and Heinz Wattie (frozen vegetables), have shifted towards organic production in an attempt to increase access to growing international markets for so-called 'clean and green' foods (see Campbell *et al* 1997). This was reiterated via interviews with representatives from Heinz Wattie, who suggested that the market for organic produce in countries such as Japan, the US and parts of Europe was unmet and was growing. In the Australian context, Lyons (1999) found that Uncle Tobys involvement in the production of an organic breakfast cereal reflected a marketing strategy to 'green' their corporate image, and thus appeal to the growing band of green consumers (Burch *et al* 1999). Interestingly, the limited success of this marketing strategy - alongside increasing costs of production - resulted in the temporary removal of this product line. The product re-entered the market two years later following increased consumer concerns in relation to food - and spurred on particularly by the genetic engineering debate (see Lyons, forthcoming).

Another suggestion is that corporate involvement in the organic industry reflects a partial shift towards 'corporate greening'. The notion here is that firms are beginning to alter production methods, advertising and marketing strategies, and other procedures and practices so as to reduce the level of pollution and/or environmental impact and damage associated with the production and/or sale of their products (see Burch *et al* 1999).

The recent entry of corporate capital into the organic industry, in concert with the tightening regulations imposed by current regulating bodies, has fostered new entrants supplying organic foods. Most of these growers have been farming conventionally for many years, and have recently made the shift to organic farming methods. Interviews undertaken with these growers revealed that this shift to organics represents a strategy to expand market access, as well as an opportunity to receive the premium paid on organic products. This is illustrated in the quotes below, which are drawn from interviews with producers from both Australia and New Zealand who currently supply food-processing firms with raw organic inputs:

> Watties were really the catalyst. No one else jumped up and said they want organic products to process and sell. Watties made us aware of the market and made us put up our hands. . . It would be fair to say if Watties hadn't been on the scene we wouldn't be doing it. Because we wouldn't have recognised the market.
> *(Brendan, vegetable grower in Canterbury, New Zealand)*

> Oh, it was economic, my whole thing stems from an economic situation. . . Knowing that you're producing something that's reasonably clean and pollution-free is important to me. But being able to produce it from a financial point of view is good. And if it wasn't financially viable I wouldn't do it.
> *(Michael, vegetable grower in Canterbury, New Zealand)*

> I'm probably more philosophically inclined that way which is probably why I thought about it in the first place, but the bottom line was the dollars coming in - which is what you've got to do. If it's not economic we can't do it. It all boils down to 'are we going to make money out of it?', and if we can't, then we can't operate as a business, so you've got to be reasonably economic about it.
> *(Melissa, pea and lamb producer in Canterbury, New Zealand)*

These quotes illustrate the significance of food-processing companies in attracting many 'new' growers to undertake organic practices, due to the premiums they offer organic producers. While some of these growers also indicate concern for broader social, health, and environmental issues, economic incentives have been the primary motive for the adoption of organic practices. Importantly, growers motivated by these economic issues are concentrated within particular locations in close proximity to processing factories. This is particularly evident in New Zealand, where Heinz Wattie and Zespri International have had success in encouraging growers (although not as many as they would have hoped!) to convert to organic practices (Campbell 1996; Campbell and Coombes 1999b; Campbell *et al* 1997; Coombes and Campbell, 1998). In Australia however, food processing companies have not entered the organic food system to this degree, and the concentration of 'new' organic growers around particular locations has yet to occur.

The companies that have recently entered the organic food system have developed a number of strategies to ensure continued provision from growers. In 1990, for example, Heinz Wattie devised a 'Grow Organic With Watties' program, to encourage producers to convert to organic practices, and so recruit a new group of producers to supply organic products for an international market. In addition, Heinz Wattie provided direct support to organic growers, including visits by field officers, the distribution of newsletters and the regular funding of field days. The company was eager both to address any problems related to organic production, and to provide new information to the growers. As a consequence, the food processing companies such as Heinz Wattie have been able to insert themselves into many stages of the organic food system. This has generated an increased interest by producers in organic production, but has also increased control by the agri-food corporations over the entire (organic and non-organic) food production system.

An outcome of this involvement by the food processors - which has occurred alongside the evolution of national and international regulatory standards - is that the organic food industry has come to resemble conventional food systems. The shifts occurring throughout the organic industry stand in contrast to many of the interests held by long-term (orthodox) organic growers. Importantly, the expanding 'corporate' organic industry - characterised by food processing, packaging and international trade - stands in stark contrast to its foundation as a grassroots movement involving the sale of wholefoods through localised and informal markets (Belasco 1993). Consequently, there appears to be a growing bifurcation between the older 'orthodox' organic growers and the 'new' entrants to the organic industry.

Resisting incorporation – the orthodox organic producers

This rift between the 'new' and 'orthodox' growers is highlighted in an assessment of the motivations for 'going organic'. While these 'new' growers indicate that the premiums paid for organic produce has led them to undertake organic practices (as discussed above), there is a large number of growers who present a range of other issues which have led them to practice organic methods. It is important to note that while the majority of 'new' growers also raised some of these issues, without price premiums these concerns were not enough to encourage conversion to organic production. Conversely, for many 'orthodox' growers, premiums are irrelevant to their decision to 'go organic'. In particular, many growers emphasise that their concerns about health effects of agri-chemicals resulted in the decision to produce organically:

> . . . We had one episode where we had guys spraying aquatic weed . . . they had respirators and stuff like that which they were supposed to be in, but didn't wear them. And ah, it was a hot day so they stripped down to their singlets and what have you. When they came back, they had got the job done, then ah, they were all starting to feel sick. They were nauseous. They had nose bleeds. Um, in the end their fingernails, their nails came out, as a result of it. All the symptoms of poisoning by paraquat diquat. *(Russ, pea producer in Canterbury, New Zealand)*

> They were spraying the air all the time. And they had 80 per cent of the kids in that area with respiratory problems. A huge percentage of primary children. Now it could have been manifest as asthma, or hay fever, or dizzy spells, or headaches, or that type of thing. Really really bad.
> *(Rachel, beef producer in Clifton, Queensland)*

Alongside the concerns of these growers, many also revealed their suspicion that the environmental impacts of conventional methods of food production had yet to be fully realised - something that promoted the adoption of organic practices. For many of these growers, organic agriculture represents an environmentally-responsible farming option:

> We are also interested in healing the earth. And Steiner was very much on about that, that's what we find Steiner is telling us, that we are ruining the earth, with chemicals and that sort of thing. And that is how we are ending up with deserts increasing and so on. And so the earth can heal, we need to use Biodynamic methods.
> *(Marchello, mixed vegetable growers in Motueka, New Zealand)*

Oh, the reason why I am organic is probably because in my youth, sort of 15 to 18 you know, you could say there was a green wave that swept Europe, which started with the oil crisis and all the environmental catastrophes. And so, just the desire of German people to have a cleaner healthier environment, that sort of brought me into it.
(Frank, mixed vegetable and kiwifruit producer in Nelson, New Zealand)

It made sense to us. Living in Sydney and just being surrounded by buildings, we thought there must be more. I guess for me I was interested in organics over conventional because while living in Sydney I had become quite interested in environmental issues, not on a..., you know, not as in waving placards. I'm not that sort of person. But I am very interested in all that sort of stuff.
(Jane, orchardist in Takaka, New Zealand)

In addition, many growers also suggest lifestyle appeals were an important stimulus to experimentation with, and adoption of, organic farming:

I prefer to live in the country. And we saw the opportunity here to start from scratch virtually. We don't have the same ideals as non-organic farmers.
(Kate, vegetable grower in Yeppoon, Queensland)

For me I would say it was the lifestyle that attracted me . . . Well you don't get much pay, but it's about much more than money, we work for ourselves. Oh it's satisfying, and much more interesting too. If you get bored with one thing you can go onto another.
(Daphne, mixed vegetable grower in Motueka, New Zealand)

My husband was an engineer, and he wanted to change his job. He wanted to do something different. He wanted to farm . . . and I wanted to be somewhere that is good for my children to live and grow up, without chemicals... You get a lot of fresh air, there are a lot of advantages, it is a quiet life.
(Naomi, vegetable grower in Childers, Queensland)

It just seemed an obvious thing to do. I think if you move to acreage and you're looking at being self-sufficient, which was the initial plan, then I think you know, some sort of natural method of farming has to be the way to go. There doesn't seem much point in going and living on acreage and pouring chemicals into the stuff.
(Mark, orchardist in Murwillumbah, New South Wales)

We love it here, and you know, I'm older, I'm 48 now, and I've got to look at what I like doing and the way I want to spend the rest of my life. I just enjoy farming.
(Rick, vegetable grower in Yeppoon, Queensland)

Many producers expressed more than one of the above concerns in explaining their own involvement in the organic industry – they are not mutually exclusive, and indeed, many growers believe health, environment and lifestyle are integrally linked. For many of these growers, the recent entry of corporate organic growers – motivated primarily by economic incentives – poses a significant threat to the future viability of the industry. Many growers in New Zealand indicate the entry of these new actors may threaten the integrity of the industry, by altering definitions of 'organic' and what might be considered 'acceptable' organic practices. For example, these growers suggest that large food processing companies have influenced national and international organic standards to reflect their own interests, and in turn, have undermined the organic industry. Growers suggested processing companies have been able to bend rules to make accessing raw inputs easier, and that some of these changes have been incompatible with the broader objectives of the organic movement. That growers in New Zealand raise this to an extent not realised in Australia reflects the significance of food processing firms – such as Heinz Wattie – throughout the last decade in shaping the New Zealand organic industry.

Reflecting opposition to the arrival of these 'new' organic growers and corporate interests, orthodox organic growers have employed a number of strategies to contest these changes. These strategies represent significant challenges to processes of institutionalisation, and suggest that the 'organic industry' - rather than being a single entity - should be considered in the plural form.

Creation of stricter guidelines

Sites of resistance are present in both Australia and New Zealand. In both nations small numbers of organic growers have recently devised their own organic certification systems to regulate organic production, and provide quality-assured organic produce to consumers. This has occurred throughout the 'Top of the South' in New Zealand, and in Australia's north-east New South Wales. Importantly, growers in these regions emphasise - as suggested in the earlier quotes - lifestyle, health and the environment in their decision to produce organically. The establishment of local and informal organic certification systems by growers in these regions marks a historic point for the organic industry, as it stands in direct opposition to the growing formalisation and institutionalisation of the organic industry by the government/industry sectors. Many producers indicate that their involvement in these alternative regulatory systems reflects their opposition to the rising costs of certification, and for a few growers it signifies their distrust of external auditing bodies. In the past these farmer-based certification systems in both Australia and New Zealand have enabled growers to ensure product quality to consumers. While organic growers certified by these programs have been unable to sell their products on the international market (where they require internationally-recognised certification), these certification systems have been successful in generating sales on the local market. Such certification systems represent the antithesis to nationally recognised formal regulatory bodies, as they are not independently audited, and therefore limit growers from accessing domestic and international markets. All 'new' organic growers emphasise the importance of meeting internationally-recognised certification standards, and many seek to distance themselves from 'breakaway' or 'fringe' groups, such as those devising their own standards. For 'new' growers, strategies such as those undertaken by long term 'orthodox' growers are perceived as unprofessional, and detrimental to the image of the organic industry.

Reinforcing networks outside the companies

In addition to this resistance, many growers have also established informal networks for the generation and exchange of knowledge. While conventional farmers have historically utilised information deriving from scientific advisory bodies and agricultural consultants, many organic growers ignore - and often directly challenge - such knowledge bases. Throughout both Australia and New Zealand, groups of growers have organised and have met regularly to exchange information with other growers. Much of this shared information is generated by growers themselves through trial and error, as well as through growers' own research and reading, rather than from the information sources available to conventional rural producers (see Lyons 1998). It is interesting to note that the majority of 'new' growers in New Zealand also engages in discussion groups with other growers, however these are organised and facilitated by processing companies. As indicated earlier, Heinz Wattie supplies much support to growers in the form of information sheets, field days and agricultural consultants. While these field days bear some resemblance to the discussion groups other organic growers participate in, they tend to draw from the expert advice of agricultural consultants and scientific advisory services. In contrast, many organic growers indicated that their participation in discussion groups reflected their opposition to scientific advisory services, of which they were distrustful.

Ignoring corporate price signals and other enticements

The orthodox growers have often entered organics for reasons of personal health, environmental integrity, and closer community participation. Maximising profits is not seen as the main ingredient to a better lifestyle. In many cases their market is small and unreliable, but they resist the 'price signals' of the corporations which are seeking their involvement. This represents a rejection not only of corporate monetary enticements to grow for a wider marketplace, but also of the wider activities of firms seeking profit from organics. Many view the entry of the corporations with suspicion - believing that the policies of such groups is antithetical to an holistic, community based, approach to farming.

Discussion: association or polarisation?

That there has been a recent entry into farming of organic producers who are very different from the core group of 'orthodox' organic growers should not necessarily be an immediate concern for consumers or producers. Are not the correct market signals finally beginning to flow through to conventional (high tech) farmers? If this latter group begins to use less chemicals and/or produces in a more sustainable manner, should this not be viewed as a highly desirable outcome? If we are also witnessing some form of corporate greening - as food corporations and supermarkets strive to deliver more wholesome, clean and green, organic, foods to appreciative consumers - is this not also a benefit for the farmers, for consumers and for the environment? Indeed, although there might be resistance, might this not best be construed as a reactionary response from a group of ideologically-motivated growers jealous that they might be beaten at their own (organic marketing) game by a more entrepreneurial cohort of producers? Finally, as Fairweather (1999) has argued, in New Zealand there are many conventional farmers who have never fully embraced high input agriculture. It might be relatively easy for such producers to adopt organic farming methods. Given the premiums on offer and the likelihood of advantageous environmental outcomes, why should they not be encouraged by government agencies to do so?

In order to understand why suspicion and resistance might be accompanying the growth in the new organics it is useful to consider recent developments in the US. In 1997 the United States Department of Agriculture (USDA) sought to develop and apply a set of legally-binding standards for organically-produced foods. The rules sought to set a standard for 'organic' well below that proposed by the organic growers' National Organic Standards Board. Under the proposed rules the USDA would not only have monopoly rights over the term 'organic', but would also have redefined the meaning of the term. It would allow for the use of food irradiation technology, the recycling onto agricultural lands of toxic sludge, the use of genetically-engineered organisms, and would condone the current practices - offensive to most orthodox organic growers - of the intensive confinement of birds and animals (see Lilliston and Cummins 1998).

The aim of the USDA legislation was not, according to Lilliston and Cummins (1998: 196), to stimulate a new group of producers to adopt the rigorous standards imposed by existing regulating bodies in the organics industry, but to reduce to the lowest common denominator that which constituted 'organic'. The latter would have the effect of promoting the continuation of a productivist agriculture, while allowing producers and corporations to take advantage of 20 per cent annual growth in the organics market in the US. It would ensure that many of the environmentally-damaging synthetic inputs supplied by agribusiness - together with the profits of those companies - would not be threatened as the organic industry expanded.

Proposed ownership of the term organic by the USDA would allow government agencies to protect the interests of the large farmers and corporations while undermining those of the orthodox organic producers - those who had spent so much time and energy legitimising the term in the eyes of the public. The Clinton Administration's attempts to redefine organic agriculture resulted in considerable opposition - although not sufficient, it would seem, to prevent a modified version of the legislation from being passed, later this year, by government (Lilliston and Cummins 1998: 200).

There is sufficient cynicism within the 'orthodox' organic industry in Australia and New Zealand to allow a prediction that any future 'incorporation' will strengthen the hand of high tech farming (and of the corporations which serve it) as pressure is placed upon governments to water down the strict guidelines imposed by the industry's current regulating bodies. Growers were suspicious of the 'green' credentials of those profiting from the new organics. But, as Lyons (1998) has reported, many of the new, corporate-linked, organic growers also sought to distinguish themselves from the orthodox growers. They were

> not influenced by an interest in the environment, but were purely motivated to undertake organic farming for economic returns, and sought to differentiate themselves from environmentalists or 'greenies' (Lyons 1998: 61).

Here, then, is evidence of what might be a growing, and irreconcilable, division among the organic producers based on particular conceptions of the 'meaning' of organic and its place in a more sustainable agriculture. The more that organics becomes a corporate-commercial reality, the less orthodox producers might be expected to accept the new parameters of production. What does it tell us, for example, if 'organic' production methods are employed on the farm, but those products are grown in a monocultural system with high water demands and which encourages the proliferation of pests and diseases? What if the organic products which leave the farm are then processed, transported long distances from the sites of production, and presented to consumers in styrofoam or other environmentally-polluting packaging? The orthodox organic growers would have good reason to be concerned that a corporate-linked organics has been 'used' to trick consumers into believing that they are receiving 'clean and green' foods, when the downstream system remains largely untouched, and the use of energy and production of wastes is no different from that of any other commodity within industrial agriculture (Burch *et al* 1999). Yet, as appears to be the case, if companies (such as UK's Sainsbury) have accepted that they must move back into the chain of production and distribution to ensure that fruit and vegetable products are grown in ways that minimise soil loss, reduce chemical use, and are transported in ways which minimise the use of fossil fuels (see Burch *et al* 1999), surely this represents a positive outcome for consumers and the environment?

If it can be accepted, as suggested in the introduction, that there is an identifiable 'greening' of western society, it is important to assess the ability of systems of agricultural production and distribution to meet the needs/demands of consumers for clean and green foods, and to contribute to sustainable modes of production. What we have identified in this analysis is the bifurcation of the organics movement as it responds to consumer demands. Orthodox organic producers believe that their 'rigorous' and 'unbiased' systems of regulation - combined with a philosophical predisposition towards small scale, 'community', and anti-corporate forms of production and distribution - are the best ways to achieve these ends. In contrast, the new organic producers are drawn to the industry for its profit-making potential - an entirely legitimate motivation within a system of capitalist production and exchange.

This tension may be resolved through the marketplace. In Australia and New Zealand, the 'new' organic producers now have ready access to distribution outlets, while the orthodox producers continue to produce for small (usually regional) domestic markets. Are we viewing a very significant division of the organic industry into those who will have access - via the intervention of the corporate food production sector - to global markets alongside those who are only able to sell to smaller (in growth and profit terms), domestic markets? If this is indeed how the division is best understood, one might predict the following:

- the corporate-linked organic sector will expand in line with the demands of overseas consumers for 'green' foods;
- pressures will be exerted to weaken the rules of regulatory bodies so as to accommodate elements of a 'high tech' approach to organic agriculture;
- the domestic-based organic sector will continue to grow, but will have an 'untidy' form - representing a combination of orthodox growers and corporate-linked growers - who are seeking the organic label to sell produce in bulk to whatever markets they can capture;
- there will be increasing polarisation between 'orthodox' organic producers and the new, more market-oriented, producers which may manifest itself in the acceptance by the latter of increasingly lax organic standards;
- the hope that many had of organic production resulting in more sustainable farming systems will largely be destroyed: 'organic' will be a name without much susbstance.

Whether or not the potential reduction of standards might lead to a rejection by domestic or international consumers is, of course, open to some debate.

In terms of the theme of this book, modern organics would not seem to herald a return to an 'old rurality'. While the orthodox organic producers could be considered close to this largely non-market form of production, and are likely to continue to produce for small-scale and domestic markets, those with capital - the corporate-linked and capital intensive producers - are on a different trajectory. They can remain rural in late modernity by accepting two otherwise contradictory ideas: the sanctity of the market, yet the need to reject productionist agriculture to fulfil the demands of that market.

Conclusion

Organic agriculture is occupying a new space in agri-food discourse, and practice, in Australia and New Zealand. 'Organic' is no longer a term that can be applied in a pejorative way to those 'alternative' producers who choose to reject high tech agriculture in the production of foods and fibres. Driven by commercial realities relating to wider tendencies of 'greening' a new group of producers has emerged who happily accept the 'organic' label. What they have done, however, is to align their interests with an agri-food sector bent upon capturing markets through the sale of organically-produced farm commodities. Those orthodox organic producers who, in previous decades, have fought for the wider acceptance of organic production, and have developed regulations to ensure the 'purity' of organics, are currently having their own approach to farming challenged by corporate firms and governments eager to re-define the meaning of organics. This is related to the imperative within existing farming systems of generating new commercial opportunities.

The orthodox growers have not only been cynical of any premature celebration of the wider acceptance of an organic ethos, but have also sought ways to resist 'incorporation'. Such actions have included: creation of even stricter guidelines - in the face of moves by corporations and government to 'redefine' (downgrade) the meaning of organics; reinforcement of their own, community-based, networks as sources of information and interaction; and rejection of the 'price signals' from companies which desire their involvement in the production of inputs to the corporate food industry.

Whether such resistance represents a long-term, successful, strategy to avoid contact with the corporate sector; whether rejection of the corporate sector is likely to dissipate; whether the temptation to produce organically - under the direction of the corporate firms and for high profits - will be accepted by the 'orthodox' organicists; whether the bulk of those resisting 'incorporation' will emerge from corporate interventions with their own regulatory system and markets intact; whether any 'corporate organics' can provide a sustainable basis for food production and distribution; and whether the state will act on behalf of orthodox growers, or the new growers and the agri-food industry, are questions which cannot be answered at this time. What *can* be said is that the organic industry in Australia and New Zealand is in a period of flux with internal divisions providing an intriguing focus for an assessment of the potential for the older 'high tech' approach to agriculture to be modified for social and environmental good.

Notes

[1] WWOOFers are Willing Workers on Organic Farms. The WWOOF guide contains a list of organic farms willing to be involved in a cultural exchange by offering WWOOFers an opportunity to work on their farm in exchange for food and accommodation. WWOOF lists are available for both Australia and New Zealand.

[2] These interviews are part of the fieldwork undertaken by Kristen Lyons from her forthcoming dissertation, *Situated Knowledges, Science and Gender: A Sociology of Organic Agriculture in Australia and New Zealand.*

Part II

Food producers and food consumers

4 The evolution of rural food provisioning in France between 1978 and 1991

GWENAËL LARMET

Following Weber (1921), the modern world can be characterised by an expansion of rationalisation in every sphere of social life. Among its many consequences, rationalisation leads to more exchanges through market mechanisms, and more impersonalised social relationships. 'The market community, as such, is the most impersonal of practical life relationships into which men can enter' (Weber 1921: 420). So the degree of market exchange and of impersonalisation of social relationships can provide an indicator of integration into modernity.

With Giddens (1990), we can distinguish between social integration, which relies on co-presence of individuals in time and space, and system integration, which implies more interactions with individuals who are distant in time and space. Earlier societies rested on social integration, but the de-localisation of social relationships in modern society requires a systemic integration. The scarcity of local and face-to-face social interactions can then be another criterion of modernity.

Sketching an ideal type of rural societies, Maget (1955) described the village as a 'group of dispersing inter-acquaintance'. Everyone living in a village knows everyone by personal experience. Belongings, character, schedules are known and allow the prediction of the villagers' usual behaviour. Opportunities for meeting between inhabitants enable the cross-checking of partial information. So everyone is at the centre of a star of interpersonal acquaintances, and all the stars form a group of inter-acquaintance.

Does the modernisation process, characterised by more market and impersonal relationships, and by systemic integration, imply the disappearance of the rural specificity, that is 'dispersing inter-acquaintance'? Rural food provisioning practices offer a focus through which to analyse the impact of modernisation on the rural world. The beginning of the 20th century saw the decrease of auto-consumption in rural France (home production of food for domestic use, using the means of a kitchen garden, an orchard, breeding, hunting, fishing or gathering), in favour of more and more purchases in small shops such as specialists or groceries (Segalen 1985). But as these shops were located next to their customers, in many rural towns, they did not imply a disruption of the rural model of social relationships. Moreover, these food shops had become meeting places, where news about village inhabitants could be exchanged between customers (Burguière 1975). Things become different with the development of large-scale retailers: these shops are not only bigger, but they are also often located far away from rural villages. Does rural specificity prevent purchases in supermarkets and hypermarkets, or, on the contrary, do these purchases show a disappearance of rural interpersonal acquaintance? How do rural provisioning practices evolve with the development of large-scale retailers?

The data used here come from Institut National de la Statistique et des Études Économiques (INSEE) surveys on food consumption, conducted in 1978 and 1991. 14,142 households, amongst which 3,972 were rural households, recorded their purchases and the origin of their food in a booklet over one week. The proportion of rural households is stable in the two surveys: 28.5 per cent in 1978, 27.7 per cent in 1991. Data have been analysed with linear models, estimated with ordinary least squares (see Annex). These models allow interpretations *ceteris paribus*. Instead of taking a reference modality for each variable, models have been centred on the mean of budget coefficients between 1978 and 1991.

We will limit our analysis to food eaten at home: the choice of the place of meals away from home does not, like the choice of provisioning places, involve the cooking of a meal, nor does it entail the same problems of distances and frequency of shopping, and forecasting of the meals. We study the allocation of provisioning between different origins. Non-market provisions have been valued at local market prices by INSEE inquirers. Of course, autoconsumption or food gifts between households have other meanings than their monetary value; they represent leisure, self-affirmation, or a way to pass time (Weber 1998). But their monetary value is the easiest way to compare them with market provisions. Our analysis rests on a budget summarising market expenditures and economies of expenditures, corres-ponding to non-market food. This fictitious budget represents the actual monetary value of food eaten at home.

The present work rests on the respective definitions of rural and urban space outlined by INSEE. A district is urban if it belongs to an urban unit, that is a territory on which there is 'a group of habitations of which none is further away from the next by more than 200 metres, and which includes more than 2,000 habitants' (INSEE 1998). Rural districts are those which do not belong to an urban unit. This statistical definition of rurality has the advantage of allowing comparisons across time. The determining criteria are the agglomeration of habitations (threshold of 200 metres) and population (threshold of 2,000 inhabitants). We have to keep this definition in mind, because we will show that, arbitrary as these thresholds may appear, they correspond to systematic differences in the observed provisioning practices.

How do rural provisioning practices differ from urban ones? How do these differences change between 1978 and 1991? We will first consider the rural population as a whole, and analyse the practices common to rural people. We will then turn to the different fractions of rural population.

Evolution of differences between rural and urban provisioning practices

The increasing share of hypermarkets and supermarkets

Between 1978 and 1991, two provisioning places saw their market shares grow significantly in rural areas: hypermarkets and supermarkets (Table 4.1). They are both self-service general food stores, the surface area of the first is above 2,500 square metres, the surface area of the second lies between 400 and 2,500 square metres. The average share of food eaten at home which was bought from butchers and groceries (general food stores where the client is served by the shopkeeper), small stores (self-service stores whose surface area is under 400 square metres), itinerant dealers, wholesalers and producers significantly decreased. The shares of non-market provisioning places also decreased: autoconsumption and autosupply (withdrawal by food craftsmen and shopkeepers from their market stocks). In 1978, the leading places for rural food provisioning were autoconsumption, groceries and butchers, with roughly 15 per cent of the food budget going to each. Thirteen years later, supermarkets represented 29 per cent of food budgets, hypermarkets represented 15 per cent and autoconsumption only came in third place with 12 per cent. In urban areas, hypermarkets and supermarkets developed, but so did bakeries, while butchers, groceries and small stores decreased in number.

Large-scale retailers are designed to concentrate in one place the most extended range of goods, to ease the grouping of purchases while offering different selling modes (self-service but also specialised departments), which explains their success (Larmet 2000). As they make comparisons and computability in economic practices easier, they are typical of modern times in Weber's sense of this. However, not every kind of provisioning place suffers from the growing market share gained by hypermarkets and supermarkets. Autoconsumption decreased in rural areas where it was higher, but not in urban areas. The share of food gifts, markets and bakeries increased even in rural areas.

Remaining differences between rural and urban provisioning

If we compare urban and rural food provisioning practices in 1978 and in 1991, they differ for the same reasons: rural people give a greater place to non-market provisioning and to local and personalised shops. In urban areas, shops are nearby and are more diversified. Rural food shops rely on long-term customers and personal relationships; rural people have possibilities, and preferences, for autoconsumption and direct provisioning.

Table 4.1: Evolution of food budget coefficients in rural and urban areas between 1978 and 1991

	% rural 1978	*% rural 1991*	*% evolution*	*% urban 1978*	*% urban 1991*	*% evolution*
gifts	2.4	2.8	0.3	1.6	1.9	0.2
autoconsumption	16.8	11.1	-5.7	3.0	2.0	-1.0
autosupply	0.8	0.4	-0.5	0.4	0.3	-0.1
market	1.9	2.5	0.6	7.5	6.3	-1.2
bakers	8.2	9.4	1.2	6.5	8.0	1.5
butchers	14.4	8.5	-6.0	12.9	7.0	-5.8
diverse specialists	5.5	5.6	0.1	8.3	7.0	-1.3
groceries	14.2	5.4	-8.7	10.1	3.7	-6.4
hypermarkets	4.5	15.4	10.9	11.0	24.3	13.3
supermarkets	11.9	30.1	18.2	24.4	33.6	9.2
small stores	7.6	4.6	-3.0	10.3	3.7	-6.7
itinerant dealers	5.6	1.9	-3.7	0.8	0.5	-0.3
producers	6.1	2.4	-3.7	3.1	1.8	-1.3
total	100.0	100.0	0.0	100.0	100.0	0.0

Hypermarkets and supermarkets are nearer and more numerous in urban than in rural areas. The variety of shops favours expenditures at diverse specialists (pork butchers, fishmongers, greengrocers, pastrycooks...). Although they may appear like peasant shopping places, markets have developed in parallel to urban growth, and are especially located in the Parisian area (Margairaz 1988).

In rural areas, small shops are used because they are nearer than large-scale retailers, but also because customers have long-term habits of buying there. When proximity is the same, rural people buy more than urban people from bakers, butchers, grocers and itinerant dealers (Larmet 1999). These long-term relationships give the habit of personalised service, and a kind of loyalty obligation, because they rest on social groups: customers, neighbourhoods, living in the same district. Moreover, small shops in rural areas have important functions in the centralisation and diffusion of news. Disloyalty to a shopkeeper exposes one to social sanctions: not to buy also means being apart from local life (Burguière 1975).

Finally, rural people get more food than urban people from autoconsumption and direct provisioning (food gifts originating from autoproduction, purchase from wholesalers or producers). This is not a constraint stemming from the long distance to the shops, because it is observed whatever the proximity of shops (Larmet 1999). It has two origins. First, households with farming connections have received know-how and values from them concerning their provisioning practices: their search for autonomy of provisioning leads them to consume more unprocessed food. Their cooking requires more time, but their production conditions are mastered or known. Secondly, detached houses are more widespread in rural areas, and the piece of ground around them allows gardening. Some urban residents have moved to a detached house in order to have a kitchen garden, which is for them like a 'productive workshop' (Weber 1998). In both cases, rural people also appreciate direct provisioning: they use their kin or neighbourhood networks to get food directly from producers, with payment (wholesale, producers) or without payment (gifts).

A reduction of the gap between rural and urban people, but a specific rural use of shops

While remaining significantly different, between 1978 and 1991, the provisioning practices of rural people have nevertheless grown closer to those of urban people. This convergence of practices has many explanations. Hypermarkets and supermarkets have moved closer to rural areas, and become more accessible (Courson 1990). Supermarkets actually grew faster in rural areas (+17.4 per cent against +9.4 per cent), while hypermarkets increased more in urban areas (+12 per cent against +10.5 per cent). Although the gap between rural and urban practices reduced for supermarkets, it increased for hypermarkets. Hypermarkets can be viewed as the symbol of modern distribution. Their bigger selling surface implies a larger range of products supplied and more impersonalised social relationships. One may then conclude that rural people are still left behind urban people in the process of modernisation. Here, however, we will argue for a specific rural use of modern food supply, linked to the extent of rural society's scale, rather then for a relative backwardness.

Of course, supermarkets are nearer than hypermarkets for rural consumers (Courson 1990). If shops move to the country, so do households. We do not have panel data, the households surveyed in 1991 are not the same as the households surveyed in 1978. But during that period, new rural people, often with higher diplomas and non-farmer parents, moved to some rural regions (Chamboredon 1980). Some of them still work in town, some are retired, but all keep part of their urban provisioning habits. *Ceteris paribus*, they buy more in hypermarkets and markets, which contributes to a convergence of rural and urban practices. But rural-born people also have more contacts with the town. First of all, a growing proportion of them work in town. Secondly they have made the acquaintance of urban people who have bought a second home, or they are still in contact with former rural people now living in town, but who come back to their native area every weekend (Chamboredon 1980). This contributes to a diffusion of new lifestyles.

Therefore, differences in lifestyles but also in the geographic space between rural and urban people decrease. In 1990, 60 per cent of rural districts are located in predominantly rural space, but 40 per cent are peri-urban districts. Peri-urban districts can be rural or urban: they have less than 5,000 jobs, but 40 per cent or more of their employed active people work in an urban pole, that is, an urban district counting more than 5,000 jobs (INSEE 1998). Between 1975 and 1990, the population of peri-urban districts has grown due to the concentration of population in urban spaces, and migration from urban poles to peri-urban districts, favoured by better conditions for purchase of detached houses (Le Jeannic 1997: 38). Daily commuters increased by 37 per cent for France as a whole between 1975 and 1990, but by as much as 89 per cent for peri-urban districts (rural or urban), and from 43 to 62 per cent for the different districts in rural-dominant space (INSEE 1998, 100). Daily commuters are certainly as important to understanding the increased share of hypermarkets and supermarkets as the movement of those shops to rural areas.

The development of some means of transport and communication particularly concerns rural people. In 1978, the rural rate of car-owners was slightly above the urban rate (72 per cent *vs.* 71 per cent), but by 1991 the gap had increased (86 per cent *vs.* 78 per cent). We could say that this widening gap is the result of commuting combined with a poor network of collective transport. But this would not account for the reasons of commuting nor for the reasons of non- commuters' trips to town. Using a car is the special way in which rural people take part in the process of modernisation: they prefer means of communication that ease face-to-face interactions. However, rural people are left behind in relation to means of remote communication, e.g. internet connections (Rouquet 2000). These figures are compatible with Maget's 1955 definition of the village as a 'group of dispersing inter-acquaintance'. But the automobile has allowed a widening of the scale of rural communities: following Mendras (1984: 319), this has increased from 1-3 km in the beginning of the 20th century to 10-30 km nowadays.[1] As a consequence, the social control of the community has decreased, and the preference for personal acquaintance has transferred to an extended space.[2]

We can contrast the increased ownership and use of cars with the decrease of purchases from itinerant dealers. Although the two provisioning strategies seem to be equivalent, they do not rest on the same customer relationships. For itinerant sale functions with periodic rounds, so that the customer, though notionally free, feels constrained to buy something each time, in order to ensure the next visit (Larmet 1993). We will see that the use of the car for shopping means a kind of liberation from the narrowness of local supply and the obligations to buy in traditional shops.

Segalen (1985: 352-363) has shown that notwithstanding the increased scale of rural areas, traditional categories of personal relationships such as kinship are still mobilised. Likewise, when located in a rural area, just as in an urban residential area, supermarkets become local attraction poles, where neighbours have great chances to meet, and the habit of purchasing allows customers and employees to get to know each other.[3] Smaller and nearer than the hypermarket, the rural supermarket has the peculiarity of concentrating social life in a space where traditional shops close and boundaries are extended. In contrast to the urban supermarket, the rural supermarket draws customers from a larger zone. Because supermarkets have come nearer, and the scale of rural communities has extended, purchasing in supermarkets does not necessarily imply in 1991 the weakly personalised relationships that it implied in 1978. Rural shopkeepers were 'local agents of sociability' (Burguière 1975). The communication functions of rural shops may not have disappeared (Segalen 1985: 320), but may have transferred to the supermarkets, where they are distinguished from the monetary side of purchase: the check-out assistants may be locally well-known, but they are not shop assistants. We may argue that rural people adopt practices that first appeared in urban areas, but only insofar as these practices meet their own idea of shopping places.

It is not only because of its material advantages, but also because it fits the rural model of social relationships, that the competition of supermarkets is so strong and has caused the decline of groceries and traditional specialised shops. After supermarkets, the gap between rural and urban people decreases most for groceries and itinerant dealers. While remaining a rural specificity, personalised service increasingly gives way to self-service. The competition of supermarkets increases more in relation to these general food stores than with respect to specialised shops, where the urban–rural gap decreases less. However, rural people living in districts without any shop do not head towards specialised shops, but group their purchases for a single journey to a supermarket. Here the rationality of travelling matters the most.

Another rural specificity, the rural-urban gap for autoconsumption and for wholesaler and producer purchases, decreases while remaining significant. Again, the renewal of the rural population and the advantages of hypermarkets and supermarkets weaken the link between autoconsumption and autonomy of provisioning in rural areas. We will come back to this point in the analysis of divisions within the rural population.

So, rural specificities remain in 1991 as they were in 1978, but at a lower level. Some of them, like the shop as a meeting place, transfer from butchers and groceries to supermarkets, enacting a rural approach to modernity.

The evolution of provisioning practices among the rural population

We have just analysed the evolution of practices common to the whole rural population. But do these evolutions reflect widespread practices among all the factions of the rural population, or rather practices by some factions only, which would increasingly differ from those of other factions? That is: does the heterogeneity of provisioning practices inside the rural population decrease or increase? For which provisioning sources and why?

Absolute and relative differentiation

A synthetic indicator of the degree of homogenisation is the evolution of the standard deviation of the mean. Average budget coefficients have increased in hypermarkets, supermarkets, markets and bakeries, and the standard deviations have also increased (Table 4.2). On the other hand, in declining shops, standard deviations have decreased.[4] This means that the decrease of non-market provisioning and purchases in personalised service shops is a commonly spread tendency for all factions of the rural population. But the choice of the preferred shops now causes more heterogeneity inside the rural population. Everyone reduces their purchases in the same shops, but each faction of population goes to different new shops instead.

However, we must be cautious of the preceding interpretation because it only takes into account the absolute differentiation of provisioning practices. But while standard deviations evolve, means also vary. If we want to control for the variation of the means, we have to reason in terms of relative differentiation via the evolution of the coefficients of variation (that is standard deviation/mean). The results are then reversed. In the rural population, the relative differentiation has decreased in growing provisioning places, and has increased in the declining ones, because standard deviations have moved more slowly than the means in both cases. The absolute differentiation allows us to interpret the heterogeneity of the actual practices whereas the relative differentiation accounts for the comparative speed of changes in practices and in their heterogeneity.

In order to explain these changes, we have to consider the different factions of the rural population. We propose to analyse the heterogeneity of the rural population according to occupation (of the head of household and an eventual spouse), income, age, type of household and car-ownership.[5] We will use an analysis of variance of the budget coefficients through regression analysis (see Annex).

Changes in the structure and in the practices of rural population

The first result of our model is to simulate two different evolutions of the means, using two ways of weighting (see Annex). The observed evolution of

Table 4.2: Standard deviation and coefficients of variation of average budget coefficients in rural areas

	standard deviation			variation coefficient		
	% *rural 1978*	% *rural 1991*	% *evolution*	*rural 1978*	*rural 1991*	*evolution*
gifts	7.2	7.0	-0.2	2.97	2.53	-0.45
autoconsumption	18.6	17.9	-0.7	1.11	1.62	0.51
autosupply	6.9	4.0	-2.9	8.18	10.46	2.28
market	6.6	8.6	1.9	3.54	3.44	-0.10
bakers	8.5	13.0	4.6	1.03	1.39	0.36
butchers	13.7	14.0	0.3	0.95	1.65	0.70
diverse specialists	9.6	12.6	3.1	1.74	2.27	0.53
groceries	20.1	14.1	-5.9	1.41	2.60	1.19
hypermarkets	15.3	28.3	13.0	3.43	1.84	-1.59
supermarkets	21.9	32.2	10.3	1.84	1.07	-0.76
small stores	16.5	14.0	-2.5	2.17	3.04	0.88
itinerant dealers	13.1	8.6	-4.5	2.34	4.44	2.11
producers	11.1	7.2	-3.8	1.80	2.98	1.17

the means can be attributed to two types of effects: changes in the structure or in the practices of the rural population. If we compare the structures of the rural population in 1978 and 1991, in 1991 we find fewer farmers and more wage earners among working men and women, more highly educated people (above the *baccalaureat*), more young people (under 45), fewer poor people, more single-parent households, fewer families with three children or more, and finally - more car owners.

These well-known facts derive from the arrival of a new rural population, which has been educated in town and has kept an urban model of birthrate as well as an urban way of provisioning. But for rural-born people, the level of education has also strongly increased. As a consequence of the increasing commuting between town and rural areas, the range of accessible occupations and incomes has also widened. The urban job market brings a higher income than the rural job market (Lagarenne and Tabard 1998). This has encouraged some young rural people to work in urban areas but to live in rural areas. What is the role of these changes in the structure of the rural population compared with the changes in its practices?

Our model allows us to simulate which average market shares would have been observed if the structure of the population alone had changed and the practices of each faction had remained the same as in 1978. For most shops, simulated changes have the same direction as observed changes but are of smaller extent (Table 4.3). Structural changes are especially important to explain the decrease of autoconsumption. The proportions of farmers, the elderly, lower-income, less well educated households, and households with many children, have decreased, and these households autoconsume the most. Conversely, a part of the increase of hypermarkets and supermarkets is due to the growing proportions of wage-earners, of the youngest and the highest income households The only contradiction between observed and structural changes concerns bakeries as the proportion of their typical customers (lower-income households and single males) has decreased. The progress of bakeries is due to a pure change of practices: commuting means more mid-day meals away from home, and so more purchases of sandwiches.

The effect of a pure change in practices, the structure of rural population being held constant, can also be simulated by our model. The resulting changes have systematically the same direction and much the same extent as the observed changes. They are stronger than simulated structural changes. So, in the observed change, the pure change of practices has dominated over the pure change of structure.

Table 4.3: Evolution of budget coefficients in rural and urban areas between 1978 and 1991 under different hypotheses

	% *1978* *average*	% *91-78* *observed evolution*	% *91-78* *change of practices only*	% *91-78* *change of structure only*
gifts	2.6	0.4	0.3	0.3
autoconsumption	16.0	-5.8	-3.6	-4.4
autosupply	0.8	-0.4	-0.2	0.4
market	1.8	0.7	0.2	0.2
bakers	8.4	1.2	1.8	-1.5
butchers	14.6	-5.8	-5.1	0.8
diverse specialists	5.6	0.1	0.2	0.4
groceries	14.5	-8.5	-7.3	-2.6
hypermarkets	4.5	10.5	6.8	3.5
supermarkets	11.9	18.2	16.6	3.0
small stores	7.9	-3.2	-3.0	0.1
itinerant dealers	5.4	-3.6	-3.9	-1.3
producers	6.0	-3.6	-3.8	1.2

Explaining the variations of heterogeneity in provisioning practices

Keeping the structure of the rural population constant, how do these pure changes in practices explain the evolution of provisioning differentiation? We will confine our analysis to absolute differentiation, bearing in mind that if we were reasoning in terms of relative heterogeneity, the social facts which affect the heterogeneity of provisioning practices would be the same but the extent of their consequences would be different (Larmet 2000). These social facts are the increase in car use, daily commuting, ageing and differences between generations, the adaptation of independent workers' lifestyles and of lower-income households to the new food supply conditions. We will focus mainly on the provisioning sources with the biggest market share variations: auto-consumption, butchers, groceries, supermarkets and hypermarkets.

The use of the car. We have seen that the high rate of car owners is a distinctive feature of the rural population. But even if this rate had remained the same, the way rural people use the automobile would have changed. Table 4.4 shows that if the structure of the rural population remained constant, car owners would have bought more from supermarkets and hypermarkets, and less from butchers, groceries, small self-service stores and itinerant dealers, which seem like proximity shops. Because in 1991 car owners used their cars more to go shopping than they did in 1978, the distance between their practices and practices of non-car owners has widened. The extension of scale that has affected rural society is not the mere result of the availability of a car, but of the will to use it, and access to a large-scale distant food supply. The disenclavement of the countryside answered a demand. This more intensive use of the car for shopping leads to an increased absolute heterogeneity in nearby as well as in distant shops.

If having a car favours large-scale retailers at the expense of personalised shops in one's vicinity, it also favours autoconsumption. This is not easy to interpret. We might suggest that using a car to buy at large-scale retailers or autoproducing a part of one's food are two ways of reducing one's dependence on local supply, hence their correlation. An idealised conception of personalised social relationships between customers and shopkeepers forgets the obligations of loyalty that it implies (Larmet 1993, 1999). Either autoconsuming or using a car to go and buy in town allows people to escape these obligations.

Commuters. The emergence of commuters causes changes in food provisioning which can be measured along many variables. Commuters own a car, they are more often younger, wage earners, especially from the middle class, and they have a higher level of income. That is the reason why these factions of the rural population bought more than average from supermarkets in 1978, and bought even more from hypermarkets and supermarkets in 1991. Unlike independentworkers, wage-earners have no professional resources, which is the reason they use more market provisioning. The development of daily commuting has increased the difference in purchasing strategies between wage earners, young and higher-income households on one hand, and, on the other hand, independent workers, old and lower-income households, who go into town less often. Therefore, the heterogeneity of purchases at large-scale retailers (hypermarkets and supermarkets) according to income and age has increased. When we group by occupation, things become more complex. The divisions between rural households have become stronger in hypermarkets, which are bigger and less personalised, but not in supermarkets, where social networks are more easily actualised. But these effects result in an overall growth of the heterogeneity of rural purchases at large-scale retailers.

Table 4.4: Evolution of the marketshares of provisioning places according to car ownership for the rural population between 1978 and 1991

	% 1978 *rural mean*	*car owners*	*non car owners*	% 1991 *estimated rural mean*	*car owners*	*non car owners*
autoconsumption	16.0	1.4*	-3.6*	12.5	1.8*	-4.8*
butchers	14.6	-0.7*	1.8*	9.5	-1.8*	4.6*
groceries	14.5	-0.6*	1.7*	7.2	-1.9*	4.9*
hypermarkets	4.5	0.7	-1.8	11.3	0.8	-2.2
supermarkets	11.9	0.9	-2.4	28.5	1.3	-3.3
itinerant dealers	5.4	-0.7*	1.9	2.4	-0.6*	1.6*
others	33.1	-0.9	2.4	29.6	0.3	-0.8
total	100.0	0.0	0.0	100.0	0.0	0.0

Coefficients centred on mean; * = significantly different from 0 at 5% level

The table can be read as follows:
In 1978, the average estimated share of supermarkets was 11.9% of the food budget. Having a car caused an increase of 0.9% of the supermarket share to 12.8%.
In 1991, assuming the structure of the rural population was constant, the average share of supermarkets was 28.5%, and the use of cars increased this share of 1.3% to 29.8%. Here the estimated average shares of 1991 differ from the observed average shares of 1991, because they are estimated on the assumption that the structure of population is constant.

Conversely wage earners, young and higher-income households bought less than the average from butchers and groceries. But in these shops the evolution of heterogeneities also reflects differences in provisioning strategies according to income and household size, in generation, and lifestyles.

The income effect. In 1978, lower-income rural people autoconsumed more than average, and spent less at the butcher's (Table 4.5). This is an effect of low income: autoconsumption allows limitation of expenditures, while meat bought from the butchers is often expensive. But the cost is not the only determinant, as lower-income rural households bought more from groceries and itinerant dealers, whose prices were higher than in large-scale retailers. Lower-income households depended more on the obligations of long-term customers (remaining loyal), for instance, because these obligations went with credit possibilities. That is the reason why in 1978 higher-income households could already free themselves more easily from these obligations. But lower-income households also found advantages or 'counter-handicaps' in their shops (Grignon 1980). Groceries and itinerant dealers had less selection than supermarkets, which limited customers to routine purchases and which also protected them from the temptation of buying more than they anticipated. So autoconsumption could be interpreted as a kind of autonomy despite the customers being captives of the shopkeepers.

In 1991 lower-income households still bought most from itinerant dealers and least from butchers, but their purchases from groceries no longer differed from the average. The decrease of groceries in French rural areas has partly freed lower-income households from their relations with local shopkeepers. They now buy more than average from small self-service shops: as these shops are self-service, one might think that the credit possibilities but also the obligation to buy locally are scarcer. As a consequence, heterogeneity related to income has decreased at butcher's and in groceries.

Although households in 1978 from every income level bought equally from diverse specialists (pork butchers, fishmongers, greengrocers, pastrycooks...), higher-income households spent more than average in these places in 1991. This is typically a distinction effect: as these shops have become scarcer and more expensive, their reputation has increased (Larmet 2000). Meanwhile higher-income households have left butchers, who appear more common.

The rationality of provisioning also stems from economies of scale. This is especially the case for autoconsumption, whose share of which increases in the budget with a greater number of consumption units in the household. The larger the size of the household, the bigger the pressure on the food budget. This effect has slightly increased over the period, which has caused an increase in heterogeneity of autoconsumption according to household size. The type of household not only indicates economies of scale alone, but also specific consumption norms. In 1978, large families, like lower-income households, favoured autoconsumption, groceries and itinerant dealers (credit opportunities and limitation in the range of available products); in 1991 they still favoured autoconsumption and to a smaller extent, itinerant dealers, but they had left groceries. So the heterogeneity has decreased in groceries and itinerant dealers.

The age effect. The probability of health troubles increases with age, which limits ability to walk or to drive a car. That is the reason why people older than 75 buy more than average in nearby shops (especially butchers) and less than average in more distant large-scale retailers (Table 4.6). These shops also allow

Table 4.5: Evolution of the marketshares of provisioning places according to income for the rural population between 1978 and 1991

1978	*rural mean* %	*first quartile* %	*second quartile* %	*third quartile* %	*fourth quartile* %
autoconsumption	16.0	1.8*	-0.4	-0.7	-2.3*
butchers	14.6	-1.9*	1.2*	0.7	1.1
diverse specialists	5.6	-0.2	-0.1	-0.3	1.0
groceries	14.5	2.3*	0.9	-2.7*	-2.9*
hypermarkets	4.5	-1.1	-1.2	1.2	3.0*
supermarkets	11.9	-1.2	-0.1	1.9	0.2
small self service	7.9	-0.9	-0.1	0.4	1.5
itinerant dealers	5.4	0.6*	0.4	-0.6*	-1.1*
others	19.6	0.6	-0.6	0.2	-0.5
total	100.0	0.0	0.0	0.0	0.0

1991	*estimated rural mean* %	*first quartile* %	*second quartile* %	*third quartile* %	*fourth quartile* %
autoconsumption	12.5	1.7*	0.1	-1.4*	-2.0*
butchers	9.5	-1.1	0.4	1.1	0.0
diverse specialists	5.8	-1.6*	0.4	0.9	1.6*
groceries	7.2	-0.4	-0.2	0.6	0.3
hypermarkets	11.3	-3.9*	1.7	2.4	2.3
supermarkets	28.5	0.2	-1.4	-0.2	2.5
small self service	4.9	1.8*	-0.3	-0.9	-2.1
itinerant dealers	2.4	1.0*	-0.1	-0.5	-1.2*
others	17.9	2.3	-0.6	-2.0	-1.3
total	100.0	0.0	0.0	0.0	0.0

Coefficients centred on mean; * = significantly different from 0 at 5% level

The table can be read as follows:
In 1978, the average estimated share of hypermarkets was 4.5% of the food budget. Being in the first quartile of income meant a decrease of 1.1% of the hypermarket share to 3.4%.
In 1991, assuming the structure of the rural population was constant, the average share of hypermarkets was 11.3%, and being in the first quartile of income decreased this share by 3.9% to 7.4%.

a daily chat (David *et al* 1992) which partially compensated for the loss in the number of conversation partners through age. Here the service provided is different from a mere facility of payment; it pertains to local integration. This might explain why elderly people's purchases in nearby shops differed more from the average in 1991 than in 1978, whereas lower-income purchases came closer to the average.

Moreover, the increase in differentiation according to age in every type of shop reveals a generation effect. Consumer competences are mostly acquired when children leave their parents and they have to take charge of their own food provisioning. So these competences are linked to the structure of the food distribution system at this time. They will then influence the consumer's behaviour all their life. This explains the reluctance of older generations to use large-scale retailers, whereas in 1991 young people preferred the grouping of purchases and the possibilities of comparison offered by hypermarkets than those who were young in 1978. Moreover, hypermarkets are often located in shopping centres where young people have the habit of buying non-food products.

The proportion of gifts coming from autoproduced food decreases with age, while the proportion of autoconsumption increases until 55-64 years of age, and then decreases. This lifecycle phenomenon becme stronger between 1978 and 1991. The share of autoconsumption is at its highest when children have left the home of their parents, and when the parents retire from their job, that is between 55 and 75. More time is available, autoconsumption becomes a pastime and a more productive activity (Weber 1998). We can assume that gifts received by the youngest respondents come from the households who autoproduce (Gojard 1994) and autoconsume the most - that is, those between 55 and 64. These gifts circulate through parent – child relationships, show the increased strength of the kin relationship and help to improve the children's standard of living. The consequences are an increase of heterogeneity according to age in every provisioning sources, large-scale retailers but also declining provisioning sources such as autoconsumption or butchers.

Farmers' lifestyle. The rationality of provisioning also stems from occupation (Table 4.7). For farmers, autoconsumption is a material resource which prevents an expenditure. But occupation does not only provide a resource, it shapes a lifestyle which shows a certain persistence through changes in the food supply. Autoconsumption requires a specific time allotment: less time for purchases, more time for autoproduction work, spread over an entire year, which gives immediately available vegetables during the harvesting season. Farmers' autoconsumption ensures a certain autonomy of provisioning, although farming is now fully integrated into the market economy.

Table 4.6: Evolution of the marketshares of provisioning places according to age for the rural population between 1978 and 1991

1978	*rural mean %*	*less than 35 %*	*35-44 %*	*45-54 %*	*55-64 %*	*65-74 %*	*more than 75 %*
gifts	2.6	2.2*	0.2	-0.7*	-1.0*	-0.3	0.1
autoconsumption	16.0	-3.7*	-0.7	-0.3	3.0*	2.0*	-1.4
butchers	14.6	-1.9	-0.3	-1.2	0.6	1.0	2.3*
groceries	14.5	-0.6	-0.9	-0.4	-0.6	1.1	1.4
hypermarkets	4.5	3.5*	0.6	-0.6	0.2	-1.3	-2.2
supermarkets	11.9	2.0	2.8	-0.2	-1.6	-1.6	-1.0
others	35.9	-1.7	-1.6	3.3	-0.5	-0.9	0.8
total	100.0	0.0	0.0	0.0	0.0	0.0	0.0

1991	*rural mean %*	*less than 35 %*	*35-44 %*	*45-54 %*	*55-64 %*	*65-74 %*	*more than 75 %*
gifts	2.8	2.8*	1.3*	-0.3	-1.2*	-1.2*	-1.0
autoconsumption	12.5	-4.7*	-2.9*	-0.1	3.1*	3.0*	0.5
butchers	9.5	-5.2*	-2.7*	-1.8	0.4	3.3*	6.7*
groceries	7.2	-1.2	-0.4	-0.3	0.6	-0.4	2.4
hypermarkets	11.3	7.2*	1.0	2.5	-3.3*	-2.8	-5.4*
supermarkets	28.5	3.0	5.2*	2.9	0.8	-5.1*	-7.6*
others	28.2	-2.0	-1.6	-2.9	-0.4	3.2	4.4
total	100.0	0.0	0.0	0.0	0.0	0.0	0.0

Coefficients centred on mean; * = significantly different from 0 at 5% level

The table can be read as follows:
In 1978, the average estimated share of hypermarkets was 4.5% of the food budget. Being less than 35 years old meant an increase of 3.5% of the hypermarket share to 8.0%.
In 1991, assuming the structure of the rural population was constant, the average share of hypermarkets was 11.3%, and being less than 35 years old increased this share by 7.2% to 18.5%.

Autoconsumption represents a compensation for an ever-reduced professional independence (Grignon 1980). Autonomy of provisioning protects people from forced purchase of food which has been manufactured for the tastes of wage-earner households (small portions, designed for low-calorie consumption, and short cooking times). Moreover, buying food from farms offends against the values of saving, prudence, and endeavour at work, which were important for 1978 farmers (Grignon 1980). In 1978 autoconsumption was practised at the expense of most shops, especially personalised ones. Farmers nevertheless favoured itinerant dealers, a traditional selling mode in rural areas (Burguière 1975). In 1991, autoconsumption was practised at the expense of the large-scale retailers: whatever the time period, autoconsumption always separates farmers from the dominant provisioning practices.

In 1991, farmers still autoconsumed more than the average, but the gap was smaller than in 1978. On the other side, craftsmen, shopkeepers, wage-earners and unemployed people autoconsumed less than average, but the gap had also decreased. This explains the reduction of the heterogeneity of auto-consumption budget shares according to occupation. Farmers abandon the more time-consuming forms of auto-production, especially those linked to breeding. The labour force has in fact decreased on farms, and spouses who traditionally took charge of the kitchen garden are less inclined to farm (Barthez 1993). That is why the autoconsumption rate was even higher than average in 1991 when the spouse was a farmer: it now depended more on the spouse's occupation than on the husband's, that is more on activities specifically aimed at food autoproduction than on the main farm production.

As autoconsumption is still the distinctive feature of farmers' provisioning, and as the variations of the autoconsumption level find their origins in the transformation of the occupation, we can assert that, *ceteris paribus*, farmers possess a lifestyle which stems from their occupation and evolves with it. A similar reasoning can be made about the lifestyle of food shopkeepers (Larmet 2000).

In 1991 farmers received more autoproduced food gifts than average, which shows another adaptation of their lifestyle. They relied on an increased division of food autoproduction between different farms. This interpretation fits with the increased specialisation of farming and the decline of polyculture. Instead of keeping it for themselves, they rely on family and neighbourhood networks to acquire autoproduced food, so that the consumption of autoproduced food remains the same, although its origin has changed. The taste for raw, unprocessed food persists, but production is split among many farms. The values of self-sufficiency and frugality have decreased for consumption as they have for work, but the method of cooking is still the same. Therefore the time spent on cooking remains the same, while the time spent on food production decreases.

So, as wage earners buy less and less from butchers and grocers, and as farmers have never favoured these shops, the heterogeneity of purchases decreases according to occupation. As much as possible, lower-income households refuse to be made captive customers of groceries and butchers, and turn to self-service small shops. This is another cause of the decrease in heterogeneity. Conversely, the more intensive use of the car for shopping reinforces heterogeneity in butcher shops and groceries. The difference between the evolution of heterogeneity in these two shops comes from the size of households (large families leave groceries, but nothing changes at butchers shops) and generation effects (which increase the heterogeneity at butchers but not in groceries). So overall heterogeneity of purchases decreases in groceries, but is stable at butchers. For autoconsumption, the decrease of heterogeneity according to occupation dominates over the increase of heterogeneity due to car use and household size.

Conclusion

First, we note the enduring character of differences between rural and urban provisioning practices between 1991 and 1978. Rural people buy less than urban residents in hypermarkets and supermarkets because they are further away from them. They buy more from butchers and groceries because they are tied to their advantages: closer proximity, credit possibilities, integration in a local neighbourhood network. They autoconsume more than urban people because they put more emphasis on their autonomy of provisioning and on raw foodstuffs.

However, provisioning practices have grown in similarity for many reasons. Hypermarkets and supermarkets have opened in rural areas, but most of all the increase in commuting and contacts between urban and rural people has developed a desire for access to large-scale retailers, especially supermarkets. They provide a more attractive supply and allow for grouping of purchases as well as the comparison of quality and prices in the same place. Buying in supermarkets takes place in widened rural areas. This explains the growth of car ownership and use. In the extended scale of rural areas, supermarkets have taken over the role of meeting places which was hitherto played by groceries and specialists, while freeing people from the obligation to buy in these local shops. This is the reason why competition from the supermarkets has so much affected traditional provisioning sources: autoconsumption, butchers and groceries. All these elements indicate the possibility of a specific rural approach to practices which symbolise modernity. Both homogenisation and permanence of differences are therefore compatible.

Table 4.7: Evolution of the marketshares of provisioning places according to occupation of the head of household for the rural population between 1978 and 1991

1978	*rural mean* %	*farmers* %	*shop-keepers* %	*middle class* %	*employees, workers* %	*inactives* %
gifts	2.6	-0.3	1.7*	0.3	0.4	-0.4
autoconsumption	16.0	11.38	-6.0*	-6.2*	-3.4*	-1.2*
butchers	14.6	-2.0*	4.4*	1.5	0.5	-0.4
groceries	14.5	-3.4*	0.5	1.0	2.4*	-0.1
hypermarkets	4.5	-2.5	-1.9	-0.1	0.6	1.3
supermarkets	11.9	-2.1	-3.7	2.5	0.7	0.6
others	35.9	-1.0	5.1	1.0	-1.2	0.1
total	100.0	0.0	0.0	0.0	0.0	0.0

1991	*rural mean* %	*farmers* %	*shop-keepers* %	*middle class* %	*employees, workers* %	*inactives* %
gifts	2.8	1.2*	0.5	-0.5	-0.7	-0.1
autoconsumption	12.5	8.4*	-5.0*	-4.9*	-2.5*	-0.7
butchers	9.5	-0.7	3.8*	1.0	1.7*	-1.6
groceries	7.2	-0.6	0.6	0.0	0.6	-0.2
hypermarkets	11.3	-4.9*	-3.6	5.9*	0.0	1.6
supermarkets	28.5	-1.6	-0.3	-1.8	2.4	-0.1
others	28.2	-1.8	4.0	0.3	-1.5	1.2
total	100.0	0.0	0.0	0.0	0.0	0.0

Coefficients centred on mean; * = significantly different from 0 at 5% level

The table can be read as follows:
In 1978, the average estimated share of autoconsumption was 16.0% of the food budget. Being a farmer meant an increase of 11.3% of the autoconsumption share to 27.3%.
In 1991, assuming the structure of the rural population was constant, the average share of hypermarkets was 12.5%, and being a farmer increased this share by 8.4% to 20.9%.

Within the rural population, these general tendencies lead to the reorganisation of social divisions. Simulations show that average evolutions are due more to changes in practices than to changes in the rural population structure. Because of the growth of commuting, the absolute heterogeneity of purchases has increased between commuters (often wage-earners, younger and higher-income households) and non-commuters in developing large-scale retailers, but has decreased in declining shops such as butchers and groceries. Intensifying car use has increased differentiation in every provisioning place; similarly, the difference in consumers' acquired competences linked to different generations has increased the heterogeneity everywhere of purchase by age-group. A distinction effect has caused a transfer of heterogeneity according to income from butchers to diverse specialists, who have gained a better reputation. Meanwhile the development of small self-service outlets has freed lower-income households from their ties to groceries through credit possibilities. Farmers have reinforced their specialisation, and farmers' wives work less and less on the farm. So farmers' autoconsumption has fallen, resulting in a lower heterogeneity of autoconsumption according to occupation.

We therefore have to distinguish strongly between the level of purchase and its evolution. Rural provisioning practices keep their specificities, while converging with urban practices. But even when they adopt urban provisioning practices, rural people keep their specificity, and particularly their preference for face-to-face interactions. This is why rural people still consider interactions in terms of social integration rather then system integration in the sense Giddens gives to these terms.

Annex

The households surveyed in 1978 and 1991 are not the same. So the variations of budget coefficients over time can only be compared at the level of social groups (for instance age groups, or income groups ...). Budget coefficients are modelled as a linear function of:

- a constant;
- a tendency parameter, which describes the average variation of the budget coefficient between 1978 and 1991;
- a set of dummy variables corresponding to the socio-demographic characteristics of households surveyed in 1978. A coefficient is estimated for each modality (for instance, each age group);
- a set of dummy variables corresponding to the socio-demographic characteristics of households surveyed in 1991;
- an error term, accounting for households' idiosyncrasy.

Instead of arbitrarily setting a reference modality to zero for each dimension, coefficients have been estimated with the following restriction: in each dimension, the sum of coefficients, weighted by the number of corresponding households, is contrained to be zero. This method is formally equivalent to to choosing a reference modality. But the constant of the model then becomes the mean of the budget coefficient estimated at the starting year, that is 1978. Thus estimated coefficients describe distances to the mean, for each year.

Practices being constant, the more a fraction of the population becomes a minority (respectively a majority), the more its distance to the mean increases (respectively decreases). So the model we use also records the effect of demographic variations. But we are able to simulate the pure effect of a change in practices, holding the relative weights of every modality constant. We obtain the coefficients corresponding to the pure effect of a change in practices if we apply to the population sample of 1991 weights describing the structure of the population in 1978. We can also reverse the reasoning, that is simulate the pure effect of a change in relative weights of every fraction of the population from 1978 to 1991, holding the practices constant: we duplicate data from 1978 and apply 1991 weights to them.

Notes

[1] See Lambert and Brémaud 1982, for an empirical study. Between 1950 and 1970, the area of inter-marriage and daily commuting widened, and the percentage of inhabitants not born in the village grew.

[2] 'Just like in past days, the *commune* is so narrowly closed in its limits that most social activities overflow. The new sociability spaces of farmers go beyond its boundaries; this holds for most of the socio-professional categories' (Segalen 1985: 327).

[3] For a description of urban conditions, see O. Schwartz, *Le monde privé des ouvriers*, Paris, PUF, 1990.

[4] The only exception are butchers: their average market share has decreased, but the standard deviation is stable.

[5] Level of education and type of region have also been introduced into the model, but we will not discuss the results here. See Larmet 1999.

5 The impact of consumer trust in the Norwegian food market

ODDVEIG STORSTAD

Introduction

In a situation where the food market is becoming more liberal, and the use of technical trade barriers is becoming less feasible, what we call *cultural* and *mental* trade barriers will become increasingly important. By cultural trade barriers we mean the preference for national, regional and local tastes, traditions and eating cultures. Mental trade barriers describe the importance of mental images – or social constructions – that consumers themselves have created around the idea that Norwegian agricultural products are safe. In this chapter, the focus is on *how Norwegian consumers regard Norwegian food production in relation to foreign food production, and how trust is created or constructed in the area of food.* The empirical material is based on a study of consumer reactions to imported beef.

A relatively extensive number of studies on attitudes has documented Norwegian consumers scepticism towards imported food, as well as their high confidence in Norwegian food production (Dalen 1997, Nygård and Storstad 1998). To a certain degree this is true even when, for example, the price of meat from foreign countries is significantly lower than Norwegian meat. This necessarily implies that many consumers find themselves in a situation where the significance of price is *subordinate* to the significance of other factors when the consumers have the opportunity to choose between different alternatives. The alternative factors that may influence their choice may be concern for the environment, animal welfare, personal health, or sensorial quality. If preference is to be given to Norwegian meat, then one of these criteria must be deemed more important than the alternatives. At the same time, there is little reason to believe that consumers are in a situation where they have extensive knowledge about these factors (with the exception, perhaps, of sensorial quality). This implies that their preferences for Norwegian instead of foreign meat must be based on some form of trust that Norwegian meat is produced, processed and/or stored in a way that ensures animal welfare, environmental considerations or the consumer's own health better than foreign meat does.

Initially there is reason to believe that while foreign meat production is associated with terms like 'industrialised' and efficient, Norwegian production remains associated with terms like 'small scale production' and 'near' or 'local'.

At the same time, one of the goals of the Norwegian authorities is to increase more efficient Norwegian meat production. To the degree that this goal is pursued, this means that Norwegian farmers will be in a better position to compete in the area of price. But when we know that many consumers choose products on the basis of other factors, this strategy for Norwegian agriculture may not necessarily be a good one on a larger perspective (Storstad 1998). The competitive edge that Norwegian meat seems to have is to a great extent based on the consumers' trust that Norwegian agriculture is less industrialised than that of their competitors. Organically produced meat has, then, an interesting position. If it is correct to assume that one's own health, animal welfare and/or environment are the factors behind Norwegian consumers' preference for Norwegian meat over foreign meat, then it is also reasonable to assume that organic products would enjoy the same advantages over Norwegian meat produced conventionally, *given that meat produced in Norway is ascribed the same characteristics as foreign meat*, i.e. that it is produced in an industrialised and efficient type of agriculture. Looking at such problems in this way is of significant importance when marketing the various product types, and also in developing policies for agriculture.

This chapter is concerned with how *trust* is constructed in this arena. With the help of interview data from a small selection of Norwegian consumers, I will first attempt to describe consumers' reactions to a concrete situation in which they are faced with having to make a choice between Norwegian and imported meat. The main focus will be on how trust is created and on which type(s) of actor(s) play the most significant roles in the construction of consumer trust. For example, is trust linked to the producer, to the control system, or to more general and abstract images, for example 'Norwegian'?

The second purpose of this chapter is to discuss the importance of trust and the construction of trust in a larger setting around the market. In such a setting, the market for organic products is of special interest. I will therefore use the same manner of thought to speculate in connection with the chances that organically produced meat may have in the Norwegian market.

The choice of meat as the 'case' in this chapter is not by chance. Meat is a product that has a unique position both regarding its use and its symbolic value.

> Meat is the most highly prized of foods. It is the centre around which a meal is arranged. It stands in a sense for the very idea of food itself (Twigg 1983: 21).

In addition to its being a product with traditionally high status (Kjærnes 1996), meat has also recently had negative associations with regard to health (Bjørkum *et al* 1997, Lien *et al* 1998). In their study of attitudes, Lien *et al* (1998) found that 1 per cent of their sample defined themselves as vegetarians, 37 per cent had misgivings about eating meat, and 18 per cent had reduced their consumption of meat as a consequence of their misgivings. Meat has played a central role in connection with the various 'food scandals' that have taken place in recent years. With the exception of scrapie, most of the aspects of risk have been connected with meat produced outside of Norway. The problems most commonly in the media have been in connection with the use of growth hormones, salmonella, mad cow disease and E-coli bacteria. Studies have also shown that meat is the product that Norwegian consumers most often regard as unsafe food (Nygård and Storstad 1998, Dalen, 1997).

It is often argued that importing products from the third world is a way of contributing towards the growth of welfare in these countries; in other words, importation is regarded as foreign aid. In the decade from 1980 to 1990, the export of agricultural products from underdeveloped countries to the western world decreased by as much as 40 per cent (Young 1997: 42). In recent years, the importation of meat from the so-called LDC[1] countries, on the other hand, has shown a moderate increase in Norway. From 2,000 tons of duty-free meat from these countries in 1996, the quantity had increased to 2,300 tons in 1997. In September 1996, the importation of beef from Namibia was stopped because of repeated findings of salmonella; thus in practice, the only LDC country from which Norway imports beef today is Botswana. In 1998, 2,500 tons of meat were imported from Botswana.[2]

Norwegian consumers' confidence in food produced in Norway

Several quantitative studies have been carried out during the last years that show that Norwegian consumers regard Norwegian food products as generally *safer* than food products produced in foreign countries (Wandel and Bugge 1993, Bjørkum *et al* 1997, Dalen 1997 and 1998). In a study from 1998, 35 per cent reported that they regarded food produced in Norway as being safer than imported food (Dalen 1998). Overstating it a bit, we can say that according to Norwegian consumers, Norwegian food *is the same as* safe food.

It is a new experience for Norwegian consumers to find meat labelled as imported in stores. The Norwegian market has been highly regulated regarding the importation of meat; while meat has been imported it has not previously been labelled with its country of origin.[3]

Studies have shown that Norwegian consumers regard imported meat as unsafe food; only 27 per cent regard imported meat as a safe purchase (Dalen 1997). An opinion poll that 4 Fakta did for the consumer report magazine 'Forbrukerbladet Tett På' showed that, assuming prices were the same as meat produced in Norway, 89 per cent of those who were asked would not have chosen beef from Africa (Forbukerbladet Tett På 1/1998). For 7 per cent it would not make any difference if they chose African or Norwegian meat, while only 2 per cent would have preferred African beef to Norwegian beef. From such a study of people's abstract choices it is difficult to conclude what they would actually have chosen had they been faced with a concrete choice in the store. It is especially difficult in this case, where the consumers were presented with a hypothetical case where there was no price difference between Norwegian and African beef. Faced by a more realistic decision, in other words comparing on the basis of a price difference that favoured African meat, there is reason to believe that the results would have been somewhat different.

Dalen (1997) found that only 7 per cent of consumers would want to purchase imported foods, while all of 56 per cent stated that they would not want an increase in importation of foods. It is worth noting that these figures come from a study that centred around food and risk – understood in this context, 56 per cent are restrictive towards importation. This result appears to be in opposition to what Kjærnes (1996) found in his study. In response to questions about conditions that would increase safety in purchasing meat, only 9 per cent stated that they would want to forbid importation. Instead, they would want better control on the part of the authorities (29 per cent) together with labeling and better information for the consumers (27 per cent).

The fact that consumers prefer food produced in their own country to food from foreign countries is a phenomenon *not* unique to Norway. Data from the Eurobarometer in 1995[4] show that 29 per cent of Norwegian consumers favour food produced in Norway. This percentage shows that Norway is almost exactly at the average for the 16 European countries that were in the study. The countries that most noticeably favoured nationally produced products are Finland (60 per cent), Austria (45 per cent) and Sweden (42 per cent). At the opposite end, we find The Netherlands (9 per cent), Belgium (14 per cent) and Italy (15 per cent). This study was carried out before the outbreak of mad cow disease, so there is reason to believe that the results would be different today.

Confidence, risk and the importance of social constructions

If we understand the concept 'food quality' to mean more than the purely sensory aspects, and extend it to include conditions surrounding the manner of production, the meaning of quality will be more in accordance with consumers' understanding of the concept. Grunert *et al* (1996) distinguish between product-oriented quality (the product's characteristics, such as fat content, etc.) production-oriented quality (the manner in which the food is produced, including regard for the environment and for animals) and consumer-oriented quality (the consumer's subjective perception of quality). According to Grunert *et al* (1996) the first two types of quality (product and production-oriented quality) are objective, in other words, measurable; the third type is subjective since it is only the consumers who are able to evaluate the quality. This is essentially correct, but as this paper will show, the consumer's perception of the product-oriented, and even more so the production-oriented quality, is highly subjective. I therefore do not see any purpose in distinguishing between the objective and subjective evaluation of quality, but would claim that the evaluation or measurement of quality has *both objective and subjective elements*. Such an understanding of the concept is necessary in any case in order to be able to analyse consumers' attitudes and their behaviour in today's food system (Storstad 1998).

In a food market that has an increasing degree of open borders and where the consumer is continuously met, for example through the media, with the idea that eating is not necessarily as safe as one has always believed, trust is important (Nygård and Storstad 1998). The consumers must be able to trust that the actors behind a product are offering a product that is safe, primarily, and that it is also what it appears to be. For example, that meat labelled 'Gilde' or 'Nordfjord Meat' is Norwegian meat and nothing else. In addition, they must be able to believe that the producers have abided by all the regulations, and that the control system does what it is supposed to do.

'To have confidence in' in plain language means 'to trust'. That means that the term confidence or trust is concrete in that it can be directed towards an object – towards *something* or *someone*; a person, a system, an institution, a country, a brand name etc. Trust is of great significance in situations that are characterised by *great complexity* and *lack of complete information*. Trust may contribute towards reducing complex situations, thus making complex social order possible (Korsnes *et al* 1997). We know from the theory of rational choice that one important condition is full information about specific products. It is often insufficient to have information on the characteristics of the product itself; the manner in which the product is produced will also be necessary information for the consumer to be able to make a choice that is in line with his own wishes. I will not spend time on arguments about why it is unproductive to have so-called 'full information' as a condition when we analyse the choices consumers make. In addition, the food scandals of recent years and the fact that less food produced locally is sold locally have not made the consumer's search for information any easier. Quite the contrary, this study builds on the assumption that the consumer does *not have full information* about the product and the production process behind it. With this as a starting point, it will be interesting to look at the importance of trust, not only in the product, but also in the actors behind the product, and at the manner in which this trust is grounded or constructed.

Luhmann (1979 and 1988) ties trust primarily to *risk*; in other words, trust must be understood specifically in relation to risk. Trust assumes that one is aware that it is possible to be faced with a situation where one's own choices may bring about risk, and trust will reduce one's evaluation of risk. I can choose to fly from Trondheim to Oslo, even though I know that there is a certain risk linked with this choice, precisely because I trust that the airline company will do what it can to make sure that there will be no risk. Trust, in other words, makes it possible to live with risk. This is especially relevant in areas where, because of time limitations and lack of insight, one must trust in the authority or competence of the 'experts'.

Giddens (1997: 32) defines trust as 'trust in a person or in a system's reliability regarding a certain set of results or occurrences, where this trust expresses a belief in the other's honesty or love, or the exactness of abstract principles (professional knowledge)'. Implicit in this definition is:

- that confidence assumes lack of complete information,
- that trust is fundamentally tied to insecurity, not to risk - in other words reliability in relationship to uncertain results, whether this concerns the individual's behaviour or the system's manner of operation,
- that trust is not the same as faith in the reliability of a person or system. Trust is that which derives from that faith. And
- that trust in the symbol or in the expert system is about the extent to which the system works as intended, not about the way it works *per se* (Giddens 1997: 31-32; 1990: 33-36).

In modern society, *confidence in abstract systems* has become more important. Our system is dependent on people's trust in the system in order for it to function. This confidence, according to Giddens (1997), is anchored to a great extent in our consciousness, i.e. we seldom reflect upon it. For Giddens (1984) confidence is the actual link between the structures of society ('the system') on the one hand, and the individual on the other. Confidence becomes, thus, a key term, since it contributes towards the negotiation between the levels of the actor and of the structure.

Giddens distinguishes between two types of relationships of confidence (Giddens 1990: 83-88):

- *Facework commitments*, which assume that the actors are in the same place geographically and have a direct relationship. In our context this is the type of confidence that we have towards the personnel in a food store and to the individual primary producers.
- *Faceless commitments*, which are relationships of confidence that are tied to the abstract systems in society. This is the type of confidence that we can

build in relationship to, for example, the Norwegian Food Control Authority.

According to Giddens, trust relationships have changed in character during the transition from traditional to modern society. This has occurred primarily because the 'world has become larger'. Previously the dimension of time/space was smaller, and face-to-face relationships were thus dominant. This is especially true in the area of food. Previously, food was for the most part produced and consumed locally; this naturally made it possible to have trust that was founded on knowledge about the individual producers. The food chain was shorter and there were fewer actors to relate to. This is part of what the Ministry of Agriculture is attempting to recreate through the competitive marketing strategy called *Norsk Gårdsmat* (Norwegian Farm Foods). They are attempting to build up a relationship of trust founded on 'face-to-face' relationships that are based on visits to, contact with, and knowledge of the individual producer.

The Norwegian experience with scrapie, which struck many herds of sheep in 1996, can serve as a good example of the significance of consumer trust. Scrapie followed in the wake of mad cow disease (Bovine Spongiform Encephalopathy - BSE) and both cases were about the prion diseases that strike animals. In contrast to mad cow disease, there is no indication that scrapie can be transmitted from animals to people to develop the human form of BSE, Creutzfeldt-Jakobs disease. Especially during the beginning of the media's coverage of scrapie in 1996, doubt was cast on whether the disease could infect people. In order to retain the consumers' trust in Norwegian mutton in particular, and Norwegian agriculture in general, a largescale slaughter of sheep was initiated. Not only infected animals were slaughtered, but also all animals that at any time had been part of a herd where there was any evidence of scrapie were slaughtered and burned. There is still no indication of any relationship between scrapie and Creutzfeldt-Jakobs disease; in addition, large quantities of healthy animals were slaughtered. Thus, it was not a 'real' risk linked to eating mutton that brought about the extensive slaughter, but the fear (on the part of the food industry) that the consumer's perception of risk, what can be described as the 'felt' or socially constructed risk, would weaken the trust in Norwegian meat products.

Considering trust in connection with risk, or perhaps more correctly, uncertainty, in the manner Luhmann (1979 and 1988) recommends, would seem to be fruitful in the area of food. In addition, scrapie is a good example for showing how the consumer's understanding of risk does not necessarily coincide with 'real' risk (MacKenzie and Berrier 1987). 'Real' risk is the *statistical probability* that a negative occurrence will actually take place (the probability that a plane will crash, that someone will get salmonella from eating meat or eggs, that scrapie can be transmitted to people, etc.). This is the form of risk that experts use, but it is not uncommon for experts and the average person to have differing definitions of risk (Margolis 1996). Experts often keep to strictly defined terms such as 'expected annual mortality', while the average person often includes other factors in his definition of risk (Slovic 1994). One such factor is the consequence of a negative occurrence, in other words the extent of the loss one would experience were the occurrence to take place. As long as the consequence of eating imported meat may be a serious salmonella infection, it matters little that the statistical probability of being infected is zero point zero something. Thus, a 'felt' risk will have consequences for the consumer's choice of action, *independent* of how great the real risk is in the eyes of the experts.

The combination of the consumer's inclusion of the size of the loss in his evaluation of risk and the fact that the consumer will often lack information about the 'real' risk means that the risk will often be socially constructed. It follows that trust will also be based on social constructions or the mental images each individual constructs about the relationship between the manner of production and risk. Positive mental images of the manner of production will create trust, while negative images may give grounds for lack of trust, not only in the actors behind the product, but also in the product itself. Since trust is often based on social constructions, trust is a *fragile phenomenon* – when the mental images change (social reconstruction), trust also changes.

Model for analysis

In the interpretation of our results, the term 'trust' will be decomposed by showing that trust that in a product living up to expectations can also imply trust in a range of underlying actors. Choosing to buy beef from Botswana may imply trust in the local farmer, in the Botswanian or the Norwegian control system, or in the abstract idea of 'Botswana' or 'Africa'. This implies that public information, for example about the quality and safety linked to imported meat, may be difficult to communicate. Likewise, trust in Norwegian meat is analysed according to the hypothesis that the actors behind the product (the producers, the control system, etc.) are of significant importance in the choice a person makes as a consumer in a concrete situation.

The empirical data this chapter is based on show that the consumer's confidence is directed not only at the product itself (the characteristics of the product); in addition, they show that *the manner in which the product is produced* is also significant. In an analysis of consumer trust it is therefore important to bring in the actors behind the product. Second, the consumer's idea of who is the 'seller' of a product varies. Since one must assume that confidence in whether or not a product is a safe purchase is related to the degree to which one trusts the actors behind the product, these actors may be regarded as those who guarantee safety, and thus are also regarded as 'sellers' of the product. Third, the 'seller' is often perceived as an abstract idea.

Figure 5.1: A model for explaining the market potential for organic, conventional and imported food

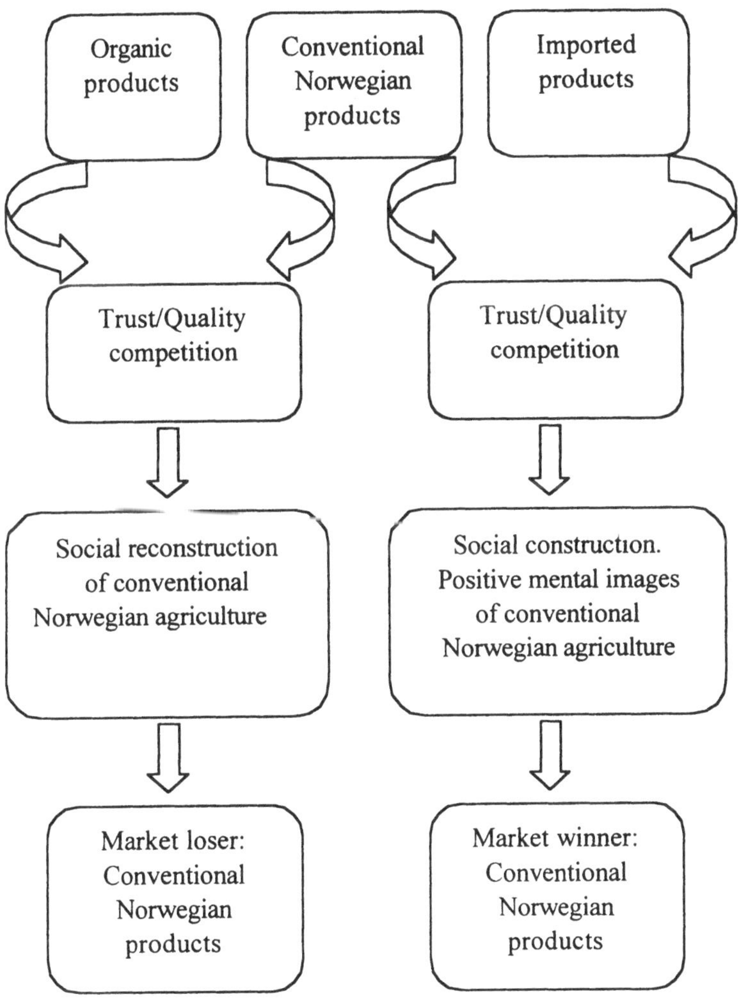

As already indicated, the price differences between imported and Norwegian meat can be significant (in this specific case, approximately 30 per cent in favour of imported meat), and from a purely economic point of view, we would explain the choice of imported instead of Norwegian meat as competition in prices, and not include qualities like animal welfare, health or environmental consequences in the model. By including trust and the manner in which consumers explain their trust, I would include an alternative way to regard this on the market level: I would propose that the competition between Norwegian and imported meat is a *competition in quality or confidence*. Only the right side of Figure 5.1 will be demonstrated empirically in this paper.

From what we already know from quantitative studies about Norwegian consumers' preferences for food produced in Norway, there is reason to hypothesise that this is largely grounded in people's positive mental images of Norwegian farm production. If we look at this from the point of view of competition in trust or quality, there is reason to claim that conventional Norwegian farming leads in this competition today. If we further assume that the same mechanisms or factors which contribute towards the consumer's understanding of which products are better than other products from an environmental, animal welfare and health point of view, then there is reason to include organic products in the model.

If the consumer's image of the way in which food is produced is included in the concept 'food quality', then organically produced food will, by definition, be regarded as having higher quality than both conventionally produced Norwegian food and imported food. This is in spite of the fact that it is difficult to find scientific documentation that organically produced food has a higher quality than food produced in the conventional way. This is true for all measurable characteristics of quality such as taste and nutrition. In other words, it is difficult to find any documentation which shows that in regard to health, organically produced food is a better product than conventionally produced food. But when 'the manner in which food is produced' is included as part of the concept of quality, organically produced foods are regarded as having higher quality. We then have the following *hierarchy of preference:*

- meat produced organically;
- Norwegian meat produced in the conventional manner;
- imported meat.

Therefore, towards the end of this article we will discuss the significance the consumer's broader understanding of quality as applied to these three manners of production may have for the marketing potential of these products.

Data and method

In January 1998, some of the 'Bunnpris' stores in Trondheim started selling imported beef from Botswana. The collection of data for this study took place in the period March-April 1998, and the difference in prices between Norwegian and Botswanian rump steak and fillets were 38 per cent and 27 per cent respectively. The imported meat was displayed in the same counter and was clearly labelled *Imported, Special Offer* but there was no information about the country of origin. The selection comprises two groups of customers:

- those who *bought imported meat*, and
- those who *clearly noticed* the imported meat, but chose to *buy Norwegian meat.* I tried to be in the store at various times throughout the day: mornings and afternoons on weekdays and on Saturdays. I conducted most of the interviews on Friday afternoon or on Saturday, since most people who bought beef bought it at those times. In Norway, beef is typically served over the weekend.

I proceeded by positioning myself where I could have a good view of the meat counter in order to see if customers chose imported meat or not. It was obvious that the customers were aware that the meat was imported, and that for many of them it was the first time they had observed such labelling. One of the things which struck me was that most of them took their time while they were making their decision. They picked up the package, squeezed it and put it back again - or put it in their shopping trolley. In those cases where more than one family member was present, it was clear that when they took notice of the imported meat, they had a discussion about it. Altogether, I had the clear impression that those who chose not to buy imported meat, as well as those who chose to buy it, all made reflected decisions.

As a rule of thumb, I would stop the first person I discovered who either bought or did not buy imported meat after giving it clear consideration; in this way I attempted to avoid selecting according to various visible criteria (gender, age, style of clothes, etc.). I believe I achieved this, with a couple of exceptions, when I did not stop customers in the oldest age group (above 75 years of age). The reason for this was that I had problems with a couple of the interviews I had done with the oldest customers. They were not able to understand what I wanted to ask about, they were hard of hearing, and it was difficult for them to concentrate while they were in the store. Of those I interviewed, there were six men, ten women and eight couples; eight persons/pairs between 20 and 39 years of age, twelve persons/pairs between 40 and 59 years of age, and four persons/pairs over 60 years of age.

I carried out a total of 24 interviews, each of which lasted from 10 to 15 minutes. Of the 24 customers I interviewed, there were six who bought imported meat, and eighteen who did not.[5]

Collecting data in a food store has its limitations. The store owner demanded that the interviews not be too long, and for practical reasons, it was nearly impossible to use a tape recorder. After having evaluated several alternatives, I decided that it would be best to make a simple questionnaire with both open and closed questions.

Initially I wanted qualitative data, but was dependent on a quantitative method (questionnaire), partly to facilitate the informants' (customers in the store) participation as much as possible. Norwegian consumers' scepticism towards imported meat has already been well documented through quantitative studies, so I wanted to use this study to take the first step towards a more qualitative approach which could contribute towards a deeper understanding of the consumers' scepticism. Subsequent use of qualitative studies to achieve a deeper understanding of quantitative results of research is a relatively little used, but interesting, research strategy (Grønmo 1996). In addition to checking off the respondents' answers to questions, I also made notes of everything the informants said. Most of their statements were explanations for answering as they did. After I made *notes of everything they said* about their answers to the questions, I retreated in order to write down what they had said in detail, since I had not been able to write more than fragments and key words during the interview itself. In this study, the analysis has been made on the qualitative data.

The construction of 'Norwegian' and 'foreign'

In response to the question about what they think about Norwegian food production, the consumers in my study answered by comparing with conditions in other countries. They had read, seen on TV and heard so much about salmonella, hormones, genetic engineering, mad cow disease, etc., in other countries, that they concluded that Norwegian farm production came out high in comparison. In general, everything that is dangerous is in foreign countries. It is the case that incidences of infections, both in animal herds and in food products, are low in Norway compared with many other countries (Skjerve *et al* 1996). Norway is one of few countries in the world with a low prevalence of salmonella (The Norwegian Food Control Authority, 2000). There have been few food scandals in Norway, and the food scandals we have seen in other countries seem to have changed the relative conditions between Norwegian and foreign produced food.

We see that large units in production and distribution are dangerous. That has become clear from the scandals during the past few years.

In other words, the various food scandals have created negative mental images, or social constructions, of how food is produced in foreign countries. That means that there is reason to believe that the *relative* opinion of consumers today is more positive towards Norwegian agriculture than it was before the 'age of mad cow disease'.

Viewing the consumers' image of that which is 'Norwegian' and that which is 'foreign' as a dichotomy, however, is too simple in describing their interpretations of which products and which manners of production take care of the consumers' demands for safe food. Many consumers have clear opinions of which countries or continents from which they would not buy food. The Netherlands, Belgium and the USA are countries that are linked with the use of hormones in their production of meat.

Those cows in the USA have pretty big legs.

There [Belgium] they go and inject an awful lot of hormones. And it's dirty there, too.

Eastern and southern Europe, Asia and Africa are also areas that are often mentioned in response to questions about countries from which they would not want to buy meat. These areas appear to be strongly tied to lack of hygiene and they do '*not have control like we do*'. The last area that is seen in a negative way includes the so-called 'EU countries'. These appear to be regarded as one and the same, and because '*we see so many strange things on TV and read in the newspapers about all kinds of strange things that are going on there*', there are many who reject 'EU-food' on the grounds that it is unsafe.

The trust in meat from Sweden appears to be nearly as strong as trust in Norwegian meat. '*Of course we have often been in Sweden to buy meat, and we are still living*'. Nineteen of the 24 persons I interviewed had very great or great trust in meat from Sweden. Because the prices of meat are lower in Sweden than in Norway many Norwegians living along the Swedish border frequently buy meat in Sweden. There are therefore reasons to believe that the trust in Swedish meat has to do with personal experience. The confidence in the meat from Denmark appears to be somewhat lower, 14 persons (of 24 interviewed) said that they had very great or great confidence in meat from Denmark. Norwegians living in the south of the country are used to buying meat in Denmark because of the low prices (compared with the prices in Norway), but negative publicity about Danish agricultural practice may be the reason that some have less trust in Danish meat: '*It's a little different there. I've seen that on TV programmes*'. '*The way they operate is a little difficult to accept*'.

Denmark and Sweden are our neighboring countries, and the closest we come to ourselves. Most of us have been there often, eaten their food and purchased considerable amounts of meat to take home with us. One can perhaps say that we have a 'face-to-face' relationship with these two countries, and on a macro level, have developed confidence in the meat from these countries that is based on 'facework commitments'. But, as we have seen for the case of Denmark, negative publicity may make some question this trust.

When we asked about meat, we specifically included countries that are further away from Norway, both geographically, mentally, and with regard to identity.

Some of those who bought meat from Botswana seemed to have a positive picture of farm production in Botswana. They pointed out that farming in Botswana is less industrialised than in the western world.

> I don't think it is so industrialised there.

> Had the difference in prices been less, I would have chosen Norwegian meat. I actually feel a bit guilty in that way. The first time I bought imported meat – that was also from Botswana – we discussed it at home that evening. We agreed then that the cows in Botswana are 'happier' than cows in the West. We had the idea that they do not have the same demands for efficiency and industrialisation there that they do in western countries. It's not like it's a factory production.

A common denominator in the consumers' evaluation of countries from which they would not buy food is linked to the manner of production – hygiene, use of growth hormones and negative effects tied to the demands for high efficiency in production. This supports our initial claim that we must understand the concept 'quality of food' as an extended definition; in addition to the characteristics of the product itself (taste, consistency, etc.) the manner in which a product is produced is part of the consumers' understanding of the quality of food.

In response to direct questions about why they said that they had a high or low level of trust in meat produced in a specified country,[6] not everyone had any exact reason. It was more a feeling they had, but did not quite 'know how to say it'.

> Don't know anything about Botswana. Am maybe more sceptical, but don't really know why.

> I am sceptical. I have more trust in Norwegian meat. That maybe isn't right, of course, but I think that's the way it is.

Confidence in Norwegian meat is undoubtedly high; none of those I interviewed gave the impression that they had a low level of confidence or they did not trust Norwegian meat. Thus, there is no reason to believe that the choice of imported meat has anything to do with lack of confidence in Norwegian meat. On the contrary, there is reason to believe that people have no reason to distrust Norwegian meat. With the exception of scrapie, which struck sheep and lambs, there have been no food scandals tied to Norwegian meat. Confidence in Norwegian meat products and production, together with scepticism towards that which is foreign appears to be decisive for those who do not choose the cheapest alternative.

> I always choose Norwegian [meat]. You hear so much about hormones and medicine and things like that.

> I don't feel sure about the quality. I feel a little uncertain – don't think it is as good as Norwegian [meat]. Maybe it tastes different, or maybe you might get sick from it.

It is also worth noting that the scepticism people express concerns their own health – they are worried about what the meat might contain that could cause disease. The explanations that were more political, environmental or ethical in regard to animals received only marginal attention on the part of the informants.

Trust in the 'system'

Trust in the Norwegian products and the Norwegian manner of production appeared to be high. Trust in what was 'Norwegian' was also apparent in another area: people have great confidence in the *Norwegian food control system*. This is supported by other studies, where 84 per cent stated that they have a very high or a high level of confidence in the Norwegian Food Control Authority (Dalen 1997). It is the *Norwegian* regulations, standards and control which appear to be a guarantee that imported meat is safe. These are the things brought up by those who buy imported meat. At the same time, several of the customers are aware of the fact that the control of imported food takes the form of spot checks and that it is not possible to trust the control system completely. None of those who were interviewed mentioned the name of the Norwegian Food Control Authority, but they used terms 'the control', or 'the control system' or 'the people who control the food'. This can be interpreted as trust in the control system in general – *an abstract form of trust*. People trust that 'the system' works as it is intended to work. A further interpretation would be that there is greater trust in that which is 'Norwegian' than there is in the concrete system. In other words, people have confidence in the control system without knowing about it in detail.

Studies have shown that compared with people in many other countries, Norwegians generally score high when it comes to confidence in various parts of 'the system'. Comparative studies which cover several aspects of political trust show that Norway is nearly always at the top (Listhaug and Wiberg 1995; Listhaug 1995 and 1998). Both regarding the Government and the National Assembly, Norwegians score higher than other western countries. Sixty six per cent of Norwegians have a positive attitude towards their government, while 69 per cent have a positive attitude towards their National Assembly (Listhaug 1998). Switzerland is the country which scores closest to Norway with 54 per cent having trust in their government and 47 per cent in their national assembly. Regarding public authorities (civil servants), the differences are smaller. Norwegians, Americans and the Swiss are all at about the same level (51 per cent confidence) (Listhaug 1998).

It is reasonable to interpret the high level of abstract trust that Norwegian consumers demonstrate towards the Norwegian Food Control Authority in connection with their generally high level of confidence in the authorities which comprise 'the system'. Putting it simply, one can say that as long as the meat is in a Norwegian store, it must have been in contact with the Norwegian control system, so people feel sure that the meat cannot be detrimental to their health. People trust that the Norwegian control system functions as intended, in other words, that no meat that is dangerous to health will be sold in Norwegian stores.

There appears to be reason to conclude that trust in that which is *Norwegian* is of key significance for the consumer's choice between Norwegian and imported meat. This is true for trust in both the Norwegian production of food and the Norwegian Food Control Authority. In addition, it is worth noting that it is the uncertainty around the aspects of imported meat related to personal health that is instrumental in the decisions people make; it is not the concern for animals or the environment. It is probably correct to maintain that consumers regard all of these aspects together, since intensive farm production is generally viewed as exploitation of both the environment and animals. But the detrimental effects such methods of production have on the meat are explicitly articulated as factors of uncertainty regarding human health.

The situation in which consumers are faced with new of choices – Norwegian or imported meat – 'activates' or forces a person to *reflect* upon trust. Studies have shown that concerns about food are not something which characterise the consumers' daily life, but when they have been asked explicitly about concerns connected with various potential risk factors, there is reason to conclude that a certain amount of uncertainty is present (Dalen 1997). As already mentioned, this uncertainty is primarily related to food that has been produced in countries other than Norway. In a study from 1998, 5 per cent state that they feel a little uncertain in regard to the contents of the food which is produced in Norway, while 72 per cent feel completely or quite certain.

Twenty two per cent state that they do not think much about whether the contents of food produced in Norway are safe or not (Dalen 1998). In addition, we know, as already mentioned, that meat is the main product that is regarded as an unsafe import product (Dalen 1997). When a consumer is then faced with the choice between Norwegian and imported meat, he or she will reflect upon the uncertainty they feel; confidence will then be an influential factor in the specific situation where a decision must be made. The consumer will attempt to minimise the 'experienced' risk or, more correctly – uncertainty – by choosing the product he or she experiences as being the least unsafe.

In the explanation of this choice there will be one or more actors which are regarded as those who guarantee safety. They may also be regarded as the 'sellers' of a product. Choosing meat produced in Norway is primarily choosing the abstract 'manner in which food in Norway is produced', which is regarded as a guarantee for the meat's qualities regarding health. For those who choose meat from Botswana, it is primarily the Norwegian control system that may be regarded as the 'seller' of the meat, but the mental picture one has of farm production in Botswana is also of significance. This indicates that in addition to the significance of the control system being Norwegian, the country the meat comes from is also of importance.

The empirical data from this study also show that the attitudes towards Norwegian agriculture are primarily based on comparisons with 'the way they produce food' in other countries. For example, the food scandals of recent years have led people to the relative view that puts Norwegian farm production in a positive light. In this type of competition in confidence and quality, there is reason to claim that Norwegian agriculture and Norwegian farm products have the leading edge. Putting it simply, it appears that the general opinion is that Norwegians do not have 'that kind of factory farming', nor do they want it. In many ways, this goes against the idea that the value of a product increases with an increase in processing. In the area of food, the opposite would seem to be true: '*Value has been added by naturalizing the product rather than industrialising it*' (Arce and Marsden 1993: 294).

Collecting and analysing the material has made a *paradox* apparent: in spite of the rejection of the idea that Norwegian agriculture should become more industrialised, there is a clear wish that food prices be reduced to a level comparable with the prices in countries with more intensive agriculture. In other words, there is reason to conclude that the consumers see price and quality (safety) in relation to each other. These results are also supported by quantitative studies (e.g. Dalen 1998).

Implications for organic products

As mentioned in the introduction, I would like to use the model of analysis (Figure 5.1) which has been developed on the basis of this study, to speculate about the implications it may have for organically produced meat. This gives rise to a hypothesis that, from the consumer's broad understanding of quality of food, there is a hierarchy of preference with organically produced meat at the top, then conventionally-produced Norwegian meat next, and imported meat at the bottom. This hypothesis has not been the object of empirical testing. All the same, it would be interesting to speculate about it.

Several empirical studies from other countries have shown that when consumers buy organically produced products they are primarily motivated by their concern for their own health (Schifferstein and Ophuis 1998). Similar studies have not been done in Norway, but we can assume that Norwegian consumers do not differ greatly from their counterparts in other European countries. We can say that the parallel is clear regarding the empirical results on which this study is based; regard for one's own *health* is decisive in determining whether the quality of one product is considered better than the quality of another. Concern for the environment and animal welfare is secondary.

We often refer to Denmark when we want to show that few organic products are produced and consumed in Norway. Why are Norwegians not as 'good' as the Danes in this area? In this debate, the factors usually used to explain the differences are poor distribution, that farm co-operatives hold back development, that there is little interest on the part of the retailer, etc. But it is also possible to interpret this through the model presented here. As this paper has shown, Norwegian consumers have a very positive picture of Norwegian food production. This picture has been developed as a result of comparing Norwegian production with foreign production. Using the same logic, it should be reasonable to present the hypothesis that the demand for organic products is less in Norway than in other countries precisely because Norwegian consumers have a very positive picture of conventional Norwegian agriculture. In other words, Norwegian consumers regard conventional Norwegian agriculture as more 'organic' than Danish consumers regard their Danish conventional agriculture.

From this frame of reference it would be natural to conclude that the best way for organic producers to profile their products would be to point out the negative conditions in conventional national agriculture: the use of insecticides, artificial fertilisation, etc. In other words, they should initiate a competition in confidence or quality with conventional Norwegian agriculture.

Using such a model for analysis (Figure 5.1) it is simple to interpret a dilemma for TINE Norwegian Dairies. At present, organic milk in Norway is part of the range of products provided by an agriculture co-operative (TINE Norwegian Dairies). The Norwegian Dairies co-operative has received a certain amount of criticism because of its lack of initiative in getting the organic milk to the consumers.[7] Currently, some of the organically produced milk is mixed together with conventionally produced milk at the dairy, so it never reaches the customer under the label 'Dalsgården' [Valley Farm] which is the name of the organic milk. The reason for this is that the demand for organic produced milk is lower than the supply.

It is difficult to imagine how TINE could advertise organic milk without simultaneously conveying the idea that the conventional milk they offer their customers is of poorer quality. When TINE advertises and sells both juice and milk, they do not compete with themselves about the quality – they can claim that both the milk and the juice they produce are of the highest quality. It is far more difficult to do the same with regard to the consumers' choice between 'regular' milk and organic milk. There is no reason to believe that the total consumption of milk will increase as a result of introducing organic milk. Along this same line of thinking, it is easy to understand why they have not used 'organic' as part of the name of the organic milk; it is called 'Dalsgården' [Valley Farm].

Conclusion

Today's Norwegian agricultural production enjoys, whether deserved or not, a high level of trust in the eyes of Norwegian consumers. The challenge for Norwegian agriculture is therefore *to hold on to and maintain* this trust. This is only possible if they 'keep their path as clean as possible'. In other words, they must make sure that the difference between the consumers' image of the 'way food is produced' in Norway corresponds *as closely as possible* with how Norwegian farmers actually operate. Let me illustrate what I mean with two simple circles. The patterned circle represents the Norwegian consumers' picture of Norwegian farmers and Norwegian agriculture - the way food is produced in Norway. This image is created or constructed in various ways and will of course vary somewhat from consumer to consumer. But on the whole, we know that Norwegian consumers have a positive impression of Norwegian farm production. They have a high level of confidence – they believe that Norwegian producers act in accordance with the consumers' interest in safe food. The other circle (white) represents 'reality' – Norwegian farmers' praxis, the manner in which they produce food. Included here is their regard for the environment, use of insecticides and animal welfare.

Figure 5.2: The relationship between the consumers' interpretation/ understanding and the farmers' practice of food production

Whether today's situation in Norway is best illustrated by situation 1 or 2 in Figure 5.2 is an empirical question. But there is little doubt that Norwegian agriculture would have a major problem if consumers were to discover a great discrepancy between what they think is 'the manner in which food is produced' and reality. In such a situation, a social reconstruction of Norwegian farming might take place. Thus the relative differences between the view of conventional Norwegian agriculture and organic production would become greater, while the perception of conventional Norwegian agriculture and of foreign production would become more similar. The result would be that conventional Norwegian farm products would not only lose part of the market in favour of organic products, but they would also lose ground in relation to imported products. This is because consumers would then reconstruct their image of conventional Norwegian agriculture by asking whether Norwegian farming is so much better than foreign farming. Or in the terminology of Beck (1992) and Giddens (1997), consumers would achieve a reflective relationship to the more or less unreflected trust they today seem to have in both Norwegian food production and the Norwegian control system.

Notes

[1] LDC stands for the least developed countries; 173 countries are classified (by the World Bank and UNDP) as LDCs. These countries benefit from the GSP which is organised under the WTO. It is up to the individual country to set the reduced duty rates for imports from countries classified as LDCs.

[2] Botswana is currently classified as a country with an average HDI (Human Development Index), but is included in the Norwegian general tariff preference system. Of the 129 consignments of meat from Botswana that came to Norway in 1998, three were rejected after the Norwegian Food Control Authority found salmonella in the meat (*Nationen*, 10.03.1999).

[3] Today there are no regulations requiring that meat be labelled to show the country of origin. From a consumer's point of view this may be experienced as problematic, but from the year 2000, requirements for labelling with the country of origin will be required in the EU.

[4] These data have been provided by the Norwegian *samfunnsvitenskapelige datatjenest* (NSD). They have been taken from the *Eurobarometer* 44.1, and the Norwegian research was carried

out by NSD. NSD is not responsible for the analysis of the data or the interpretations that have been made here.

[5] Since the method used does not assure a representative selection, it is not possible to conclude that 25 per cent of the consumers would buy imported meat, given the price difference in this case.

[6] In the questionnaire I had questions asking the customers to state the level of confidence they had in meat from the following six countries: Denmark, Belgium, Great Britain/England, Sweden, Belgium and the USA.

[7] Only 1 per cent of the milk produced for consumption and sold in Norway today is organically produced. The difference in price between conventionally produced milk and organic milk is considerable (20-30 per cent).

6 Towards a consumer images approach – exploring the quirks of modern food consumer behaviour

J.C. DAGEVOS and H.J.M. HANSMAN

Consumption blooms

The food market is increasingly becoming a consumer market. The development of the food market from a product-driven to a consumer-driven orientation is firmly embedded in, as well as consistent with, a process which characterises the wider socio-economic context. Affluent states and economies have shown an unprecedented rise recently. Both set off consumerism. The consumer society has taken root and flourishes in these fertile – or, affluent, for that matter – grounds. This profound and far-reaching tendency resonates in recently published scholarly literature. Nowadays, strategic management books declare time and again that consumer-mindedness is a *sine qua non* to maintain economic growth; competitiveness and commercial success thrive to the extent that enterprises are able to meet the fast-changing preferences and demands of individual consumers.

In sociological discourse, consumers and consumption are also gaining growing attention. Although it remains difficult for sociologists to study consumption and consumerism in a non-normative, i.e. objective, way, it is acknowledged that 'it is clear that contemporary consumption is too complex to warrant a simple good/bad judgement' (Belk 1995: 73; see also Bocock 1993: 110 for a similar conclusion). Whereas scholars who severely criticise consumerism and ('*over*') consumption are not difficult to find, their objections to modern consumption practices, in which consumerism as a way of life reigns supreme and commercialisation is a deeply ingrained feature, do not reduce the importance of consumption. On the contrary, many sociological scholars are convinced that consumption is the name of the game. Nowadays it is possible to detect a nascent contemporary sociology of consumption (as a proof of this claim, the following might be cited: Baudrillard 1998; Bocock 1993; Corrigan 1997; Featherstone 1991; Fine & Leopold 1993; Firat & Dholakia 1998; Gabriel & Lang 1995; Miles 1998; Miller 1995; Ritzer 1998; Slater 1997; Statt 1997, to name but a few).

Many sociologists endorse the perspective that 'it is time for sociology to throw off the old production-dominated paradigm in favour of a new, consumption-oriented one' (Campbell 1995: 96). George Ritzer sheds light on the way that the socio-economic centre is changing from production towards consumption in the course of *The McDonaldization Thesis* (see Ritzer 1998: 68, 70, and 118). In the same vein, Peter Corrigan (1997: 182) speaks of consumers as 'the perfect creatures of capitalism', and proposes that 'perhaps it is time to stand Marx on his head and claim that consumption, and not production, is the central motor of contemporary society' (Corrigan 1997: 1). Even Jean Baudrillard, who walks in the footsteps of such founding fathers of sociology as Max Weber and George Simmel, and who cherishes the opinion that consumption is a function of production, nevertheless notes: 'The system needs people as workers (wage labour), as savers (taxes, loans, etc.), but increasingly it needs them *as consumers*' (Baudrillard 1998: 83).

These examples illustrate that much of the prevailing sociological literature understands the shift that is going on. The shared presumption of various economists and sociologists mirrors the direction in which modern economic and social life is developing. *Consumer* capitalism is unfolding as consumption increasingly invades the human landscape. Its relevancy to research ('theory') as well as business and marketing ('reality') becomes more and more visible. This shift from a production-led way of thinking and acting towards a demand-led one, which is picking up real momentum, marks a break with the socio-economic perspective of the Industrial Age which focused on the supply side of economies and the means of production labour, nature, and capital.

The advent of the emotion economy

A mere economic or materialistic idea of consumption and the consumer society is today insufficient and unrealistic. In *addition* to this perspective, so it is thought, consumer society and its inhabitants should be regarded from socio-cultural and socio-psychological points of view.

In this respect, one is tempted to suggest that there is some sort of broader socio-cultural renaissance to be found. It appears that topics such as trust, cooperation (shared) values, arouse greater interest in professional as well as public circles. Probably it is not an exaggeration to perceive a kind of 'softening' or maybe even 'spiritualisation of economic life', to use the title of the final chapter of Francis Fukuyama's book *Trust* (1995). This sign of the times is found in particular in two recent thought-provoking books.

The first is Joseph Pine and James Gilmore's *The Experience Economy* (1999), and the second is *The Dream Society* (1999) by Rolf Jensen. These authors envisage that we are on our way to a change of logic in consumer behaviour. To give readers a quick flavour of their character and contents, we offer some quotes:

> No longer do customers purchase goods merely for their functional use but also for the experiences created during purchase and use.
> The emerging Experience Economy opens up possibilities for new strategies of moving beyond goods and services as the primary source of profits.
> We are, in fact, living more and more in an intangible economy, in which the greatest sources of wealth are not physical. We aren't yet used to an economy in which beauty, amusement, attention, learning, pleasure, even spiritual fulfilment are as real and economically valuable as steel and semiconductors (Pine & Gilmore 1999: 100, 143, and 187).

> [F]or much of the world, the focus of acquiring more and more of the material aspects of life is drawing to an end. Instead, we have a growing interest in humankind's emotional side.
> In other words, you may say that the Dream Society marks the end of humankind's multimillenial epoch of material domination; at the same time, it is the first postmaterialistic era – the beginning of something new.
> In the Dream Society the emotional aspect of consumption will play an ever more important part (Jensen 1999: 7, 50, and 112).

Leaving their differences aside, these books have in common the idea that important aspects of social interaction in private and day-to-day life should not be neglected when considering professional or product-consumer relations. Explicitly or implicitly, the authors stress the value of the different types of glue that bind relationships, such as reliability, loyalty, dependability, communicative and collaborative skills, 'emotional intelligence', or creating a good atmosphere.

The latter in particular brings us to another item that can be deduced from this literature: de-materialisation. This refers to the idea that neither the price nor the intrinsic qualities of products exclusively or automatically influences the decisions modern consumers make. Their behaviour is also triggered by less obvious and calculable factors which affect their attitudes, habits or needs. De-materialisation refers to what is *behind* or *around* a product, rather than the physical product itself. In other words, the 'atmosphere' of a particular product or brand – or, to use Jensen's phrase, the story side of a product – is of paramount importance. Quality and price are not sufficient to pursuade consumers to pay attention to and eventually, pay for, certain goods or services. Consumer behaviour is a very subtle phenomenon indeed – or rather, a very *sensitive* affair. The (post)modern world of today is, at least partly, an *Erlebnisgesellschaft* [Experience Society] (Schulze 1997). According to Pine and Gilmore (1998, 1999) what we call a 'sensationalisation-development' will establish itself in a societal setting in which a vibrant experience economy prevails.

Inspired by Pine and Gilmore's as well as Schulze's and Jensen's line of reasoning, we prefer to announce the advent of a so-called *emotion economy*. This is, in short, largely based upon the satisfaction of nonmaterial needs, that is, the fulfilment of emotional needs. In terms of food consumption: consumers feel drawn to the *taste* of *feeling good*, or have an *appetite* for *sensual* fulfilment (see also Dagevos & Hansman 1999). Food consumer research, in turn, needs to find an empirical basis for testing whether it is in fact true that consumer purchases are increasingly made for emotional reasons – which is anything but a trivial problem.

Despite the ongoing process of 'rationalisation', we would caricature human beings if we do not treat them as 'coloured' by emotions – as the portrayal of *homo economicus* usually does. It is, in this respect, reassuring that emotions are receiving growing attention from economists lately (see Elster 1998). Nobel prize laureates Herbert Simon and Amartya Sen have always proclaimed that the complexity of *homo sapiens* should not be confused with the simplicity of *homo economicus*. These and other scholars never stop cautioning against blurring the difference between the consumer as a *homo economicus* and the actual reasonings of flesh and blood consumers with their attendant inconsistencies and uncertainties. Fukuyama, for instance, provides an up-to-date vision of *homo economicus*. He entertains the thought that this idea of human behaviour is not terribly realistic in its depiction of how people actually behave, saying that 'we call people who calculate their self-interest with absolutely cool rationality psychopaths, not normal human beings' (Fukuyama 1999: 185). This kind of clockwork rationality may be interesting from a strictly utilitarian *theoretical* point of view; however, when we are interested in exploring and explaining *empirical* human behaviour, we should try to have a look inside this black box. Inspired by recent reflections on consumer society and behaviour, our suggestion is that future studies try to find out if and to what extent consumers' food choices are influenced by their own non-materialistic needs as well as by the attributes of consumer goods.

This does not mean that we wish to ignore the material basis of consumption. Our point is that serious attention is – and should be – given to the fact that human behaviour is swayed by cognition and reason as well as by passion and emotion. The long neglect of the latter is largely due to the supremacy of the idea that behaviour, at least in academic research since the Enlightenment, is simply and solely a matter of a head-led satisfaction of material needs and desires. However, real life pre-eminently indicates that head and heart are inseparable. And consumers are sensitive to messages, associations, images, lessons or experiences that pluck their heart strings.

In an affluent world in which the appeal of many consumer goods does not depend on mere functional purposes and values, one has to look for added value for goods and services in the realm of the non-material. We call this *access value*. Assessing the consumer's hopes and fears, desires and dreams is a precondition to generating access value. Getting into the heads and, preferably, hearts of consumers appears to become more important when an emotion economy is expected to grow in the immediate future. Roughly, the way to reach the food consumer's stomach is through his heart.

The emerging emotion economy, as noted above, is based on the logic that functional reasons and values are waning and emotional, or non-materialistic or post-materialistic (Inglehart 1990) reasons will be increasingly decisive in the buying decisions of consumers. In the network society (Castells 1998) and network economy (Kelly 1998), economic life is envisaged as largely 'socialised' (involving commitment, honesty, responsibility, togetherness, reciprocal altruism) and 'personalised' (invoking image and imagination, empathy and enchantment). Competition, in this context, will be about excess value (product quality, price, functionality, availability) as well as, and above all, access value (feel-good, enjoyment, beauty, fairness).

Whether or not one likes to speak of a 'full belly syndrome', there is reason to believe that enrichment of the spirits and souls of consumers is taking over an increasing part of the food market. In the emerging emotion economy, consumers are expected to choose products on their emotional appeal, which leads inevitably to producers being purveyors of emotions and experiences. This trend in which non-material-values-are-objects-of-(measurable)-value is already underway, as the following quotations show:

> While the economic offering becomes more and more intangible (...) the *value* of that offering becomes more and more *tangible* (Pine & Gilmore 1999: 190).
> Trust and the development of a shared ethical understanding are key components of success with the reward dimension in which price follows value (Preiss *et al* 1996: 147).

Figure 6.1: Briefly on competition and emotion economy

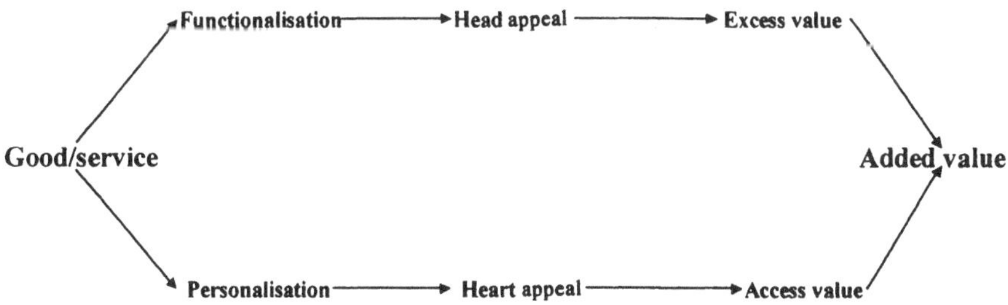

This perspective in which 'virtues become economic assets' (Fukuyama 1999: 256, see also p. 14) is strengthened by situations in which consumer goods have taken over functions which previously belonged to monolithic political ideologies or religious rules. That is to say, goods are used to tell others – and yourself – who you are, they hallmark the group to which you (want to) belong, or help you to live the life you want to live. Consumer goods are embodiments of a lifestyle, of success, luxury, integrity, wisdom, audacity, etc. We are dealing here with non-material consumption in which goods or services are secondary and more or less devoid of utility value. Contemporary consumption has much to do with personality ('express your true self', 'show who you are'), and personal well-being ('self-empowerment', 'self-esteem').

The emotional or non-material factor in current economies is a driving force behind the consumer society – the attraction of (big) brands and their logos is just one manifestation of this phenomenon. In this way consumer goods as appliances of lifestyle, identity etc., also contribute substantially to a materialistic world and world view – an aspect that will not be overlooked by critics of consumer society:

> To have is to be. For many western young people at the end of the twentieth century the belief 'I am what I consume and what I have' remains strong. To misquote Descartes 'I shop therefore I am'. To this extent material possessions are part of self-identity (Gunter & Furnham 1998: 43).

In their search for youth cultures and the ways in which these are creating new markets, Lopiano-Misdom & De Luca express themselves in a less materialistic way. With an eye to 'Generation X' in particular (the age group between sixteen and twenty-nine) they write:

> [T]hey'll be looking for products that have 'soul'. What do we mean by soul? In the year 2000, the only objects that will survive are those that people relate to mentally, spiritually, and emotionally (Lopiano-Misdom & De Luca 1998: xii).

We have argued here that in analysing consumer behaviour, we should not neglect emotional influences on the choices consumers make. By implication, then, reaching consumers requires keeping abreast of shifts in the tempo of the times, and, more specifically, a thorough sensitivity for the ways consumers feel and think.

In the affluent western world, a true and genuine relationship between supplier (producer) and buyer (consumer) is not just a matter of exchanging products and money. To enhance relationships of trust with consumers, producers at all levels of agri-chains should be(come) receptive to changing consumer preferences and (moral) concerns, to their (secret) wishes and (vital) needs. Otherwise they will lose their credibility and licence to produce, with serious repercussions for their market position and competitiveness. In the competitive world of today and tomorrow, interactivity and responsiveness in connection with transparency are the keys to success and depend satisfactorily serving upon establishing and strenghtening bonds between suppliers and buyers, the needs of consumers, and two-way communication, i.e. a dialogue with consumers (see e.g. De Bruijn & Dagevos 1999; Goldman *et al* 1995; Preiss *et al* 1996).

Coping with two main problems

The two previous sections pose critical problems. The first section raised the problematic idea that there seems to be some kind of linear relation between recognising the importance of consumers and difficulty in understanding their foibles, fads and fancies. Recognition of the essential role consumers have when we are talking about food chains and systems coincides with these creatures turning out to be whimsical and not easily grasped By the time visionary people in agribusiness have become accustomed to the perception that '[e]very business competing for the future is customer-centric, customer-driven, customer-focused' (Pine & Gilmore 1999: ix), they have reached the conviction that consumers are by no means easy to 'catch', let alone to pigeon-hole.

Affluent country consumers are becoming increasingly 'eclectic', individually and collectively. All this perfectly accords with an evolving process of fragmentation in the food market. Niche markets are apparently becoming more regular and integrated parts of the food market as a whole.

The 'floating', 'unmanageable' or 'butterfly' consumer is acknowledged. Consumers often move in complex and seemingly chaotic ways, as is confirmed by recent recognition that classifications of them on the basis of classic segmentation criteria are becoming hard to make. Socio-economic and demographic criteria traditionally used in (quantitative) consumer research appear out of step with today's complicated and fragmented food markets. Their exploratory power is declining. This does not mean that quantitative consumer and marketing studies are either impossible or useless, as Lopiano-Misdom and De Luca (1998: 163) contend. Rather, it calls for the development of *additional* methods to track and crack contemporary consumption patterns. This coincides with Colin Campbell's view:

> Whilst it is reasonable enough to stress that the sociological approach to consumption seeks to break with that utilitarian focus on the satisfaction of want which has traditionally characterised economics, to replace it with a paradigm that reduces consumption to merely a process of indication or signification is surely to throw out the baby with the bath-water (Campbell, 1995: 117).

So, although our consumer images approach might be thoroughly sociological, and is also partly a reaction to the limitations of classic variables in explaining modern food consumer behaviour, it has no intention of throwing those 'old' variables away. However, the fall of traditional segmentation criteria suggests that the time has come to analyse consumers in a more encompassing way than just in terms of income, educational level or age. This is linked to the second problem posed above in the second section.

The second problem focuses upon the idea that the food consumer market is only partly a tangible world in which food products, functionality, prices and monetary possibilities are the only driving forces. We also live in a non-material world. In the world of affluent consumption, exchanging money and material goods is just one way consumers have of dealing with their preferences, intentions, needs and desires. Therefore we need an approach that is able to cope with intangibilities influencing consumer behaviour, one which, in other words, incorporates 'surrounding' aspects of behaviour such as consumer concerns, variety-seeking, distinction, the culinary interest or emotional significance of food.

Our suggested solution brings the two problems together: to apply an approach that aims at using traditional segmentation criteria ('objective') and socio-cultural (or 'subjective') criteria *simultaneously*. This is an attempt to pay attention to the well-known economically or materially defined perspective on consumption, and at the same time, to emphasise consumption as a socio-cultural and psychological phenomenon, to be understood from an emotional perpective.

Ideas concerning identity, lifestyle or postmaterialistic values, are additional tools in our ongoing attempt to understand and predict human behaviour. In our changeable, insecure and complex world of today all kinds of tools and perspectives are needed to forecast future demand for certain products and services, and to detect reason in consumer behaviour. More specifically, our perspective aims to be in tune with the overall 'message' of modern consumer research, and studies on consumer culture and modernity, in which considerable attention is paid to approaching the modern world socio-culturally, and its inhabitants psychologically, to uncover their belief systems, thoughts and actions.

Two dimensions underlying consumer images

Our contribution to understanding modern food consumer behaviour is the introduction of a so-called consumer images approach. Four different consumer images have been developed by the Agricultural Economics Research Institute (see Dagevos & De Meere 1997; Diederen *et al* 1996). Here we introduce these relatively new images and outline their contours. Later we reflect briefly on some of their origins in social scientific thought.

We focus on food consumer behaviour from a two-fold perpective. The four consumer images are given interpretation in both a 'material' and 'non-material' manner. Non-materialistic behavioural influences such as animal welfare concerns or, for instance, the need to feel pampered, are as important as influences of a materialistic origin, such as price or product quality. Consumer images are archetypes and portray possible consumer preferences and behaviour. They depart from assumptions about key orientations which trigger (future) consumer behaviour. Two dimensions are expected to underlie consumer behaviour. The first dimension ranges from materialist to non-materialist, the second from individualistic to collectivistic.

The individualistic angle points to consumer behaviour which is directed to the satisfaction of personal needs. It is to a large extent independent of any social constraints. A self-centered attitude is given a prominent position. Collectivistic consumer behaviour is much more tuned to the social or cultural structure – in order to get social approval or to protect cultural inheritance, for example. Collectivism refers to the influence of behavioural considerations in which the social or physical environment is taken into account. Materialistic objectives influencing consumer behaviour are product- or price-oriented. Non-materialistic objectives are basically intrinsic - to achieve an emotional state (fun, happiness, sense of security, etc.). Non-materialistic value can be process-oriented too, when it concerns emotional, ethical or ecological aspects of production processes. This places (food) consumption in a broader context. Based on these two dimensions, then, four different consumer images are distinguished in a four-fold analysis of types of consumer behaviours, labelled *calculating*, *traditional*, *unique*, and *responsible* respectively (see Figure 6.2 for overview).

Calculating consumers

Figure 6.2 shows that the *calculating* or rational consumer is supposed to act in ways which are primarily focused upon materialistic benefits. On the individualistic-collectivistic dimension, the orientation is predominantly individualistic. Calculating consumers are product-oriented and their preferences for certain outcome is 'pro-self', implying that they are individualists 'who tend to maximize their own outcomes with little or no regard for others' outcomes' (Van Lange *et al* 1997: 733).

A further salient characteristic of calculating consumer behaviour is that it is without much light and shade. The calculating consumer's behaviour pattern is mainly 'grey' and mainstream. Variety and diversity are for utilitarian reasons made subordinate to efficiency and availability. In this respect calculating consumers are, in Ritzer's words, strongly influenced by the McDonaldisation of society: their consumer behaviour searches for efficiency, calculability, predictability, and control. An overall result is homogenisation. The so-called Americanisation of the (western) world, and westernisation of the global village as a whole, are inextricably bound up with standarisation tendencies. Calculating consumers voluntarily step into and stay in the cage of McDonaldisation. For them, the 'McDonalds' Golden Arches (...) iconically summarize the McDonalds' promise – to provide a sense of security and inclusiveness, and to be included in the McDonalds family' (Finkelstein 1999: 76). This particular fast-food mammoth is emblematic of calculating consumers who are in favour of a consumer society in which a great deal of their decisions are prepared for them on a mass-market scale - the IKEA or Coca Cola customer, with Pepsi's New Generation, or the aficionados of one-stop (mall) shopping.

Figure 6.2: Consumer images

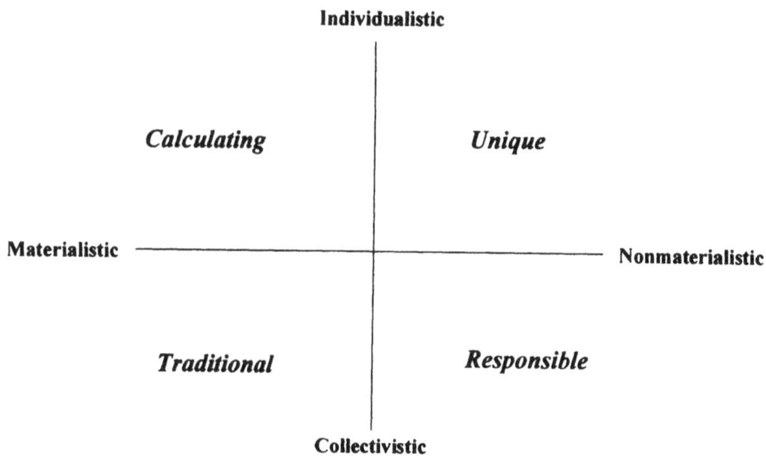

The behaviour of calculating consumers is oriented first of all to price and product, and motivated by immediate and personal satisfaction of their materialistic needs. However this does not necessarily indicate that their behavioural intentions are insensitive to social pressure. Though their consumer behaviour may depend less on convictions, commitments or traditions and more on practical motives (e.g. time and/or money), they feel comfortable if they behave and buy within the social contraints that exist at a particular time and place. They are willing to adhere to the prevailing mores, modes, mentalities, and mass-manufactured goods. Conformity and uniformity, not exclusivity, are pivotal. By and large, the calculating consumer has a 'fuel' approach to food. Foods and drinks are appreciated in terms of quantity, nutritional value, cheapness or easiness to prepare and eat ('presto food').

Calculating consumers are thus less interesting in relation to qualitative changes in the food market. However they are of great importance in quantitative terms. They undoubtedly represent a large part of the consumer population. This power of numbers is responsible for the fact that calculating consumers are a driving force behind changes in the food market in the framework of 'the reign of quantity'.

Traditional consumers

Traditional or conservative consumer behaviour is commensurate with the behaviour of calculating consumers so far as the materialistic dimension is concerned. It is product-oriented and even more pre-occupied with price and price differences between products or points of sale. Saving time is relatively less important – traditional consumer behaviour is largely immune to all kinds of convienence food such as fast food, pre-prepared meals or oven-ready and microwave dishes.

Psychologically, traditional consumers resist change more strongly than calculating consumers. Traditional consumers cling to what is familiar, to their assumptions, prejudices, rules, and routines. Their habits are not easily dislodged and they are unwilling to supercede convention. This is also reflected in the sphere of food consumption, where culinary traditions – possibly passed on from generation to generation – are harboured. Food choices and tastes are fairly conservative as a result.

Traditional food consumption behaviour is not to be neglected as it helps to provide stability and continuity in the food market. In spite of the idea that we are living in the midst of an ever-accelerating pace of change and that the dynamic and ephemeral nature of the contemporary world is frequently mentioned, we should not forget that many (mass) food habits and food choices remain largely unchanged over a considerable period of time. In this respect food consumer behaviour parallels other social processes, in which change is approached with suspicion or denigration. Modernism asks for the development of new mindsets and perspectives, the cultivation of new lifestyles and practices. Thus, it demands flexibility and a drive to innovate in thought and feeling as well as in acting. But as we are all more or less unwilling to take risks, change is frequently considered a dangerous thing which should be resisted. Human beings would rather take the road of evolution than revolution. This pertains to conservative consumers *par excellence.*

On the individualistic-collectivistic dimension, traditional consumers are pro-social: 'Prosocials tend to maximize outcomes for both themselves and others (i.e. co-operation) and to minimise differences between outcomes for themselves and others (i.e. equality)' (Van Lange *et al* 1997: 733). This implies that they contribute to a levelling of differences and tastes. Even more strongly than in the case of calculating consumers, the development of sameness is explicitly and deliberately targeted. Likewise, traditional consumers take the conventions and customs of their social groups seriously. Uniformity and routine characterises the sphere of conservative food consumption, although traditional consumers are less apathetic and more active than calculating consumers, and more sensitive to the idea that stable patterns of consumption are part of the common good. In this respect their conformism differs from that found in calculating consumer behaviour. Whereas calculating consumer behaviour is basically triggered and adjusted for pragmatical and practical reasons based upon functional calculations of economy and effectiveness, conformism as an ingrained feature of conservative consumer behaviour originates from grounds of principle. It is more strongly related to what Durkheim calls *conscience collective* - the collective consciousness which is the basis for social solidarity. The mundane area of traditional patterns of food consumption, as a result, is more firmly fixed in the (unwritten) 'laws' and 'spirits' of a particular time and space-bound community. Traditional consumers take into account that individual rationality might cause collective irrationality and consider their doings as part of the whole set of traditions, habits, codes and conventions existing in the country, county, or 'culture club' they belong to.

Their proclivity to take core values of their (local) socio-cultural systems into account reveals the pro-social attitude of traditional consumers towards their living environment. Commitment to the community and its constraints is present, implicitly as well as explicitly. In sharp contrast to calculating consumers, whose main concern is with immediate satisfaction of their personal needs, traditional consumers are more frequently inclined to take (long-term) consequences of their consumer behaviour into account. In this and in other ways, traditional consumer behaviour is the opposite of what we have labelled unique consumer behaviour.

Unique consumer behaviour

Unique and *responsible* consumers, the two consumer images we turn our attention to next, represent post-materialistic tendencies. This means that the significance of materialistic, object-centred items, as dealt with in relation to calculating and traditional consumers, is lessened. A post-materialistic need such as self-expression is a central motivation in unique consumer behaviour, while quality of life influences responsible consumer behaviour.

Unique consumer behaviour stands for modernity in another way too. It is a result of a social process that marks (post)modern times: individualisation. Unique consumer behaviour thrives on individuality and an urge to independence. 'Freestyle' is adopted as a lifestyle (see also Lopiano-Misdom & De Luca, 1998: 84-6). Unique consumers are also competitive in orientation, 'and competitors tend to maximize their own outcomes relative to others' outcomes, seeking relative advantage over others' (Van Lange *et al* 1997: 733, italics omitted). Unique consumer behaviour is self-centred – it is about ego, even narcissism sometimes, about self-awareness and self-expression.

Unique consumers are genuine children of late twentieth-century consumer society. They stand for modern, hedonistic, variety-seeking man. Although it is a century ago that Thorstein Veblen introduced the notion of conspicuous consumption, it is a typical feature of contemporary unique consumers. The cognate notion of distinction (Bourdieu 1984) is also relevant to unique consumers. They are searching for differentiation. They are driven by a curiosity about the curious, they venture into the unknown and have an exploratory spirit. Behavioural motives are found in a search for exclusivity and identity value of material goods. In contrast to calculating and traditional consumers whose prime behavioural drives are devoted to use value and exchange value, identity value dominates and triggers off unique consumer behaviour. In other words it is the identity or symbolic value of material goods which appeals.

Inasmuch as unique consumer behaviour is profoundly influenced by perception and (self)expression, products and product qualities are means rather than ends in themselves. Personal belongings which reflect one's identity, which 'appeal to the heart, bespeak a certain lifestyle, status or adventure', are favoured (Jensen 1999: 34). This links back to our earlier discussion of Schulze's (1997) notion of 'lifestyle victims' in search of customised goods and services that match a particular and personal *Erlebniswert* [experiential value]. Note again that experience or identity values are usually not offered by a (food) product *per se*, but value is attached to it by the consumer on the basis of what this product is expected to do for him or her personally. It is the value that lies within the consumer (access value), rather than the value within the product (excess value), that produces the behaviour.

Unique consumers belong to the 'pleasure' camp: they fancy fun, and find it in following consumer patterns directed at novelty, surprise, variety or exclusivity. They are devoted to finding and buying something which is (in the eye of the beholder) 'unique' and contributes to feeling or showing that one is special. Not surprisingly then, they favour a pluralistic 'anything goes' market which has a bewildering variety of products and product images in abundance. Unique consumers do not feel lost amidst the plethora of new and distinctive food products routinely and continually made available by food manufacturers. This illustrates, incidentally, that we should always remind ourselves that a consumer-driven foodmarket by no means implies that the days of producers as a crucial driving force behind the development of rampant consumerism are over.

Because unique consumers dare to be different, trendwatchers or marketeers are particularly interested in trendsetting consumer behaviour that fits their image. Quite often unique consumers will belong to the early adapters or the dedicated followers of fashion.

Responsible consumers

Engagement typifies *responsible* consumers. Responsible consumer behaviour is strongly motivated by social and non-materialistic considerations. Their social value orientation is pro-social. Responsible consumers' behaviour is in touch with their thoughts and feelings about the (future) world. Like traditional consumers, they tend to take their own actions into consideration in thinking about the world they live in and the things they believe in. In contrast with traditional consumer behaviour, however, their scope is frequently wider than the world of their own group. Responsible consumers also do not normally cling to traditions. They are willing to embrace unconventional lifestyles and take the resulting consequences seriously. Whereas feelings of nostalgia and (rural) romanticism of some kind might play a role in their ideas and opinions, generally they put their minds to the future.

Responsible consumers' political and social commitment relates to what has recently been described as consumer concerns. This refers to the fact that consumers take a wider circle of reference into consideration when consumption is at stake. They see both food consumption and food processing in much more moral terms. Responsible consumers are concerned about some of the practices used in the production and consumption of food, and the ways in which these appear to be evolving: industrialisation of food production, insecurities about the application of modern biotechnology, recent 'plagues' on the livestock of England and The Netherlands, environmental pollution or adverse effects on the landscape from agricultural activities. Quality of life is defined more broadly than product qualities and personal satisfaction. It is associated with the well-being and sustainability of plant, animal and human life.

One of today's most famous trendtrackers, Faith Popcorn, identifies this as a forthcoming style of behaviour. She claims that a trend called 'save our society' is becoming more and more ubiquitous (Popcorn 1991). Lopiano-Misdom and De Luca (1998) refer to youth cultures in which vegetarianism and veganism (total refusal to consume or wear any animal (by)products) are going mainstream. This particular example may illustrate a very strict form of responsible behaviour, and responsible consumers normally follow a less consistent mode of consumption. Nevertheless it is repeatedly declared that we are at the start of a '"green" wave of recycled, organic and responsible products' (Lopiano-Misdom & De Luca 1998: 145). In the food market this consumer trend is already on the horizon. Some organic food products are strengthening their market position, and responsible consumer demand is already enticing on-trend retailers. A further proliferation of consumers demanding 'feel-good-products' will increase the importance of responsible consumer behaviour, sensitive to (non-materialistic) norms and values.

It is readily understandable that conscious consumers want to know more exactly what they consume and how their food is produced. This relates directly to process-oriented quality, as defined by Klaus Grunert *et al*:

'Process-oriented quality is concerned with characteristics of the production process (...) like the fulfilment of ecological and ethical production standards' (Grunert *et al* 1996: 76). Responsible consumers, therefore, are willing to support companies that are environmentally concerned, whose business ethics are (politically) correct, and who can be trusted with respect to the purity, integrity, authenticity, and fairness of both their products and their information. In these days of critical and well-informed consumers – as well as influential non-governmental organisations like Greenpeace to represent them – consumers should never be underestimated when it comes to honesty and humbug. It would be a mistake to assume that 'idealistic' consumer concerns lack moral fibre, are mere sentimentality or will fade away automatically in the course of time.

Building relationships with your customers starts with access value – in which subjective matters, and even idiosyncratic styles of behaviour and fashions in conduct, have a central position. In the expanding network economy food products and production processes must also coincide with the ethical principles and moral values that circulate in socio-cultural spaces and motivate consumers' conduct. In a 'socialised' emotion economy, producers must understand their customers, in order to recognise and anticipate what consumers are willing to stomach.

A note on Weber's four ways of social action

To conclude discussion of the four consumer images, we note again that they are founded in sociological theory. To illustrate this briefly, we associate the four images with the types of social action identified by Max Weber.

Calculating behaviour, to begin with, relates directly to the orientation of social action which Weber labels as *Zweckrationalität*. Weber's profile of our calculating consumer is about social conduct which is 'instrumentally rational' (*zweckrational*), that is, determined by expectations as to the behaviour of objects in the environment and of other human beings; these expectations are used as "conditions" or "means" for the attainment of the actor's own rationally pursued and calculated ends' (Weber 1978: 24). Traditional consumers thrive on – *nomen est omen* – tradition, 'that is, determined by ingrained habituation' (25). The relentless pursuit of pleasure and hedonistic individualism by unique consumers is, in Weber's words: 'affectual (especially emotional), that is, determined by the actor's specific affects and feeling states' (25). Their conduct is mainly driven by spontaneity and impulsiveness - anathema to responsible consumers whose actions are well-considered and primarily motivated by a sense of conviction. Principles prevail. Weber, in this context, speaks about behaviour as 'value-rational' (*wertrational*), that is, determined by a conscious belief in the value for its own sake of some ethical, aesthetic, religious, or other form of behavior, independently of its prospects of success' (24-25).

Table 6.1: Consumer images and Weber's types of social action

Consumer images	Weber's types of social action
Calculating	*Zweckrationalitä*
Traditional	*Traditional*
Unique	*Affectual*
Responsible	*Wertrationalität*

The relevance of Weber's ideas is enhanced when we bear in mind that in real life *every* consumer is, in principle, more or less able and inclined to act in a way that is consistent with *all* of the four distinguished consumer images. It should always be remembered that consumers are many-headed monsters. One moment they long for fast-food or convenience food, the next they are especially interested in organic food, etc. We have to develop ideas and research tools which are able to cope with the fact that consumers act in different ways which are by turns consistent with the behaviour as defined with respect to different kinds of consumer images. As a consequence, consumer behaviour and consumer images must be regarded as a *continuum*. It is particularly interesting to hear what Weber has said about this problem. He emphasised that:

> It would be very unusual to find concrete cases of action, especially of social action, which were oriented *only* in one or another of these ways. Furthermore, this classification of the modes of orientation of action is in no sense meant to exhaust the possibilities of the field, but only to formulate in conceptually pure form certain sociologically important types to which actual action is more or less closely approximated or, in much the more common case, which constitutes its elements. The usefulness of the classification for the purposes of this investigation can only be judged in terms of its results (1978: 26).

In the next sections, therefore, the results of empirical research are presented. The consumer images approach is deployed to 'catch' food consumers in The Netherlands. The images were tested for the first time amongst a sample of several hundred consumers in The Hague.

Figure 6.3: Conceptual model
Explanatory drives are enclosed by circles / Explanatory variables are enclosed by squares

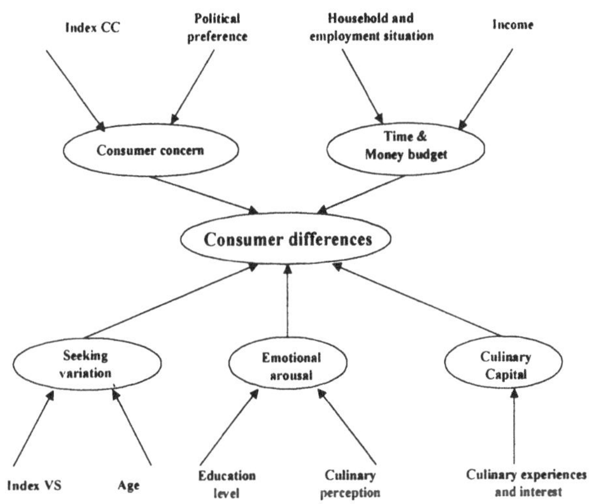

Modelling food consumption

The mechanisms or drives which account for (food) consumer behaviour within our framework are presented in Figure 6.3. The explanatory mechanisms or drives are enclosed by circles. The variables or indicators used in our analysis as explanatory variables are enclosed by squares. These include both demographic and cultural attributes of consumers. The conceptual model is a first step towards the idea that consumer behaviour is influenced by economic or material motivations as well as by socio-cultural ones.

To avoid misunderstanding, it is important to state that the two latent dimensions presented above have not been directly operationalised. It seems obvious that both should be incorporated as independent or explanatory variables in our model, following for instance such scholars as Inglehart (1990) or Triandis (1995), but the usefulness of their scales is questionable with respect to our research with its focus on food consumption. Many such (scale) measurements are applied to research areas such as political orientation and ethnocentrism. The model of food consumption used here is less abstract and concentrates on finding indicators for these latent dimensions. Using this approach enables both traditional consumer attributes such as time and money budgets, and non-traditional attributes such as culinary capital, to be incorporated into the model. We assume that these attributes reflect the individualistic-collectivistic dimension as well as the materialistic-non-materialistic dimension.

Beginning with a few considerations concerning motivations of a material origin, even if people were willing to eat trout or lobster every day, they are faced with certain restrictions which determine, to some degree, whether they are in a position to do this. First of all, they need to be able to afford the products. Secondly, they need to have time to prepare the food and sit down to have dinner. Money and time budgets affect consumer decisions and consequently their behaviour. Both are often considered the most determinant of behavioural conditions. According to economic utility theory, income and prices play an important role in consumer behaviour. Given their available income and the prices of different goods, consumers choose those options which maximise their utility (Wierenga & Van Raaij 1987). This also applies to time budgets (Linder 1971). We use employment and household situation as an indicator of time restrictions. Financial restrictions are reflected by the household income situation. With regard to the consumer images, time restriction is an important drive for calculating consumer behaviour. Traditional consumer behaviour is characteristically cost-conscious.

The next mechanism is 'consumer concern', in relation to health, environment, technological intervention, bio-industry, ecological production, etc. In general, consumer concern reflects anxiety on the one hand and risk perception on the other, due to technological developments which result in alteration of existing food products or introduction of new food products. We expect that those who are more 'concerned' prefer vegetarian or organic food. Two variables are taken as indicators of consumer concern: political orientation and an index of attitude items (index CC). Consumer concern is supposed to be a motive for responsible consumer behaviour in particular.

Variety-seeking expresses openness to the unusual, or willingness to try something new on the menu. It represents people's attitude towards unfamiliar food. The level of variety-seeking appears to be a prominent factor in explaining why people have different tastes, and consequently, differ in consumer behaviour (Van Trijp 1995). Two variables are considered here as indicators of variety seeking: an index of attitude items (index VS), and age. We assume that younger people are less traditional and more open to new developments than older people, resulting in a higher level of variety-seeking and a more diverse consumption pattern in yourger age-groups. This motive is thought to be typical of unique consumers.

Socio-cultural motives are crucial in understanding modern consumer behaviour. Unique and responsible consumer behaviour are particularly influenced by non-material needs and wants. Before we elaborate on the two final motives and their indicators, we would like to bring the materialistic/non-materialistic dimension back to mind, since these drives originate from this realm.

Information theory suggests that people differ in information processing capacity and receptiveness to complex information. This results in different levels of attention, interest or excitement and explains why some people turn to less complex stimuli, while others develop a preference for more complex stimuli (Berlyne 1960).

Table 6.2: Explanatory variables

Driver	Indicator	D/C*	Operationalisation
Time budget	Household situation	D	Single, youngest child 0-6; youngest child 7-12; youngest child 13 years of age or older
	Employment situation	D	No work, <15 hours/week; 15-29 hours/week; ≥ 30 hours/week
Money budget	Nett household income	D	≤ f 2500,-/month; 2501-4500; 4501-5500; >f 5500,- per month
Consumer concern	Political orientation	D	Liberal, social, Christian democratic
	Index consumer concern	C	11 Likert items (α=.77)
Variety seeking	Age	D	≤ 24; 25-34; 35-49; 50-65; ≥ 65
	Index variety seeking	C	8 Likert items (α=.91)
Emotional arousal	Education level	D	Low (primary education); medium (secondary education); high (higher education)
	Culinary perception	C	Orientation toward cooking: 4 Likert items (α=.84)
		C	Orientation towards eating: index of 4 Likert items (α =.73)
Culinary capital	Culinary experiences and interest	C	Number of cookery books (1), recipes (2) and regular reading of magazines about cooking and food (3)

*D = demographic attribute; C =cultural attribute

Following early aesthetic theory, Holbrook and Hirschmann (1982) adopted a hedonistic view which emphasises that consumers act not only rationally but also non-rationally, pursuing feelings of fantasy and fun. They argue that consumer behaviour cannot be fully understood without looking at the symbolic, hedonistic and aesthetic nature of consumption. Both the information-processing and experimental nature of consumption determines the level of emotional arousal, which consequently influences the diversity of consumer behaviour. Two kinds of variables are considered to be an indicator of emotional arousal regarding food: culinary perception and education level. Culinary perception refers to one's enjoyment of preparing and eating meals. Education level reflects the capacity to deal with complex information. However, level of education might reflect different kinds of behaviour. It can be taken as an indicator for information processing, but also as a proxy for information-search behaviour, or the use of particular information sources.

The final explanatory motive is culinary capital. In our model, culinary capital connects to culinary knowledge, skills, competence and interest in food and cooking. The theoretical basis is provided by human capital theory. An application of this body of thought can be found in the status theory of Bourdieu (1984) in which he ascribes lifestyle differences to the accumulation of cultural and economic capital. Three variables function as indicators of culinary capital: the number of cookery books people have at home, the number of recipes they possess and their media behaviour, operationalised by the number of times magazines about cooking and food are read.

In our conceptual model, variables such as age, level of education, political preference or household situation are considered as explanatory. Although one might be inclined to characterise them as backgound or proxy variables, given methodological limits on measuring theoretical dimensions – the variables are meaningful in explanatory analysis. They are used as indicators for certain sociological, psychological or economic stages of personal development, and for this reason defined as explanatory variables, both conceptually and empirically.

As the transformation of motives or mechanisms into indicators has been explained, we can now proceed to their operationalisation. Table 6.2 gives an overview of the way in which the explanatory indicators are made operational. Note that our model, at this stage, focuses primarily on the main effects, not on the inter-relationships between explanatory variables and drives. Note too that the 'domain' of the empirical research is a sample of 1,000 inhabitants of the city of The Hague in The Netherlands. We asked a number of questions about monthly food consumption in general, and such items as eating different kinds of meat, chicken, fish and vegetables, favourite meals and dishes (Italian, Dutch, vegetarian, ready-made food, casserole, fast food etc.), in particular. A total of 380 people returned the questionnaire.

Classification of consumers

Before getting to the question of how demographic and cultural attributes contribute to differences between consumers, we classify consumers on the basis of their behaviour. We asked people what they usually have for dinner, along with questions about the number of times they eat certain sorts of food and dishes.

Since we have nominal and ordinal data, we applied a non-linear principal component analysis, known as Princals (Gifi 1985) to identify different consumer groups. Princals conducts a correspondence analysis by fitting a two-dimensional solution on the data. The four-fold matrix of Figure 6.4 contains the Princals solution.

Figure 6.4: Princals solution of behavioural variables

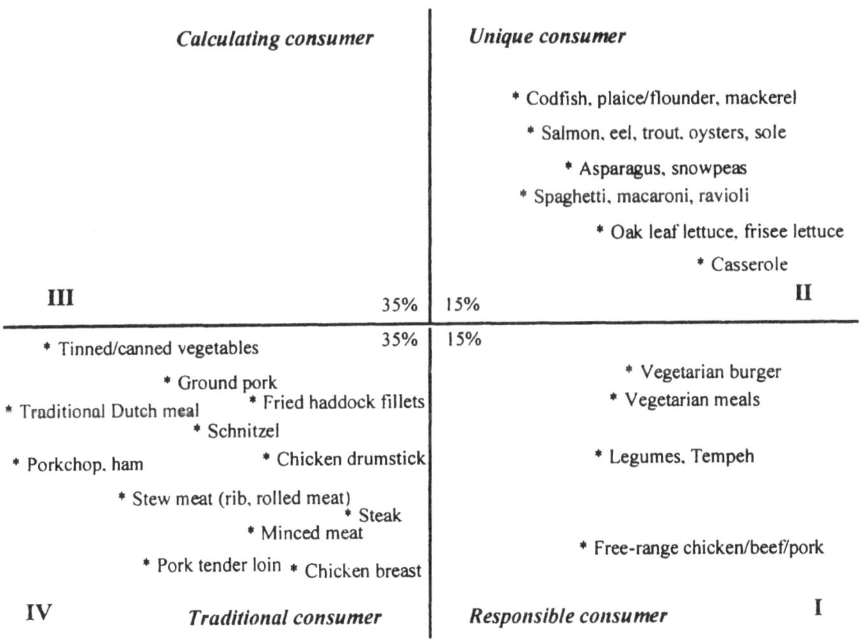

The first quadrant (I) shows a clustering of organic and vegetarian food products. Here we are dealing with a segment of responsible consumers, who prefer healthy and organic food. Responsible consumers represent about 15 per cent of our total consumer population.

Quadrant II reflects a concentration of consumers who like to eat fish – particularly the more expensive types such as trout, oysters and sole. They also seem to enjoy Italian food instead of typical Dutch food. They give preference to non-traditional vegetables such as asparagus, snowpeas, oak-leaf lettuce and frisee lettuce. In general, they seem to seek exclusivity and variation. Their consumer behaviour corresponds with the profile of unique consumer behaviour. About 15 per cent of the population may be regarded as unique consumers.

The third quadrant (III) is home to calculating consumers. It appears to be an empty house. Although this might seem odd at first glance, closer scrutiny reveals that this is perfectly consistent with the presumptions made. As the results show, the calculating consumer does not differ in any particular way from other consumers. Here, we are dealing with an outcome that is in order with what was presupposed. A calculating consumption pattern follows neither vegetarian nor traditional consumption patterns, nor do these consumers seek variety. Their consumer behaviour probably depends less on certain convictions, commitments or traditions and more on practical motives (e.g. time and/or money). This kind of reflection is partially supported by the results of our multivariate analyses as presented later, which suggest that calculating consumers neither seek real variety nor possess much culinary capital. Calculating consumers represent about 35 per cent of the consumer population.

Quadrant IV reflects the traditional consumer, who likes to eat typical Dutch meals (potatoes, vegetables and meat). The choice of meat is particularly conventional (pork chops, minced meat, chicken breast, and the like). About one third of the consumer population seems to fit into this category.

Finally, although 30 per cent of the variation is explained by the two dimensions, we do not consider this, at this stage, a serious problem, since our primary goal is to identify a consumption structure, and that does manifest itself. There seem to be two main reasons for the meagre explanation score. First, consumer images are defined at quite an abstract level, so they are not easily connected to concrete consumption patterns and products. Second, we have to take into account that we are dealing with results of an early attempt to test the developed consumer images empirically. There are many kinds of errors in our questionnaire which we hope will be 'debugged' as the work proceeds and more empirical tests are carried out.

Consumer images explained by demographic and cultural factors

This section discusses the outcomes of a multivariate analysis. Both demographic and cultural factors have been modelled simultaneously. Using the four modes of behaviour identified above, these factors help us explain how consumer segments differ in terms of personal attributes. Figure 6.5 contains the mathematical specification of the logistic regression model.

Figure 6.5: Model specification, logistic regression

$$\text{Ln}\,(P_{type(k)}/1\text{-}P_{type(k)}) = \beta_0 + \sum_{i=1}^{n} \beta_i X_i$$

K	=	1,2,3,4 (consumer types)
P	=	Probability of being a consumer type K
1-P	=	Probability of not being consumer type K
β_0	=	Constant
β_i	=	Parameters which reflect the effect of explanatory variables
X_i	=	Explanatory variables

This estimates the effect of one unit change in the independent variables on the logarithm of the 'odds' (log-odds), such that the dependent variable takes one value rather than the other when controlling for the effect of the other independent variables. The odds are the ratio of two probabilities: the probability of being in one category of the dependent variables (e.g. being an responsible consumer) divided by the probability of being in the other category, 1-P (e.g. being a nonresponsible consumer). To apply logistic regression, some of the quantitative variables were transformed into qualitative, discrete, ones. This approach was adopted, even though it results in some loss of information, because we are dealing here with a limited number of respondents, which increases the risk of unreliable parameters in case of quantitative modeling.

Table 6.3 shows the outcome of investigation (for further details see Hansman 1999). Four different models are presented, one for each type of consumer. We use 'forward selection' to determine the effect of each variable, given the presence of other variables in the model. Only the parameters which are statistically significant are shown in the table. The parameters reflect the relevance of every variable, in terms of explaining (predicting) differences between consumer images. In technical terms, they reflect the importance of every category (of the independent variable), compared with the average effect of all categories of the independent variable.

Looking at the model chi-square, it turns out that adding both demographic and cultural variables results in a significant improvement. This applies to all four models.

An example is given to clarify how the information presented in Table 6.3 is interpreted. The relationship between educational level and calculating consumer behaviour (last column of Table 6.3) indicates a negative effect of high education, which is statistically significant (indicated by an asterisk).

Table 6.3: Differences between consumer images

Drives	Responsible consumer		Unique consumer		Traditional consumer		Calculating consumer	
	b	se	b	se	b	se	b	se
Educational level								
Low			-.07	.30			+.31	.21
Medium			-.77	.32*			+.35	.21
High			+.84	.24*			-.66	.23*
Political preference								
Social							-.68	.23*
Christian Democratic							+.93	.33*
Liberal							-.25	.23
Variety seeking								
Low	-1.02	.38*			-.41	.25	+.81	.25*
Medium	+.80	.28*			-.04	.23	-.19	.24
High	+.22	.25			+.45	.18*	-.62	.21*
Consumer concern								
Low	-.54	.35			-.21	.24		
Medium	-.15	.25			+.57	.18*		
High	+.69	.24*			-.36	.20		
Culinary capital 2								
<10 recipes			-1.36	.42*			+.69	.24*
10-19			+.67	.33*			-.00	.30
20-39			-.16	.44			-.25	.39
≥40 recipes			+.85	.28*			-.43	.28
Culinary capital 3								
0-1 magazines	+.40	.23						
2-3 magazines	-.52	.24*						
≥ 4 magazines	+.12	.30						
Constant	-1.79	.27*	-1.79	.27*	-.94	.17*	-.43	.22*
-2 Log Likelihood	267.5		242.5		360		338.3	
Model Chi-square	23.7;df=6;p=.0		38.3;df=5;p=.0		18.1;df=4;p=.0		61.4;df=9;p=.0	

parameters logistic regression; only significant ones; * p=<.05

High education turned out to have a negative effect on being a calculating consumer. By taking the exponent of the parameter, we are able to calculate the odds. For the higher educated, the odds are exp(-0.66) or 0.52 times the average for all educated groups, and 0.52/exp(+0.31) or 0.38 higher for the lowest educational group. Since the parameters of low and medium education are not significant, their odds of being calculating consumers are considered to be average.

Responsible consumers emerge as having an average score with regard to variety-seeking and culinary capital, and a high score on consumer concern. The latter supports our earlier discussion and accords with our findings that we are dealing with a segment of the consumer society which highly favours healthy food and organic food. The traditional demographic factors do not seem to be of much importance, and are apparently of no help in identifying (future) responsible consumers.

Higher educated people and those with a lot of culinary capital (recipes) stand out among the unique consumers. Again this supports our discussion of consumer images. We are dealing here with consumers who like to eat exotic fish, would rather Italian food, and prefer non-traditional vegetables. Based on their consumer behaviour we would expect them to score high on variety-seeking. This, however, is *not* the case.

Although the traditional consumer likes to eat typical Dutch meals (potatoes, vegetables and meat), and traditional consumer behaviour is fairly conventional, this apparently does not exclude variety-seeking – within the array of conventional food products, that is. The influence on traditional consumer behaviour of consumer concern is average.

The calculating consumer emerges as more Christian Democratic in political orientation, with a low or medium level of education, and low levels of variety-seeking and consumer concern. His calculative approach to daily life is presumed to result in a limited need for variety and diversity, and a greater concern for efficiency. The results support these presuppositions.

What can be concluded from these results? Cultural variables, along with demographic variables, are indeed important for identifying consumers. As far as demographic variables are concerned, education level turned out to be the most discriminatory. With regard to cultural variables, variety-seeking, consumer concern and culinary capital are important attributes. Generally our empirical analysis supports the consumer images approach and helps us to identify consumer segments on a more fundamental basis. Although the results are promising, however, we have to keep in mind that differences among consumers are partly only images. For instance, on the basis of images we would expect unique consumers to be highly variety-seeking and traditional consumers to be more highly involved with ethical or environmental concerns. Neither assumption is supported by this analysis and this calls, among other things, for a more extensive elaboration of the underlying mechanisms or drives which contribute to the origin of consumer images.

Concluding remarks

The last two sections end with a brief reflection on some of the results. *Including* the ifs and buts, we hope this study has revealed that there is reason to continue working along the lines of the consumer images approach as introduced above. We believe it is safe to assume that further development of methods and concepts to improve our ability to understand and foretell what makes food consumers tick is particularly worthwhile, now that it is clearly recognised that older types of analysis are less adequate today. Besides, the capricious conduct of consumers has to be addressed within the context of a fast-moving food market where products and prices are not the only triggers of consumer behaviour. Modern consumers are increasingly interested in, and critical about, what happens to their food as it travels through the food chain. Instead of taking things for granted, many people nowadays question the integrity of food systems. We have also paid attention to relationships between food consumption and the personal life or lifestyle of consumers. Their consumption pattern is part of their identity. These kinds of concerns and considerations have an influence on modern food consumer behaviour. The consumer images approach aims to provide insight into this.

The consumer images approach conveys the idea that the clientele of the food market is quirky. Further empirical tests and theoretical development will hopefully improve our capacity to understand food consumer behaviour and to provide further clear and unambiguous answers to what food consumers want and why they want what they want.

The other theme of subsequent enquiries will be to say more, and more accurately, about the extent to which the expected emergence of the emotion economy is actually occurring. Are parts of the food market arena becoming emotionally driven, prospering by growth in consumption of a non-material origin, and if so, how fast and in what ways is this trend taking over? Further empirical validation - preferably by longitudinal study - is needed.

Part III

Living with nature?

7 Changing social representations of the countryside: rurality, nature and heritage within a French National Park

EILEEN O'ROURKE

Introduction: French rurality

The rural has traditionally been defined in opposition to the urban. Mormont (1996) observed that in France the term rurality emerged during the joint industrial and urbanisation process of the mid-nineteenth century, which not only led to labour migration towards the city but, with the aid of improved communications, to the integration of rural areas into a larger economic and socio-political system. Neither was the pre-industrial French countryside solely an agrarian space, rather its relatively autarchic agriculture was inserted within a larger social system of landlords, merchants, artisans, labourers and industrial workers. What Jollivet (1971) aptly referred to as the 'ghetto paysan' emerged after the post-World War II period, when European agriculture was mobilised around the central theme of food self-sufficiency, and the industrialisation process was essentially grouped around urban poles. However, the political agenda has once more evolved. Europe is today confronted with the cost of agrarian over-production and the non-competitivity of much of its agriculture within what is increasingly a global market system. There may be a future for productive agriculture in optimal regions, but the so-called 'Less Favoured Areas', which cover over 50 per cent of both French and EU territory, are increasingly encouraged to make the transition from a singular focus on food production to farm diversification and pluriactivity, with emphasis on off-farm employment, agri-environmental work and agri-tourism (Agenda, 2000; Fuller, 1990; Cork Declaration, 1996).

At the beginning of the twentieth century, half the active French population was engaged in the agricultural sector. This figure fell to one third after the World War II and today agriculture employs less than 7 per cent of the active French work force (Hervieu 1994). The 1990 population census indicates that

74 per cent of French rural inhabitants are salaried workers, which implies that farmers are today a minority among other rural inhabitants (Institut National de la Statistique et des Études Économiques (INSEE) 1992). French rurality can no longer be seen as synonymous with agriculture. As the boundary between rural and urban spaces seems increasingly porous, the question of what constitutes the rural (and the urban) is once more problematic. The former demographic definition of rural spaces is seen as static and outmoded.[1] The question increasingly posed is 'who is rural?' (Mormont 1996; Murdoch 1996; Marsden *et al* 1993). In France, Eizner (1996) reminds us that despite the diversity within rural France one thing is sure, the old rurality whose rhythm was dictated by the church belfry is fast disappearing. The urbanisation of France (and Europe in general) and the fact that in less than 150 years (1850-1990) it made the demographic transition from being 80 per cent rural to 80 per cent urban has not only redrawn its political and socio-economic map, but arguably the urbanisation process also represents one of the most profound changes of consciousness that has occurred in the twentieth century. During the early urbanisation process the city was associated with technicity, progress, modernity and culture, whereas the rural was synonymous with tradition and anti-progress. However, since the 1970s the above representations have evolved. The city is increasingly associated with stress, pollution, unemployment, exclusion and delinquency, and the representation of the rural is shifting from a site of primary production to an association between countryside, nature, leisure, health, patrimony and above all else the all-encompassing, but elusive, concept of 'quality of life'.

It has been argued that French rurality is simply following what happened in Britain a few centuries earlier, in that it is being transformed into countryside synonymous with landscapes and aesthetics (Mathieu 1998). Either way, the old opposition between the archaic countryside and the modern city is less pronounced in the contemporary so-called postmodern epoch, rather one is seen more as a continuum or complement to the other. Many modern city dwellers have a commuter lifestyle of weekends and holidays away, frequently in the countryside. The majority of French people still have rural roots, and almost 10 per cent of them (the highest proportion in Europe) possess a holiday home, which is generally to be found in the countryside (Perrot 1998).

The recent interest in all things rural is often interpreted as a counter-tendency to the alienating urbanisation process, and is ultimately seen as an existentialist quest for identity and roots (Fischler 1993). Within this context it is not surprising to find that peasant identity is also being reinterpreted. The portrayal of the French peasant has oscillated between images of Zola's (1887) brutish and backward agrarian proletariat of the nineteenth century, to a certain romanticisation of the peasant to be found in the writings of people like George

Sand. During the Vichy period the high moral value of the peasant as opposed to the dangerous urban classes was highlighted. This ideology was reflected in General Petain's famous statement that 'la terre ne ment pas'. However, the backward image of the peasant was once more in vogue during the modernisation drive of the 1950s and 1960s, which was in many ways presented as a form of liberation from the brooding soil. This sentiment was captured in a film of that epoch titled 'Les Damné de la Terre', which was set on the desolate Causse Méjan, which now forms part of the Cévennes National Park. The film depicted the enslavement of man to the un-giving soil. That image was once more revised in the 1990s with the release of the highly popular film 'Le Bonheur est dans le Pré', which portrayed the Gersois Peasants of the south of France as living in harmony with nature and as having preserved the highly regarded French 'l'Art de vivre et du bien manger', along with strong family ties. In short the social values of the past. The reworking of peasant symbolism along with the increasing demand for rural heritage, implies that the rural is no longer just a spatial entity, it is also a temporal category frequently used to mediate between the past, the present and the future. Hervieu (1994: 95) captured the contemporary sentiment when he declared that French rurality is slowly being transformed into a 'tabernacle of symbols', a place where images and symbols are created, consumed and transformed. This chapter explores the contemporary representation of French rurality, with special reference to the links between rurality, nature and heritage.

Methods

The interpretation which follows is centred on a case study, that of the inhabited Cévennes National Park, where nature, heritage, tourism and agriculture are closely entwined. The field data was collected during what is most aptly described as a year's anthropological field work in the area. Along with the collection and collation of historical, natural resource and statistical data, over a hundred semi-structured interviews were conducted among the diverse local actors, including farmers, tourists, holiday home owners, conservationists and local administrators. The interviews were set around the agriculture, environment, society *problématique*, but they were sufficiently open for the people to tell their own story, what they considered as relevant. Throughout the interviews short notes, generally key words or phrases, were taken in order to help recall.[2] On returning home the interviews were written up in full in a field diary.

The analysis of the qualitative diary material, in common with all forms of analysis, is a search for patterns in data and for ideas that help explain the existence of those patterns (Bryman 1988). On re-reading the diaries, in the light of insights gained from participant observation (living in the area) and other documented data sources, certain themes and patterns began to emerge. The diary material, which was typed into a word processing package, was subsequently organised under different categories and sub-categories, which corresponded with the emergent themes. The categories and sub-categories were not rigid structures, other people may have come up with different ones. Their function was simply to help me organise the amorphous field diary data, and thus aid conceptualisation and analysis. There can be no set analytical formula for analysing such ethnographic work, only basic guidelines sensitive to contexts. The interpretation which follows arises from an interplay between the classified field diaries, participant observation and supporting statistical and documented evidence for the area. In what follows I essentially want to compare and contrast the different representations of the rural nature-heritage linkage held by conservationists, farmers and tourists within the National Park.

Introduction to field site: Cévennes National Park

The Cévennes National Park (PNC - Parc National des Cévennes), situated on the southern edge of the Massif Central, covers an area of approximately 300,000 ha (91,417 ha in central zone and 237,000 ha in periphery zone), most of which lies in the Department of the Lozère, one of the most rural and sparsely populated French Departments. It is at least a two-hour drive on bad roads to the Mediterranean coast and to Montpellier, the nearest large urban pole. Clermont Ferrand, the capital of the Massif Central, is approximately three hours by road from Florac, a small town of 2,000 inhabitants which hosts the National Park's headquarters. The area contained within the National Park is best described as a peripheral hilly region located at some distance from the main markets and urban centres.

The major attraction of the Cévennes National Park is its rich cultural heritage and human encultured landscapes, essentially composed of the limestone of the Causse Méjan, the granite of the Mont Lozère and the rugged schist hills of the Cévennes (see Figure 7.1). The Park has the rare distinction of being inhabited throughout and most of its land, including that of the central zone, is privately owned. When the National Park was created in 1970 its dual - and what may appear like a contradictory - objective was nature conservation and socio-economic development in this naturally disadvantaged, but culturally and ecologically rich rural territory (Crosnier 1996; Monjauze 1971). The first Director of the PNC, Mr Monjauze (1971: 91) described his vision of the Park as a conservatory of 'provincial aesthetics', an 'art form consecrated to the glorification of the past and its environment', while at the same time being a 'motor of socio-economic development for its inhabitants' (author's translation). In 1984 the whole area was classified as a biosphere reserve and it is estimated that over 800,000 tourists visit the Park every year (Calvet 1995).

Figure 7.1: Cévennes National Park

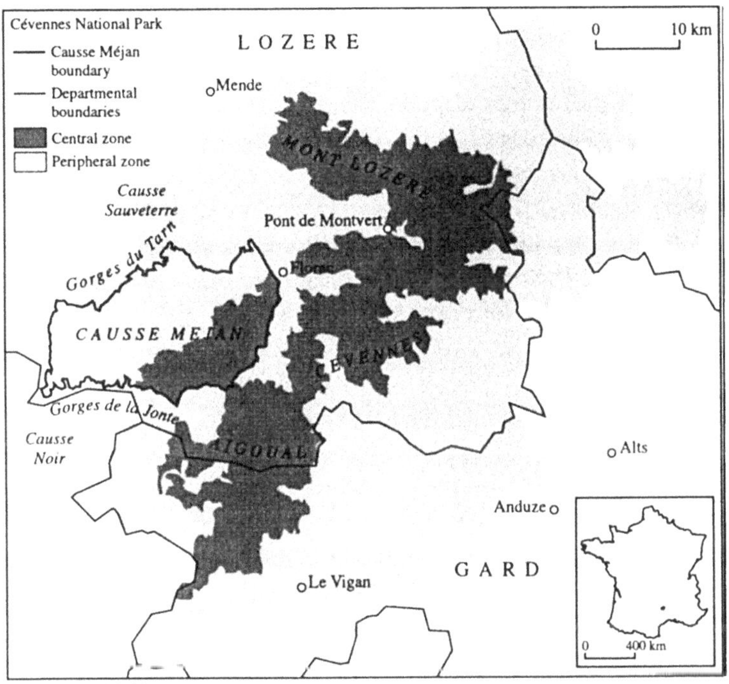

Source: PNC–BDCOM90

The area contained within what is today the National Park has known very little industrialisation, apart from coalmining around the Alès basin and the once thriving Cévenol silk industry. Both activities have now practically ceased. In keeping with many other isolated rural areas it essentially made the transition from a pre-industrial to a post-industrial stage, which is part of its present-day attraction. In the past the inhabitants of the Cévennes, Mont Lozère and Causse Méjan lived essentially from semi-autarchic agriculture, which was frequently complemented by a cottage textile industry and outside seasonal work in the vineyards and olive groves of Languedoc, and fruit picking as far afield as Provence. In the course of the nineteenth century the Cévennes lost approximately 80 per cent of its population, due to rural exodus and the quest for a better life in the city (Crosnier 1996). The population density within the National Park today varies from approximately 10 people per km^2 in the Cévenol valleys to as low as 1.4 people per km^2 in its depopulated Méjan plateau. The Cévennes National Park provides an ideal setting in which to study the new environmental and cultural heritage interests being projected onto the rural, even though they did not originate there.

Conservation within the Cévennes National Park

The 1960 French law concerned with the creation of National Parks confers on them a fundamental role in the domain of the protection and conservation of nature. However, nature and agrarian culture are closely entwined within the PNC, and it was apparent from the outset that in order to protect the Park's landscapes it would have to protect the traditional farming practices that had fashioned them (O'Rourke 1999). For example, the ecologically and aesthetically highly valued steppe of the Causse Méjan is the product of the hoof and teeth of its ovine grazers, and the cultural landscapes of the Cévennes are synonymous with terracing and chestnut cultivation. The Park's stated objective today remains that of its founding mandate - to combine the protection of nature with sustainable development (PNC 1994). In terms of socio-economic development the farmers within the Park have profited from agri-tourism, with grants of up to 80 per cent of costs for the construction or renovation of high quality gîtes within the Park's central zone. Tourism along with the Park's high quality imagery is also important for the sale of territorial farm produce (*Produits Agricole de Terroir*), such as cheeses, sausages, hams, honey and pâté, either directly from the farm yard (*vente direct*) or in local markets. In summary, ecology, farming and tourism are closely knit within this local rural economy.

However, it is not a closed symbiotic system and its symmetry is liable to be disrupted by internal conflicts of interest and different representations of the area, along with external policies. For example, the sheep farmers of the Causse Méjan and the beef farmers on the Mont Lozère are dependent on agricultural subsidies for over 80 per cent of their available revenue (Lhuillier 1993). Those essentially European subsidies are calculated on the number of animals owned by the farmer rather than on their management of space. Headage subsidies along with subsidies for farm buildings and fencing combined with changing social expectations, such as the rejection of shepherding, have led to a general abandonment of difficult pasture lands and terraced agriculture. There is a trend towards farm intensification and indoor (*hors sol*) practices, including the buying in of foodstuffs from outside the area (Benoît & Valéro, 1997). This is especially true of the Roquefort sheep milk producers on the Causse Méjan and the Pélardon goat cheese producers in the Cévennes (Benoît & Valéro, 1997). The retreat of traditional labour intensive agri-pastoral systems and the decreasing grazing intensity in certain sections of the Park have in turn provoked landscape and biodiversity changes, most notably the uncontrolled spread of scrub vegetation (essentially boxwood and juniper) resulting in the closing-in of open landscapes.

The National Park, whose terrain is already almost two-thirds wooded, perceives the advancing scrub as a threat to its remaining biodiversity-rich open landscapes, as well as adversely affecting landscape diversity (PNC 1994). The Park in collaboration with the European LIFE programme launched a project to 'recreate' the highly valued steppe of the Méjan (PNC 1996). They propose to do this by the mechanical clearing of the scrub followed by traditional style ovine grazing. In an attempt to render this system attractive the National Park launched a new mutton label, known as '*Agneau de Parcours*', which is a traditional lamb raised outdoors, as opposed to the majority of lambs produced on the Méjan today who in the name of economic and subsidy rationality are raised indoors (Benoît & Valéro 1997). It remains to be seen if the Park's '*agneaux de parcours*' are economically competitive as well as socially acceptable, since this system is more labour intensive in that it requires shepherding and the management of poor pasture lands, which in turn restricts the number of animals and consequently income from subsidies. This example not only highlights the close links between biological diversity and farming systems, but the fact that the Park is now talking about 'recreating' nature brings into question the common perception that nature provides a solid bedrock of how things really are. One of the major attractions of 'green tourism' is the perceived 'naturalness' and 'authenticity' of nature and the countryside, whereas it is increasingly evident that desirable nature and landscapes vary according to cultural and historical determination. The notion of recreating nature also tends to imply that it was complete in the past, but the anthropic Méjan landscape has had several pasts (Vernet 1995). What criteria do we use in deciding which past we want to 'recreate'/'revisit': should it be its glacial past, its Gallo-Roman past, or its agri-pastoral nineteenth-century past? Having chosen the nineteenth-century past for aesthetic, biodiversity and nostalgia reasons, how can we be sure that those fashions will not change in the future, and besides who is the arbiter of today's tastes?

The farmers' point of view

The local farmers tend not to share the conservationists' vision of nature or the countryside. The indigenous rural population, although not hostile to the Park (except for isolated pockets frequently linked to hunting rights) see it as something from the outside, a Parisian bureaucracy which was imposed on them. Not only does the Park escape local control and the well developed patronage system, but it represents an alien way of thinking that has its roots in what Leynaud (1985) referred to as urban post-industrial societies. Within the difficult topography and harsh climate of the Cévennes there is a strong local history of protecting oneself from nature, rather than protecting it. Even though the local people are aware of the previously referred to advantages of being within the National Park, the Park along with the large scale Government afforestation projects which preceded it are associated with a form of dispossession (Mousset, 1992). It infringes on the strong local Occitan desire to be '*maître chez soi*'. Besides, the Park is perceived to favour the interests of nature and tourists over those of the local agrarian population. The farmers within the Park believe that their interests and world vision are better served by the Ministry of Agriculture, which is deeply entrenched within the fabric of rural France. It was under the guidance of the Agricultural Department (DDA) and the '*Chambre d'Agriculture*' that the local farmers, following the 1960 Agricultural 'Loi d'Orientation', set about modernising agriculture in the area and raising farmers' standard of living. Even if today they, in keeping with European farmers in general, are heavily dependent on subsidies (Bazin 1994), there can be little doubt that overall the farmers' standard of living has greatly increased from what is universally seen as a poverty-ridden past.

Accompanying the fruits of progress is a change in the farmers' own self perceptions. Today, the majority of them (with the exception of the newly arrived *Néo-Ruraux*)[3] want to be seen as modern professional farmers, and their social status is attached to such things as the size of their tractors and other farm machinery, along with the possession of modern sheepfolds, cowsheds and dwelling houses. This conflicts with what the 'outsiders' -

tourists and conservationists - are looking for, and the farmers have to be careful about how they play their dual identity, that of romantic shepherds and pastoralists for the 'outsiders' and modern, professional stock breeders for the 'insiders'. This tension is aptly reflected in the symbolism of the shepherd who in the past was depicted as essentially someone 'who could not do any better', and was often a child or the village simpleton, but today the shepherd is presented as a romantic figure. However, the latter image is essentially for the tourists. The one remaining salaried shepherd on the Causse Méjan describes himself as a social outcast because he did not modernise and become an '*éleveur*', someone who owns his own herd.

Despite the diversity in the farming systems practiced within the National Park, there is an underlying commonality in the farmers' perception of and interaction with the landscape they inhabit. Ingold (1992) reminds us that the way people think about the environment is shaped by their mode of engaging with it. The farmers within the PNC and elsewhere engage in an essentially utilitarian way with their land and the countryside, even though they may also have sentimental, even aesthetic, attachments to their ancestral land. Above all else the land represents a means of livelihood and the farmers are preoccupied with economic survival. They have a lived relationship with the countryside and its landscapes, rather than the disengaged aesthetic or ecological vision of the tourists and conservationists. Whereas the farmers accept, even highlight, their traditional role as the guardians or caretakers of nature and rural landscapes, they insist that this is something they do unselfconsciously via their work. The landscape is seen as a product of their agrarian activities rather than being an object in its own right. However, the farmers within the National Park today admit that they are no longer in control of the evolution of their landscapes, which they attribute to the lack of labour and changing farming systems, in particular mechanisation. Ironically those semi-abandoned rural landscapes are increasingly seen as part of the nation's cultural and environmental patrimony, to be preserved for future generations.

What the farmers describe as environmental degradation is not a reduction in biodiversity or landscape aesthetics, but rather the undoing of their agrarian control over nature, reflected in the naturalisation of the landscape, the spread of non-edible shrubs and weeds, the disrepair of stone wall terraces and irrigation systems and the diseased and unkempt chestnut groves in the Cévennes. They may describe such things as degradation, but they have no intention of cutting back bushes, repairing terraces and mountain irrigation systems no longer in use, unless they are well paid to do so. Some farmers within the PNC have joined agri-environmental initiatives such as 'Article 19', which arose from the European Directive 797/85. However, the vast majority of them are adamant that they are not doing such work out of environmental concerns, rather their motivation is the obtaining of subsidies for something they were already effectively doing. The farmers mistrust outside environmental experts, and ecologists are invariably presented as naive and bad managers of nature. They frequently point out that it is the biodiversity 'unknowingly' created by their agrarian ancestors that is today so highly valued. The fact that today's farmers have at their disposal far more powerful technology is not commented on.

However, the farmers also realise that times are changing, and whether they like it or not the rural discourse is no longer solely an agrarian one. They are confronted by 'outsiders' and other users of rural spaces, essentially in the form of holiday and retirement home owners, passing tourists, and people in pursuit of leisure activities such as hunting, spelaeology and hang-gliding. There are also people who come to the area in search of alternative life styles. The indigenous population may resent these intrusions into their world, the inevitable clash in mentalities and the disruption of local power structures,[4] but they also realise that the outsiders may provide a valuable and increasingly necessary complement to their agricultural incomes and rural livelihoods in general.

An increasingly important condition for rural development in France is the valorisation of a product by linking it with territorial (*terroir*) identity (Wachter 1987). Philo and Kearns (1993) argued that territorial marketing and territorial identity construction is a device to attract people to places and to 'sell' the area and its produce to the extra-local. In effect it is an attempt to sell identity and contextual meaning, including the social values of the past. The farmers within the Cévennes National Park are well placed to exploit its high quality environmental and cultural heritage imagery and symbolism. For example, the sales image of the locally produced famous Roquefort cheese places heavy emphasis on its links with tradition and continuity with the past. The image of wild, windswept Causses, sheep and shepherds along with traditional stone arched sheepfolds are all designed to give a strong emotional, territorial identification with the product. In reality the Roquefort cheese production process, including the milk production system, is today a highly modern mechanised system (Barthe 1995). Similarly, many of the people producing and marketing Cévenol territorial produce such as Pélardon cheese and Cévenol pate, are recently installed *néo-Cévenol* outsiders without local territorial roots and whose production system frequently has little in common with the so-called traditional farming model once practiced in the area. The image of authenticity may be more important than actual authentic links with tradition and territory.

The tourist gaze

The majority of the tourists who visit the Cévennes National Park are of urban origin and the type of tourism practiced there is best described as 'Green Tourism'. Research by the Park indicates that over 60 per cent of its visitors fall within the professional socio-economic class, and they describe their visitors as 'cultivated, independent people who love nature and are fleeing from cities and crowds' (Calvet 1995). Regardless of their socio-economic background, the tourists that I interviewed were united by a love of nature and the countryside, and they frequently make the rather romantic connection between nature, purity, health and well being. In opposition to the artificiality of the city, the rural has come to represent the 'authentic', the 'real' - authentic nature, authentic human contact and authentic nourishment. MacCannell (1973; 1976) argued that today's quest for the authentic is a modern version of the universal human concern with the sacred. He describes the tourist as a kind of contemporary pilgrim seeking authenticity in other 'times' and other 'places' away from their everyday life (see also Urry 1990). The tourists interviewed described their everyday urban lives as stressful, work oriented, competitive, impersonal, polluted and artificial. Their holidays in the countryside are about restoring a balance via contact with nature, authenticity and what they frequently refer to as more genuine human contacts. The words they use to describe this purification process include - 'detoxification, physical and spiritual renewal, equilibrium, leisure, freedom, calm, silence, isolation, a return to source'.[5]

Tourist perceptions of nature and local farming practices

The tourists' perceived naturalness of nature and the countryside was translated into their frequently expressed desire that nature should be allowed to take its own course. True nature has come to represent something wild, free and independent, all highly charged symbols. The tourists' desire for naturalness and unencumbered nature not only contrasts with the farmers' desire for a domestic controlled nature, but it also contradicts their own frequently mentioned appreciation of nostalgic agrarian landscapes and an active agrarian presence in the landscape. From the interviews it is clear that in general the passing tourists did not have a profound knowledge or understanding of the functioning of the local environment and society, and they frequently erroneously believe that the landscapes of the National Park had not changed since time immemorial. The perceived invariance of the landscape and society comforts and reassures, but the majority of the tourists fail to see that 'order' is not permanent, it has to be maintained. The tourists' stated attraction to the 'wildness' and 'isolation' of the countryside carries over to their fascination with certain emblematic wild birds and animals, such as vultures and a herd of archaic Przewalski wild horses that were recently introduced by the Park on the Causse Méjan. Not only do these wild creatures animate the landscape and embody it with naturalness, a highly sought commodity, but they are also symbols of resistance, freedom and wildness. However, the imagery conjured up by such powerful symbolism is often less real than what it stands for. In effect the wild horses are contained behind high electric fences and their survival and that of the vultures is heavily dependent on the Park's intervention and protection. The tourists on the other side of the fence can experience the thrill of 'wild otherness' while at the same time they know that their own personal safety is assured.

The tourists interviewed were invariably of the opinion that the agriculture practiced within the National Park was traditional and unpolluting, and that the local farmers respected nature unlike their industrial counterparts in places like Brittany. The Park's farmers may perceive themselves as modern, but the tourists insist on seeing them as 'wild and close to their customs and traditions'. Some tourists interviewed went so far as to describe the place as 'underdeveloped like parts of Africa'. Others referred to it as a 'living museum', a place mythically infused with timelessness. In the tourist dialogue the link between nature, purity and tradition is obvious, as if the past and those who remained true to it are 'pure' and only the present is 'polluted'. The myth of the purity and naturalness of the landscape is carried over to the food produced in such places, which the passing tourists invariably perceive as wholesome, unpolluted and traditional. Eating farm fresh territorial produce not only represents a biological quality, but as Bessière (1998: 25) points out, satisfies a desire to go back to culinary roots, 'as if eating were a quest for nutritional truth and essence hidden in a dish'. Luginbuhl (1991) argued that eating such food is a nostalgic attempt to access the rapidly disappearing rural peasant identity. The French 'paysan' in an attempt to shake off his often derogatory image spent the last thirty years trying to be modern, but under the contemporary tourist gaze frequently finds himself faking 'peasantry', while in private he expresses the fear of becoming an anachronism in the 'modern art museum' under construction. It is interesting to note that many of the holiday home owners, especially those of local origin, who undoubtedly have a more profound knowledge of the area than the passing tourists, frequently bemoan the fact that they can no longer find the quality and flavour of the mutton, bread, chestnuts and cheeses of their childhood. Some of the holiday home owners declared that they now have to return to urban centres for their food provisions.

Tourist interaction with the local society

Along with perceptions of natural wildness the tourists interviewed invariably perceived the local communities as isolated and cut off from the outside world. They frequently added that they are looking for this isolation while on holidays, and some even dream of having it all year round.[6] The tourists often wonder how the locals hear about what is going on in the outside world, which is probably more a reflection of their own sense of disorientation than representative of the local reality. Perceptions of isolation, space, distance and naturalness are relative. I did not meet a single local inhabitant who considered themselves to be either physically or socially isolated. Today the vast majority of the indigenous inhabitants have cars and telephones and they tend to judge things in terms of time rather than distance. Over the winter they stock up in food supplies, and besides snow ploughs maintain most roads open all year round nowadays. The less mobile, such as the elderly, who frequently live alone in isolated mountain hamlets, explain how in fact the world comes to them. Mobile traders, who include a baker, general grocer, butcher and fishmonger pass once or twice a week, and the postman will deliver bread and medication if necessary. One is frequently reminded that one can be far more isolated in a big city, like the cousin of a man from the Causse Méjan who was dead for a few weeks in his apartment in Paris and it was his mother from the Lozère who raised the alarm when his letters stopped coming. It is believed that such incidents could not happen in this sparsely populated rural area, within a day the postman or a neighbour would know that something was wrong. Practically all these apparently isolated rural communes are inundated with requests for houses and access to land for agricultural installation purposes, which they cannot meet.

Despite the tourists' stated desire for local human contact, on questioning it is clear that for the passing tourists (as opposed to holiday home owners who may or may not be socially inserted within the local community) that contact rarely extends beyond the gîte owner and business contacts with shopkeepers, restaurant and café personnel in local towns. On the whole a rather strict divide is maintained between the locals and the tourists. Many of the locals who claim never to have had a holiday in their life (even though some of them could well afford it) see the tourists as privileged, whereas in fact some of those tourists complain of working all year round in big ugly cities, with the sole motivation of being able to afford a week's holiday in a 'rural paradise'.

Even though the majority of the tourists are described as respectful and some even talk of a cultural exchange, the other side of the coin is that the tourist gaze, and that of their cameras makes some of the locals feel marginal and particular. A sheep farmer on the Méjan, whose family also run a profitable

'*table d'hôte*' (restaurant), declared that he sometimes feels like something out of a 'Disneyland cartoon'. He does not mind the mass July-August tourists taking photographs of his sheep, but he does not like it when they turn the camera on himself and asks him 'stupid' questions like why and how he lives on this barren, deserted plateau? The one remaining traditional salaried shepherd on the Méjan, who has become a major tourist attraction, and who would appear to have a bit more insight into the tourist mentality, believes that the tourists who plague him with similar questions are in fact in quest of their own identity, which they hope to discover via the Méjan pilgrimage. The tourist fascination with the identity and 'real lives' of others can create an obvious intrusion into that world. MacCannell (1973) argued that the people being observed, both in an effort to protect themselves while at the same time profiting from the tourist opportunity contrive back-stages, or tourist spaces for 'staged authenticity'. In time such representations come to constitute a closed self-perpetuating system of illusions (Urry 1990). The general clash in mentalities and world visions between the local rural population and the urban tourists leads one to speculate that the people most apt to capture the quality rural tourist market in the future may well be outsiders, professionals in the tourist business, who set themselves up as 'farmers' – keep some animals and a good vegetable garden. In this way they are able to provide the all important rural agricultural ambiance, while at the same time meet the tourists' culinary and other holiday needs, including that of personal contact. This trend is already observable within the National Park where some of the most successful '*ferme-auberges* and *table d'hôtes*' are run by recently installed outsiders, who via international marketing manage to remain open for most of the year, whereas on average gîtes within the Park are rented for between 14-16 weeks of the year (Comité Departmental du Tourism 1996). These may well be the 'professional outsiders' most adept at exploiting local heritage and imagery, not least that of authenticity.

Heritage or outdoor museum?

The conservation of the past conceals the present, or as George Orwell (1949) put it, he who controls the present controls the past. The tourists and holiday home owners interviewed invariably saw the National Park in a positive light and associated it with quality, taste and the conservation of local natural and cultural patrimony. The local farmers are also seen as the guardians of local landscapes and rural heritage; but there is also a fear (especially among the well informed holiday home owners) that the farmers and local people in general may not always be able to appreciate the value of their own heritage. Some tourists claimed that today such heritage forms part of humanity's global patrimony, and is strictly speaking no longer just a local affair. There is a feeling that luckily the Park is there to keep an eye on the locals, and contain such atrocities as the widespread construction of ugly modern sheepfolds, cowsheds and fences which dissect the landscape. The disrespect for traditional stone architecture and the placing of tarmacadam on country dirt roads were also frequently mentioned tourist 'eyesores'; as if stone is 'authentic', cement and tarmacadam are 'modern' and 'inauthentic'. Nevertheless, the tourists did not express the desire for mud motorways joining their urban homes to this rural idyll. It would appear that the patrimony of the past that is to be preserved is selectively chosen. Authentic remnants of the rural past, such as fly-infested agricultural installations and piles of dung commonly found in the middle of these mountain villages, are to be sanitised.

The indigenous rural population tend to have a different understanding of their heritage, and they frequently express the belief that the 'outsiders' are not only trying to appropriate their space but that they are also distorting their past, their heritage. A classic example of contested heritage is the National Park's Eco-museum, opened in 1983 at the Pont de Montvert, on the Mont Lozère. The local population is adamant that it does not contain their patrimony, even though they supplied the museum with many of the objects on display there. Initially the locals refused to visit 'their Eco-museum', in whose creation they had no say. They designated it as a place for the tourists, until one day the Curator of the museum hit on the idea of organising a free dinner there for the members of the large local retirement club. After the initial breaking of the ice, the senior citizens admitted that the museum was interesting and tastefully done, but they were still adamant that their patrimony and identity was not contained there. Is this strong aversion to the museum due to the fact that theirs is essentially an oral patrimony (Stephani 1972) that cannot be fossilised or contained in a rather extra-terrestrial saucer-designed building, which paradoxically does not conform to the Park's own architectural standards?[7] Or is it as Bessière (1998) argued, that a possession or knowledge that one gives up or wastes cannot be called heritage, since we ascribe no particular value to it? One gets the impression that the authentic '*paysans*' of the Mont Lozère, Cévennes and Causse Méjan are too close to their past to see it as a form of art or heritage. Rather they are preoccupied with making a living there today, and the main advantage they see in the Eco-museum is that it created a bit of much-needed local employment Once more it is an active local presence in the landscape and community that the local people prioritise rather than what they perceive as the fossilisation of some fictitious past for the pleasure of economically and culturally dominant 'outsiders'. Ironically it is the rural inhabitants' sense of belonging and perceived rootedness in tradition that renders them an important living element of today's heritage industry. However, the 'insiders' insist that their lived authenticity involves moving with the times, rather than the 'outsiders' learned authenticity which prioritises preservation and un-change.

This association was highlighted at one of the local 'Association Méjan' meetings I attended, where a debate concerning the commissioning of a new video of the Causse was discussed. There already exists a video on the Méjan, but the local people are not very happy with it, because it concentrates solely on its historical monuments, archaeological sites, caves and natural history. Whereas they want to show the contemporary life of the place - its modern farms, buildings and tarmacadamed roads which they perceive as symbols of success and progress. But the fear was expressed that such a video might turn off the tourists. It would appear that contrary to the tourists' authenticity discourse, it is more often than not some preconceived image of nature and the countryside that they are after rather than its contemporary reality. It would be wrong to present the contested rurality as simply a struggle between 'insiders' and 'outsiders', today those categories are ill-defined and frequently interdependent. For example, in one of the Causse Méjan villages, Hurs-La-Parade, it is essentially retired local sheep farmers along with some holiday home owners of local origin who are objecting to the free passage of sheep through the village. The local agricultural Mayor perceives this situation in a very negative light, that of people who have forgotten their origins.

Inventing tradition

The indigenous rural population, especially its elderly members, frequently expressed the fear that nowadays everything is mixed up, and that their past is being turned into a folklore that never existed. In the Cévennes the numerous local *fêtes de transhumance* were often given as typical examples of such practices. Those who took part in the (now practically extinct) transhumance trails of the past are adamant that it was far from a *fête* in the past, rather it was a very difficult and anxious time of the year. The tourists who flock to such festivals today think that they are participating in something authentic, an authenticity that a former transhumance shepherd from the Mont Lozère described as '*bien du cinéma*'. But the tourists have to pay for their escapism, such festivals are important to the economic survival of many local traders. Besides, is not the indigenous Caussenard's recently created *fête du mouton*[8] (sheep festival) on the Causse Méjan, along with the local reworking of the sheep and shepherd symbolism equally cinematic? Both the Caussenard sheep farmers and the tourists are using the past to justify and legitimise the present.

Hobsbawm (1983) argued that inventing traditions is a process of ritualisation characterised by reference to the past, often a fictitious past, and that we should expect it to occur more frequently when a rapid transformation of society weakens or destroys the social pattern from which old traditions had been designed. Or is the study area simply engaged in an ongoing process of reproduction and transformation bounded by structure and history in what Rogers (1991) referred to as the human capacity to creatively invent or manipulate novelty in old terms and continuity in new circumstances? Besides, 'novelty is no less novel for being able to dress up as antiquity' (Hobsbawm 1983: 5). Neither can heritage be seen as simply a link with the past, it has also come to include the present and the living, such as peasants and landscapes.

Conclusion

The contested representation of the countryside, nature and rural heritage in the study area brings to the fore questions of identity and imagery. It has been argued that within traditional societies identity and meaning were contained within a moral code of conduct, and that in the epoch of late modernity that code has given way to a form of subjective reflexivity and an attempt to construct an ethical code of behaviour (Ray 1999). Within the ethical code of conduct the power of agency in the construction of identity and meaning is enhanced, rather than being 'handed down' as within the traditional model. Within the so-called postmodern epoch, meaning and social identities are increasingly constructed through the exchange of sign values. Everything, including heritage (imagined or real), landscapes, architecture, peasants and food becomes grist in the mill of identity construction and exchange. However, it is difficult to estimate how much freedom the individual actually has in the construction of new identities and lifestyles and how much is predetermined by mediatised imagery. Are the tourists, farmers and conservationists of the Cévennes reading the rural for culturally pre-established notions or signs of French rurality, culminating in the emergence of what Cloke (1997: 370) referred to as a 'virtual rurality'? Or does the fact that imagery and symbolism is open to multiple meanings and interpretations guarantee a plurality of choices underpinning the contemporary construction of multiple ruralities and multiple identities?

The iconography and staging of the rural today, along with the profound sense of loss and powerlessness when confronted with the irreversibility of time (exacerbated by the contemporary lack of faith in technological progress), and the tendency to turn nature into a mirror of society, are all bound up with the postmodern condition which tends to **problematise** the distinction between representation and reality. Western culture may well have taken up residence in Baudrillard's (1983) famous world of simulations, within which reality is defined as that of which it is possible to give an equivalent reproduction. The former fear of nature is overcome by simulating it, and the fear of uncontrolled reality is alleviated by creating a 'virtual reality'. In the final analysis the conservationists' and tourists' interests in the preservation of rural natural and cultural heritage is not about a return to the past, rather they – along with the local agricultural population – are constructing new ruralities. Future ruralities risk being centred more on 'lifestyles' rather than on 'a way of life'.

Notes

[1] In France rural communes are defined as those which are not urban: urban communes being those which belong to 'urban units', which are defined by their size of population (more than 2,000 inhabitants) and by a criterion of spatial contiguity.

[2] I did not use a tape recorder, because it was felt to be too intrusive and would put my informants ill at ease.

[3] The term *néo-ruraux* is used to describe a generally young urban people influenced by the 1968 'back to nature' movement who arrived in the semi-abandoned Cévennes in the early seventies. They were attracted to what they perceived as the liberty and freedom offered by rural life, communal living and working close to nature (Léger and Hervieu, 1979).

[4] In many of these apparently rural villages the election of the local Mayor, who in the past was automatically a local farmer, along with the election of the 'conseil municipal', have become in many villages a fight between 'insiders' (who do not necessarily live or work locally) and 'outsiders', such as *néo-ruraux* and holiday and retirement home owners.

[5] The perception of the countryside as a refuge continues. For example, in the Cévennes there is an important drug rehabilitation centre for young urban addicts. One also finds a new wave of unemployed urban people looking for refuge there. Their profile is very different to that of the initial well educated *néo-ruraux*. Today's refuge seekers tend to be unskilled and they generally lead an itinerant nomadic lifestyle with no worldly possessions or attachments apart from a canine companion.

[6] Some gîte owners like to tell stories of inexperienced 'green tourists' who were frightened by the silence and emptiness of places like the Causse Méjan, and could not even bring themselves to spend a night there.

[7] The reason being that the traditional material was too expensive and since the museum is situated outside the Park's central zone, its architectural restrictions do not apply to it.

[8] The 'fête de mouton' on the Causse Méjan was created in 1970 by the then agricultural technician, who wanted to give the Caussenards pride in their new-found identity, that of 'éleveurs' (sheep farmer). Prior to the arrival of Roquefort dairies on the plateau in the early twentieth century, sheep, who were kept essentially for their manure and wool, were secondary to cereal cultivation on the Méjan. Similarly, the emblem of the recently created folklore club is two dancing sheep.

8 Farm families as local environmental managers?: a case study on wider farm development

BARELD VAN DER PLOEG

Introduction: narrow versus wider farm development

There is a growing interest in the multi-functionality of rural space, in which both agricultural production and other functions are considered to be important. An OECD publication (1994) refers to these other functions as follows: 'Rural areas are home to a wide range of natural and man-made features - also called amenities including wildlife and flora, ecosystems of special interest, recreational areas as well as cultivated landscapes, unique settlement patterns, historic sites, and social and cultural traditions that cannot be transferred or recreated elsewhere'. The title of this publication, 'The contribution of amenities to rural development', is at the heart of the case story that will be told in this chapter. One of the points that will be brought forward is that it is not at all easy to bring about positive contributions by amenities to economic development. In fact Blöchliger in this OECD publication mentions, besides (target) option 4, three other possible interactions between amenities and rural development (illustrations and 'back-lash' mine):

- development leads to the destruction of amenities; (e.g. standardised farming practices develop at the expense of unique regional characteristics)
- non-development leads to the destruction of amenities; (e.g. farming fades away and takes with it cultural landscapes)
- preservation/ promotion of amenities leads to non-development; (e.g. severe constraints on farming such as prescribed high water tables)
- preservation/ promotion of amenities leads to development; (e.g. the creation of a financial spillover from city people who are 'consumers' of the countryside to farmers who welcome the role of being a 'local environmental manager') (Fuller 1990).

(back-lash: from the fourth interaction back to the second)

The reality of rural development in the Netherlands, in particular in the peat-meadow areas

In the Netherlands the first option was most clearly the case in the era of scale enlargement and specialisation in farming, when there was an emphasis on adapting sites to allow for application of uniform farming systems. This emphasis on site reconstruction was probably stronger in a man-made country like the Netherlands than in most other countries. Particularly in the period 1950-1980 most rural regions were, in the Dutch expression, 'taken on the spade' during land consolidation and re-allotment projects (*ruilverkavelingen*). Policies for water management provide an example. Agricultural production goals in most cases determined at what level water tables should be managed and in consequence, how the rural area should be reconstructed (system of waterways, dams, pumping stations or mills and drainage at field level).

After about 1980 other goals besides agricultural production became more important as factors in the design of land consolidation plans. New plans were often the outcome of clashes of interests in the rural area, although officials preferred the label 'integrated planning'. Plans were adapted to the wishes of farmers, tourists and stakeholders of nature and landscape values. For some regions the view even prevailed that farming systems should be subservient to nature and landscape goals. This applied to the so-called 'relation policy areas' (Ministry of Agriculture) and especially to some small marginal peat-meadow areas (wetlands or fens) which were brought under EU mountain farming regulations. In these areas physical handicaps for conventional farming were conserved or even brought about – e.g. high water tables in the peat-meadow areas - if this was considered necessary for nature and landscape qualities. On the other hand farmers were eligible for subsidies if they were willing to participate in a management agreement scheme (EU regulation 2078/92). This income subvention was to prevent the preservation of amenities from leading to farm non-development (Option 3) and finally to farming fading away and taking with it - as a back-lash effect - the typical meadow landscape and natural species like meadow birds and botanic grassland (Option 2).

Recently awareness is growing of nature and landscape values in rural areas outside the management zones ('white zones'). Besides the management agreements based on acceptance of constraints by farmers, now there are also schemes which reward positive actions by farmers for nature or landscape qualities, especially in field margins but also all over the fields (e.g. protection of birds' nests). In these schemes, grass-roots farmers' nature organisations mediate between subsidising authorities and individual farmers who conclude contracts about 'nature production' (Option 4).

'Development' phrased in terms of internalising amenities in farm economics

Option 4 is a special case of internalising amenities in farm economics. It represents the kind of internalisation in which economic decisions refer to amenities as *benefits* (economic resources) for the farm business. Internalising amenities in farm economics takes a different direction if negative effects to amenities from farm development/functioning are counted as a *cost* factor by the farmer.

Both kinds of internalisation presume changes in the context for economic decision making compared with a former situation in which so-called externalities existed, sometimes of two different kinds: (a) positive externalities such as cultural landscapes and (b) negative externalities such as the pollution of natural resources (van Kooten 1993). An example of a context change that allows for internalising a former positive externality is mentioned above in Option 4 as the creation of a financial spillover from city people who are 'consumers' of the countryside to farmers who accept the role of being a 'local environmental manager'. An example of a context change solving former negative externalities in the Netherlands is the so-called System of Mineral Balances which imposes a fine on farmers who exceed norms with regard to the physical 'input minus output' rate (Mineral Surplus) on their farms. In our case story, an even more relevant illustration is that governmental regulations for the management of surface water can be designed in favour of amenities. This is achieved by prescribing high water tables which are needed for sustainable organic (peat) soils and thus for the continuity of the typical peat-meadow landscape and nature. In zones with very high water tables (up to the grassroots), income compensations are given to farmers ('mountain' area regulation), but in zones with moderately high water tables (60 centimetres below surface), farmers have to live economically with yield reductions and relatively high production costs.

From a farmer's perspective one kind of internalisation is described in positive terms (benefits) whilst the second type is stated in negative terms (costs). From a non-farmer's perspective the two might be seen the opposite way round.

In the view of non-farmers, the first can be associated with expenditures, assuming their willingness to pay for amenities. The establishment of markets is crucial here, following Tomlinson (1996) who described them as institutions which enable exchange to take place, in this case between (a) farmers who commit themselves to the 'cultivation' of rurality and rendering services to visitors to the countryside; and (b) urban people who derive satisfaction from their stay in a rural area or simply from the idea that farmers are taking care of rural amenities. Bryant and Johnston (1992) describe this type of 'consumption countryside' with its active role for farmers, in areas which are close to urban centres, as 'agriculture in the city's countryside'. This implies that farmers in wealthy urbanised societies can make a living out of the 'cultivation' of rurality (nature, landscape and a clean and relaxing environment) and its economic exploitation (e.g. agri-tourism). This type of agriculture in densely populated regions implies a delicate combination of: (1) articulation of differences between rural and urban space, (2) establishment of functional linkages between rural and urban systems (e.g. a public footpath passing farms), and (3) economic linkages to create the financial flow (spillover) from urban people to farmers.

As a consequence, farmers might have to abstain from adoption of what is labelled progress in mainstream farm development, especially if this would destroy amenities. A complication is that amenities are not always just there (in the countryside) but sometimes exist primarily as a 'social construction' (concerning rurality) in the eye of the beholder. Farmers may try to produce 'counter-social constructions' for instance by suggesting that the countryside is most attractive if it is alive and not an open-air museum. A striking illustration of contested social constructions is the introduction of automatic milking systems in dairy farming. This can be presented as a symbol for industrial farming systems clearly at odds with the idea of rurality, but in our study area it is presented by an organic farmer as a way to give back to cows the natural situation of being sucked (milked) more then twice in 24 hours.

The second way to internalise amenities in farm economics *for non-farmers* has environmental benefits (including the prevention of environmental harm) without financial costs. If society wants farmers to behave in an environmentally friendly way, it can use a 'soft mechanism' (convincing them about standards of good farming practices) or a 'hard mechanism' (over-ruling by collective decision making, obligatory standards and eventual charges on those 'negative externalities' which are not strictly forbidden). The 'soft mechanism' appeals to countryside stewardship, the 'hard mechanism' leaves little choice to farmers other than to calculate differently in farming than before. For broad-minded farmers there is a middle course, named 'licence to produce'. These farmers are searching for positive action (the creation or conservation of amenities) as a possible substitution for a future more painful 'hard mechanism' which would be imposed from outside. A major incentive might also be the desire to improve the image of agricultural products. Being broad-minded in conventional farming is not far away from Wider Farm Development, in which 'environmental services' and quality products are perceived as means to realise additional value for the farm, especially when a farmer's orientation to the world market is (partly) replaced by an orientation to niche markets.

For non-farmers a 'free ride' to amenities might depend on their power or influence to reduce farmers' free property rights over land to conditional property rights. The reader should remember the example of the Dutch system of Mineral Balances and the conflicting interests between farmers and non-farmers around policies about water tables. It is highly relevant that non-farmers in recent years have become much more influential in the Dutch Water Boards (*Waterschap*), because these Boards can prescribe water tables also in ditches between farm fields. Farmers in an urbanised society like the

Netherlands are in the position of a small minority. Yet real possibilities for 'free rides' to rural amenities are limited for the urban majority as long as farmers cannot be replaced as 'environmental managers', as in the case of peat-meadow landscapes. Because farmers have to make a living in a market economy, there is always the danger of Option 2 above, the fading away of farming taking with it the highly valued cultural landscape.

Problem formulation and delimitation of wider farm development as a possible way out

The most striking change seems to be that the countryside is no longer regarded as the property of agriculture, to be adapted to the production needs of farmers, but is increasingly claimed as the property of a range of mainly urban groups (nature conservationists, tourists etc.), who believe that it is agriculture which has to adapt to the needs of amenities in the local countryside and not the other way around. Farmers, particularly in urbanised areas, are faced with a double-bind dilemma, involving conflicting external claims: (a) an urban society asking for rurality, and (b) a world market urging cheap production. Internalising amenities as additional costs in farm economies can easily transform this double-bind into a deadlock for farm development.

Against this background, the positive way to internalise amenities into farm economies is taken as a goal in this case study. The combined 'cultivation' and economic exploitation of amenities is named Wider Farm Development (WFD). Three main types are distinguished:

- deep farm development, e.g. organic farming and regional quality products such as on-farm cheese making, where amenities are tied up in agricultural products provided that consumers want to pay for the additional labour, care and out-of-pocket-costs to the farmer compared with conventional products;
- ecological widening, especially the management of nature, landscape and/or physical environment features such as 'clean water ditches', as an economic activity carried out alongside agricultural production;
- social widening, particularly farmers rendering paid services to visitors to rural amenities, e.g. on-farm tourism and educational/recreational excursions to farm fields with nature management.

Description of the case study in terms of farmers framing endogenous development

The peat-meadow district in the western part of the Netherlands will be described as an example of WFD. This is presented as an endogenous process (born from within), from the perspective of farmers who try to maintain or enlarge their room for manoeuvre.

The description can most easily be understood by keeping in mind the keyword *framing.* Framing is embodied in the pre-occupations of the farmer which determine his *sense of direction,* or in negative wording his *blind spots.* We will look for 'framing' in different fields.

The first field might be called institutional framing. Emerging organisations for exchange between farmers who take care of amenities and non-farmers who are interested in this ('the social organisation of wider farm development') are discussed later.

The second field might be called personal pre-disposition framing, and we focus on 'leitbild framing' in which the farmer's occupational outlook is analysed in relation to the choice between narrow (conventional) and wider farm development. Is WFD for me as a farmer in line or at odds with my basic ideas about what farming should be? Next, the way in which male and female farmers perceive their occupational capabilities is considered, especially those needed for wider farm development. Can my personal capabilities as a farmer be employed more fruitfully in narrow compared with wider farm development, in terms of both economic success and self-realisation?

Whilst we then analyse how farmers ideally define their occupational preferences, we then discusse the real choices made by them when facing the total set of alternatives for farm and income strategies, including types of pluri-activity without any direct relation with rural amenities such as second branches of agricultural production and off-farm jobs ('dynamic analysis at farm level'). This is particularly interesting if WFD and other farm and income strategies are perceived as competing alternatives, as neutral or in a relationship of mutual re-enforcement. Which role does WFD, for instance nature management, fulfil in my overall strategy as a farmer?

Finally, we look at WFD as a case of collective learning. This learning is related to the challenge of finding a sustainable solution for the future of this vulnerable spatial system called the peat-meadow district or city's green heart, implying that 'water' should be the main steering (framing) factor in its development. Is it possible for us farmers as a group in this area to make a long-term contract with the surrounding urban world in which we are paid to be environmental managers of the city's green heart?

The City's Green Heart: a case of wider farm development

The peat-meadow district in the western Netherlands covers about 100,000 hectares of land and water. At some places the peat layer can be found from the surface to 10 metres below it. The land surface in this area has been sinking continuously since the time of reclamation around the year 1100. Locally the peat has already almost vanished, due to oxidation of the soil. For the conservation of peat soils it is essential to maintain high water tables in the broad ditches between the meadows. Water Boards (*Waterschappen*) have legal power to prescribe such high water tables to farmers. Depending on the level at which the water tables are fixed, peat-meadow areas represent seriously disadvantaged farm locations (wetland zones) or moderately favourable farm locations, zones in which water tables are at least 60 centimetres below grassroots, for conventional dairy farming.

Figure 8.1: Dutch Green Heart Metropolitan (*Randstad*) Area

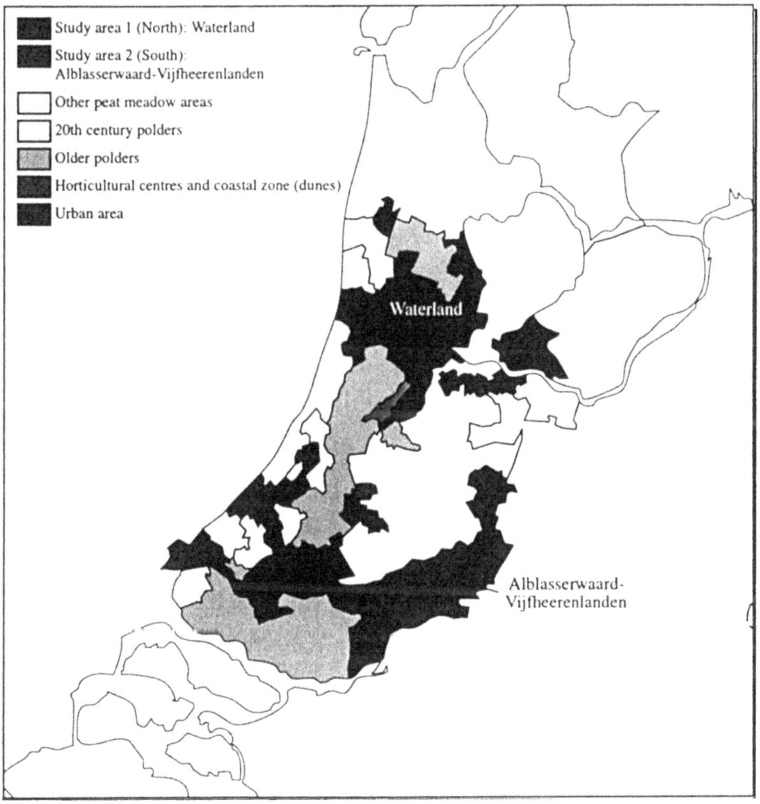

The peat-meadow district is surrounded by several urban centres. The area is popularly called the cities' green heart (*Groene Hart van de Randstad*). The green heart was protected with some success from urbanisation by physical planning and also by physical local conditions (high cost of building houses on soft peat soils). The green heart is a powerful symbol in Dutch politics because it is thought to prevent the Randstad cities from growing together. Besides this, the meadow landscapes like other polder landscapes are seen as a part of the national identity. The high bio-diversity of the peat meadows, especially on extensive grasslands, is an additional reason to protect this green heart. The district also has a function in international eco-systems, especially for migrant birds.

Farming in the district in recent history reflects some of the dominant trends in this sector during the era of industrialisation. Most important here are two trends towards uncoupling of farming:

- from its natural environment (relatively) and
- from direct interaction with its nearby urban markets (rather drastically).

The relative uncoupling of farming from the natural environment can be illustrated most clearly by the case of fertiliser. Initially the peat-meadow district was known for its highly productive grasslands, due to the 'natural nitrogen' set free by the process of oxidation in peat soils This comparative advantage became less important with the introduction of chemical fertilisers. On the other hand the disadvantages of the soft soil became more clearly visible when farming was mechanised, including the replacement of boat farming by tractor farming, and because of higher cattle densities induced by the use of chemical fertilisers.

The rather drastic uncoupling of farming from direct interaction with nearby urban markets can be illustrated with the disappearance (to a large extent) of some typical farming systems in the area. The most interesting farming system was found in the centre of the current Green Heart. There, until the 1960s, many farms had an integrated system of cheese-making based on the output from dairy farming and providing the input (skimmed milk) for pig production. These pigs were called 'butchers' pigs', as compared with 'bacon pigs' for the export market. The farming system of cheese production + dairy farming and pig production was prosperous and extremely laborious for farmers and farm wives (in charge of the cheese making). It was also marked by high land use intensity (cattle density). However there was a large diversity in land use within each farm. Fields at the back of the farm properties - deep in the polder - were often managed in an extensive way which allowed for a rich bio-diversity.

Until about 1980 both uncoupling tendencies were stimulated by the prevailing forces in agri-business and governmental policies. Since then there has been a counter-movement directed at recoupling farming to its natural environment. In some parts of the peat-meadow district, the so-called relation policy areas, extensive grassland farming was promoted by (a) restrictive policies concerning management of water tables, and (b) financial support policies or management agreements. In recent years this was followed by a grass-roots movement (farmers' nature organisations) backed by government subsidies, aiming at higher nature qualities in zones where circumstances allow rather intensive agriculture production practices.

In recent years there has also been a counter-movement aimed at recoupling farming to nearby urban centers. In government initiatives - e.g. the Valuable Landscape Project (WCL) in Waterland - and in farmer initiatives such as the 'city's garden' (*binnentuin*) of the regional farmers union WLTO, a restoration of linkages between urban and rural systems was put on the agenda. There is a close connection between this 'social recoupling' and the 'ecological recoupling' mentioned above. Farming is to be made an ecologically and socially friendly business. Some of the proposed changes can be classified as ecological widening of farm development (the uptake of nature conservation) whilst others can be characterised as the social widening of farm development (e.g. the uptake of farm tourism). The orientation to special quality products with high value added is here considered deep farm development.

The information base of this chapter

The main sources of information for this paper are:

- A survey of wider farm development in 1996 among 105 cattle farmers in the peat-meadow area North of Amsterdam (Waterland). This survey was part of an international multi-disciplinary research project to support EU Agri-Environmental Programs (regulation 2078/92). The sociological perspective on individual farmers (attitudes) and social organisations (support system for wider farm development) in Waterland was elaborated in Van der Ploeg (1999).
- A follow-up survey in 1999 (first results) on 33 dairy farms among young farmers (up to 40 years) and their wives in a peat-meadow area between Rotterdam and Utrecht (Alblasserwaard/ Vijfheerenlanden).
- Desk research by two research institutes (Agricultural Economics Research Institute, Staring Centre for Research on Green Space), for four provincial governments in the Randstad area into future possibilities of Land Based Farming in an Urban Environment. This identified several factors in the development of conventional and wider farm development:
  - factors internal to the farming sector, including changing EU price and market policies;
  - factors external to the farming sector, especially the urban market for wider farm development, and
  - factors at the interface of rural and urban, especially the urban influence on prices for agricultural land (relatively high) and the limiting consequences of this for scale enlargement in conventional farming.

The study areas Waterland and Alblasserwaard cover the northern and the southern edge respectively of the total peat-meadow district. Both are front runners in the grass-roots movement for WFD. The two farmers' nature organisations here (Waterland, Den Hâneker) are among the oldest and largest organisations of this type in the Netherlands.

It should be noted that farming in both study areas seems to be more compatible with WFD because of the *relatively* low cattle density compared with the central part of the peat-meadow district.

Social organisation in regions of wider farm development

Waterland

The social organisation of WFD in Waterland was recorded via an inventory in which each farm field was checked for existing contracts between farmers and other parties in favour of nature in 1999 covered about one third of the total area. Two years later, this area has shrunk considerably.

A second category is fields with (only) voluntary protection of meadow birds. These farmers are members of the Waterland association for voluntary protection of meadow birds, founded in 1983. Membership implies that farmers give free access to their fields to volunteers (often with an urban background) in the protection of meadow birds and commit themselves to protecting the marked nests as far as possible, especially when cutting the grass. Many farmers in the area are enthusiastic bird-watchers themselves. However, the breeding season coincides with a busy period on the farm and especially in cases of scale enlargement farmers are not able to supervise the total farm area themselves.

A third category is fields with a contract for 'paid nature production' between the farmer and the farmers' nature organisation, founded in 1995. Farmers who agree to such a contract are obliged to follow a course in nature management. They must also record the 'amount of nature' as part of the contract. The level of payments depends on the number of animals (nests) or species (plants). This is checked at random by the organisation. The tendency is towards more comprehensive contracts. Many farmers started with voluntary protection of meadow birds, then converted to the protection of meadow birds and extended this to nature management in field margins or even ditches. Nature management in field margins has consequences for farmers' activities in the fields (adjusted cutting of grass, no fertiliser), but also for the yearly dredging of ditches (spreading the sludge over the fields). Farmers with rather comprehensive contracts get average payments of about 100 Euro per hectare. However, this is not just additional income because the payments are largely compensation for higher costs and yield-reductions.

A fourth category is fields with a management agreement between farmers and the government. These are found mostly in zones with obligatory high water tables. Some fields have been under management agreements for 20 years. The most important contractual constraint is postponing grass cutting until mid-summer. Payments to farmers for fields under management agreements may rise to about 750 Euro per hectare. Recently the task of concluding management agreements with farmers has been partly delegated by the government to the farmers' nature organisation. Also the previously strict zoning of fields on which agreements are possible has been relaxed (the so-called 'wide coat' approach).

A fifth category is fields which have been bought from farmers by nature conservation organisations with the help of government subsidies. These fields are located in zones with a high water table. Some are managed by employees of the owners as 'real nature' (swamps) or for botanic grassland and peat-meadow wildlife. However, most are let out rent-free to farmers who accept management constraints comparable with those in management agreements. A new development is 'positive management of nature' on these fields.

Some types of WFD are not directly land based. Waterland has new production chains for regional quality products, especially Waterland beef (which was served at the 1998 EU summit meeting in Amsterdam). There is also a service chain for recreational activities such as renting a room or a boat on a picturesque farm. The most important of these other types of WFD is organic farming. In Waterland this option was embedded in institutions specialising in organic farming. A short distance from Waterland (in Limmen) there is a small dairy company specialising in the processing and marketing of organic milk. Supervision of organic farming standards is done by a national institution. In the peat-meadow areas, however, organic farming is not a solution for all problems of sustainability. Under wetland conditions, organic dairy farmers have severe difficulties in achieving self-sufficiency in cattle feeding.

Alblasserwaard/Vijfheerenlanden

In this sub-area on the southern edge of the peat-meadow district the same types of WFD are found as were mentioned above for Waterland; however, a smaller land area and fewer farms are involved. So Alblasserwaard/ Vijfheerenlanden fields without any contract would occupy a larger area, especially in the west near the windmills of Kinderdijk.

Embeddedness of wider farm development at the farm level

Guiding images in farming

The Waterland survey allowed farmers' practices and farm changes to be classified as friendly or unfriendly to ecological and social qualities of the countryside.

The classification of *practices/farms* was related to a classification of *farmers' attitudes*. Sensitising concepts in the classification of occupational attitudes were:

Table 8.1: Basic preferences appearing from factor analyses of farmers' answers[a] to a question about characteristics of their ideal farm

Characteristics of an ideal farm	Factors with factor loading by 'characteristics' *Factor 1* Alternative *vs.*Conventional Farmer	*Factor 2* Entrepreneurial Farmer *vs.* Yeoman	*Factor 3* Family Farmer	*Factor 4* Adventurous Farming	Frequency of appreciations (indication)
Continuous farm growth		+75			++ -
Thrifty farming (low expenditure)		-72			++ -
Yields as high as possible	-35	+47	-31		+ - -
Automatic milking system		+46	-42	-48	++ - -
Second branch: agricultural production		-40	-42		++ -
Second branch: on-farm tourism	+90				+ - -
Second branch: nature managment	+60			-38	++ -
Organic farming	+75				++ - -
twin purpose cows: milk and beef	+70				++ - -
High risk farming: highly gainful but risky				+82	+ - - -
A real family farm: farmer, farmer's wife and farm successor			+79		+++ -

[a] Multiple choice: definitely not, not, yes, definitely yes. A fifth alternative was added for farmers who 'really are between yes or no'.

Source: Survey 1999

- Entrepreneurial Farmer versus Yeoman (Austin *et al* 1996) and
- Conventional versus Alternative Farmer (Beus and Dunlap 1991).

An illustration of these dimensions can be found in Table 8.1.

The results from the survey indicated:

- an unfriendly pattern, found mainly on expanding dairy farms managed by farmers with an entrepreneurial-conventional outlook;
- a compensatory pattern (positive and negative 'externalities'), found mainly on expanding dairy farms managed by farmers with an entrepreneurial outlook, varying from 'alternative' to a need to obtain from society a 'licence to produce';
- an indifferent pattern (no positive or negative 'externalities'), found mostly on stable or contracting beef or sheep farms, often former dairy farms, managed by farmers with a yeoman-conventional outlook;
- a friendly pattern, found most on non-development farms whose farmers are marked by a yeoman outlook.

The study tried to evaluate farmers' attitudes as a factor in farm development. However, in many interviews - especially on former dairy farms (mostly shrinking or part-time enterprises) farmers indicated that their occupational attitudes had changed following their disengagement from future-oriented full-time farming: 'If you had asked me this fifteen years ago, I would have given quite a different answer'.

The follow-up survey in Alblasserwaard/Vijfheerenlanden was limited to farmers and their wives aged less than 40 on dairy farms. In the interviews one question was about characteristics of the ideal farm 'that may be quite different from your current farm'. After the first run of factor analyses (varimax rotation) the following outcome for male farmers appeared (Table 8.1).

Alternative farmers who are positive about WFD (farm tourism, nature management, organic farming) may be entrepreneurial farmers or yeomen. This conforms with field impressions during interviewing. Figure 8.2 presents a systematic overview of the possibilities with regard to the attitudes of farmers. This typology will be referred to in the descriptive analyses which follows.

Human resources in farming

Dairy farmers' and farm wives' views about occupational skills

In Alblasserwaard, farmers and farm wives were asked about the importance of specified occupational capacities, and to evaluate these capacities in terms of their own strengths and weaknesses. The specified occupational capacities were:

- craftsmanship in working with crops and livestock;
- entrepreneurship in determining future policies for the farm;
- organisational skills in running a business;
- social skills in human relations;
- administrative skills in data management;
- technical skills in handling machines and equipment.

Cluster analysis of 24 variables revealed two clusters with a different participation rate in wider farm development (see Table 8.2).

Figure 8.2: Systematic overview of possible attitudes of farmers

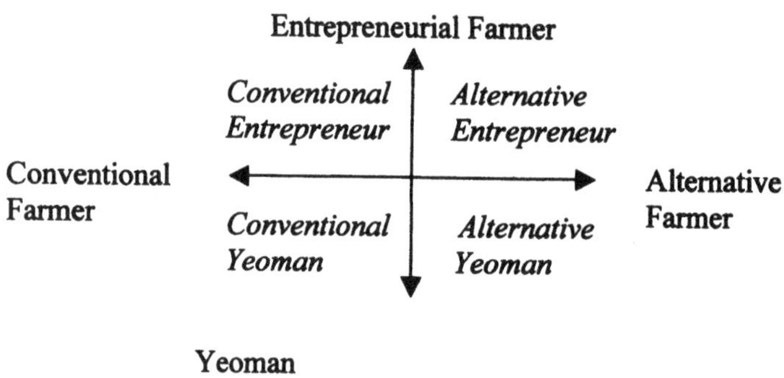

Couples who put a high value on organisational and administrative skills (cluster 2) are more often involved in wider farm development than couples who valued craftsmanship and social skills more highly (cluster 1). However, this difference is not found if the most frequent type of WFD (nature conservation) is left out. Social wider farm development and also deep farm development (including organic farming) are most often found in cluster 1. Farm women have a large influence in this type of WFD.

The higher participation in nature management found in cluster 2 (organisational and administrative skills) seems in line with findings by Van den Ham (1998) about differences between dairy farmers with and without contracts for nature management. According to Van Den Ham, farmers with nature management contracts are less focussed on detail (craftsmen?) and have a better view of the farm as a whole. Van den Ham found that such farmers generally achieved higher farm results compared with other farmers with a similar enterprise type and size, even if the additional income from nature management is excluded.

Craftsmanship appears from the survey to be a male farmer's domain, whereas social and administrative skills appear to be the stronghold of female farmers. Social skills could also be labelled 'social craftsmanship'. The label 'craftsmanship' refers to an emphasis on informal knowledge, especially on the role played by intuition and experience. A qualitative impression is that as a

Table 8.2: Perception of occupational capacities related to the participation in wider farm development

	Number of farm families (M/F)		
	perception of occupational capacities		
Participation in WFD: (ecological, social and/ or deep farm development)	Cluster 1: Craftsmanship Social Skills	Cluster 2: Organisational Skills Administrative Skills	Total
No activities, definitely no plans	36	21	27
No activities, unlikely	14	5	9
for the future	14	53	37
Present, or likely for the future	36	21	27
Present, plans for expansion	100	100	100
Total	14	19	33
N			

factor in the expansion of WFD, entrepreneurship is also strongly marked by informal knowledge including from personal discussions with fellow farmers (e.g. in meetings of the farmers' nature organisation) and non-farmers (e.g. visitors to the farm and co-residents of the village).

Formal knowledge is most clearly present in administrative skills. This is the domain of female farmers because in general they take care of the book-keeping. However, male farmers enter this domain when strategic external relations are involved: contacts about large investments, contacts in circles studying mineral balances, etc. A professional who is in touch with the farmers' nature organisation said that the strength of this organisation is in innovation based on local knowledge of male farmers, while the administrative skills of female farmers are not fully used. Here it should be noted that payment for nature results is based on a recording of these results.

Dynamic analysis at farm level (Albasserwaard)

In the world of farmers there is more pluriactivity than wide and deep farm development alone (Table 8.3). For farmers in our most recent survey the most common types of pluriactivity are nature management (ecological widening) and/or a job off-farm for the farmer or the farmer's wife.

Table 8.3: Frequency, economic importance, task divisions (M/F) and plans with regard to pluriactivity

Types of pluriactivity	% of farms/ families having this type of pluriactivity	% contribution to family income (estimation by respondents)	Who is responsible for this activity? (M/F)	% of all respondents considering tostart or to expand in this
Second branch of agricultural production e.g. sheep or pigs besides dairy farming	34	10-19 (varying from nil to 30%)	MF	?
Deep farm development, e.g. Organic Farming or cheese making	19	10-19 (two thirds in this income category)	MF	9
Ecological widening: e.g. contracted nature management besides dairy farming	47	< 10 (all cases!)	MF	22
Social widening: e.g. on farm tourism or educational excursions on farm field with nature management	9	<10 (two-thirds in this income category)	MF	19
Off-farm Job: e.g. Farmer who is a lorry driver between Rotterdam harbour and German Ruhr or farmer's wife who returned to her former job in a hospital	41	≥50 (two-thirds in this income category)	MF (F = most frequent, M= most important income)	6

Source: Survey 1999

The income generated by external jobs is much larger than the additional income from nature management. External jobs often contribute more than 50 per cent of family income. This applies especially to external jobs held by farmers, and not by farm women. Income from nature management is largely a compensation for reduced yields or additional expenditures.

In our analysis of the city's farming, the options 'second branch of conventional production' and 'off-farm jobs' are interesting as possible functional equivalents for our central theme 'widening + deep farm development'. The ongoing study for the total peat-meadow district reveals that WFD can probably only compensate for a minor part of the income loss that is expected over the next decade from changes in EU prices and market policies and increased competition in conventional farming. Outlooks for second branches of conventional production in the peat-meadow district are gloomy because the soil is only suitable for permanent grassland, and buildings for intensive livestock and horticulture would be costly. So, the main alternatives for additional income, besides WFD, are off-farm jobs, and improvements in farm structure and functioning in dairy farming.

Farmers in our most recent survey reported many changes in the structure and functioning of dairy farming on their holdings (Table 8.4). Farm expansion in terms of acreage, herd size and total milk production occurred on most holdings. This last element is partly based on a rather general gradual increase in milk production per cow. Decrease in the use of external inputs (fertiliser, concentrates) follows from more severe environmental regulations, and improved methods of grassland and roughage management For the future most farmers expect a further reduction of the mineral surplus. Most expect a further expansion especially in terms of the quantity of milk they produce.

Many of these younger farmers started as entrepreneurs on rather small- or medium-sized farms. Their continuous expansion in farming is the result of hard work and a relatively low level of household spending. The study in the total peat-meadow district showed that farm investments are at the same level compared with other Dutch dairy districts, although on peat soils the economic results are lower.

High entrepreneurial aspirations, modest personal aspirations

In our recent survey farmers and farm wives were asked about their most and least important goals in farming. Having as many hours for leisure as in most other occupations was generally classified as one of the least important goals, as was the goal of having 'personal expenditures comparable with other occupations'. Qualitative impressions from interviews tell us that working hours for many respondents are extremely long. This applies especially to farmers with an off-farm job and to most of the farm wives. Some female farmers have a triple task: household, farm work and an external job. Role divisions between male and female farmers are generally rather traditional.

Table 8.4: Structural and management changes in conventional farming (recent past, expected future)

Characteristics of dairy farming at farm level	Changes in the last 5 years –	0 (lt 5% ch.)	+	Total (100) n= 32	Expectations for the next 5 years –	0	+	Total (100) n= 32	1999 level
-Workers (SLU)	10	67	23	100	-	93	7	100	2.0
-Acreage (HA)	3	27	70	100	-	55	45	100	32
-Milking Cows (number)	17	40	43	100	7	43	50	100	50
-Milk Production (tons)	3	28	69	100	3	13	83	100	433
-Milk Production/ Cow (KG)	-	37	63	100	-	30	70	100	7,129
-Milk Production/ Hectare	30	23	47	100	-	38	62	100	10,718
-Artificial Fertiliser (N)/HA	58	42	-	100	3	90	7	100	?
-Yields (nutrition value)/HA	7	71	22	100	3	90	7	100	?
-Concentrates bought (KG)	54	33	13	100	10	80	10	100	?
-Mineral Surplus (N/HA)	68	29	3	100	52	48	-	100	?

Source: Survey 1999

Farmers generally only have minor tasks in the household. On the other hand, farm wives have seldom been instrumental in the core business of these farms, the management of dairy cows and the production of roughage. Their task is often to take care of young cattle, attend to book-keeping and to render general assistance. This division of roles can also be found in nature management and organic farming (see overview, Figure 8.3). However, for social widening indications were found that the leading role was most often taken by farm wives.

Motives for wider farm development and other forms of pluriactivity

Figure 8.3 presents a systematic overview of the motives for pluriactivity. At the top, farming is close to conventional full-time farming notwithstanding some widening. Options at the bottom are closest to stable part-time farming and shrinking full-time farming. In the middle are fully integrated forms of pluriactivity and options of so-called development pluriactivity (Wilma de Vries 1995) which in the Alblasserwaard means that earnings from a second activity are used to invest in dairy farming as the core business. One farmer was very clear about this: 'Everything my farm yields remains in the business because our cost of living is covered by other activities'. On this basis he was able to buy land and milk production rights from other farmers who had left dairy farming.

On the other hand, we found no examples of 'development pluriactivity, farming subservient'. Several farms (all organic) could be labelled 'fully integrated PA'. But only one of these was close to the ideal typical situation of the city's farming. This was a farm with organic production, cheese-making, a small farm shop and educational excursions for visitors.

Pluriactivity in the peat-meadow district is often marked by the wish to have an opening to the world. For farmers, this applies especially to the most frequent type of pluriactivity: nature management. For them this activity is a form of public relations and a way to acquire social recognition. It is the best way to show your enterprise and agricultural region to the outside world. If you want to communicate with urbanised society it is better to stress aspects of farming with nature than aspects of high-tech farming.

For farm wives, horizon-widening pluriactivity is present in external jobs (meeting people) and in (other) types of wider farm development in which social contacts are essential, e.g. farm tourism and on-farm selling of products.

Only one case of 'substituting PA' was found: a farmer who had land heavily constrained by high water tables, land-rent contracts (on fields owned by nature conservation organisations) and management agreements. The farmer did not like this situation but did not see any realistic alternative. The basic idea behind the label 'substituting PA' is that net production yields on these fields are so low that the reason for existence of these farmers is 'nature conservation', not food production.

Older farmers, and those without milch cows, were excluded from the Alblasserwaard sample. This is why no clear representatives of the option 'survival PA' were found. However, some farmers and farm wives spoke about their future in terms of possibly shrinking farming. They are *not* in the current movement towards wider farm development. Off-farm jobs are more important to them.

Learning for wider farm development

Learning at three levels

Earlier, we discussed the development of WFD for individual farmers and for regions with special constraints and opportunities; here, both are addressed together.

Figure 8.3: Different contexts for pluriactivity

Motives for pluriactivity[a]	Types of pluriactivity Countryside activities other than agricultural production Ecological widening: e.g.nature management	Social widening: e.g. on-farm tourism	New activities in agricultural production Economic deepening: additional value added in agr. prod.	Economic widening: second branch of agr. prod.	Off-farm job
Horizon widening PA The primary motive is non-economic, but to 'meet people' and to show the farm to the outside world	X-------- : mostly	-------X farmer :			X Mostly farmer's wife
Licence to produce PA The motive is indirectly economic: public relations for farm, region or sector	: X------- :	: ------ X :			: : :
Development PA (agriculture) Second activity supports investments in core-business (dairy farming)	: x----	: ----x			: X Farmer, F's wife, Successor
Complementary PA Second activity is based on resources (esp. labour, land) that cannot be used in the core-business	X Field margins and surplus fields because of milk quota	These options for dairy farming are less relevant compared with arable farming (having a labour surplus outside peak periods)			

Integrated 'PA' Farmer widens his occupational task without a distinction between first and a second activity	....	....	: X organic farming and regional quality products		
Development PA-2 Farming is subservient to development of a second activity	....	X..... e.g. farm management is adjusted to recreational needs of tourists	....X e.g. farm management is adjusted to cheese making		
Substituting PA Farming is largely replaced by a second activity	Extensive grassland (high bio-diversity): most of the income is from management agreement X				
Survival Strategy PA Second activity enables the farmer to maintain a non-development farm as a source of identification and as a place to live	: X ------- Marginal full-time farm dependent on subsidies	----------	----------	----------	X Part-time farming

1) Classification partly from Wilma de Vries (1995)

The learning involved in wider farm development seems to include a number of different dimensions: an occupational paradigm switch, especially from 'conventional farmer' to 'alternative farmer' (see above) and including 'entrepreneurs' who change their orientation from anonymous world markets towards nearby personalised urban markets; new skills, especially know-how about social services, ecological services and high quality products establishing new (or reformed) support systems for WFD, especially organisations such as farmers' nature organisations for marketing new countryside products, but also systems for training and information supply related to 'new skills'.

Integrated picture from Waterland

Waterland illustrates learning processes that take many years. Current WFD originated here from a conflict between organised farmers and nature conservationists around 1980. In 1983 a paradigm switch by both parties resulted in the foundation of the Waterland organisation for the voluntary protection of meadow birds. In return for concessions by nature conservationists in a project for land reconstruction, farmers gave free access to their fields to volunteers in the protection of meadow birds. They committed themselves to full co-operation in protecting nests and young birds especially by cutting the grass carefully. Farmers started to realise that 'cultivation of nature and other amenities' was appreciated by society no less than food production. In the years following, many farmers found personal satisfaction in protecting meadow birds and improved their skills in nature conservation.

The farmers' nature organisation Waterland, founded in 1995, aims to professionalise these new tasks. One side of this professionalisation is that environmental performances are remunerated indirectly by governments and other subsidisers of the organisation. Another is systematic training of farmers in their new tasks. At present most farmers in Waterland have followed a course in nature conservation.

The surveys in the two peat-meadow areas suggest a future mosaic situation. There will be a mixture of conventional farms, organic farms and other farms with WFD. Most of the organic and other farms with WFD will be marked by an internal mosaic as well. Most fields are used for intensive grassland production, but this is generally combined with some form of nature management (especially protection of meadow birds with the help of volunteers) and with extensive nature production in field margins and possibly also in ditches. Some scattered fields are managed by farmers as botanical grassland with a low agricultural production value. Farms which are entirely in high water zones will withdraw from dairy farming and will become almost fully dependent for their income on payments from management agreements.

The desk research for the total peat-meadow district revealed that, within the context of current 'organisations for exchange' between farmers and non-farmers, additional income from WFD will not be high enough to compensate for the decline in income that can be expected from developments in EU policies and world markets. Even with maximum acceleration in the process of scale enlargement and closure of smaller farms, the income lag on the remaining farms compared with other Dutch dairy districts will continue or will worsen.

Thus there is a serious need to look at possibilities for enlarging the economic reach of wider farm development.

Society as a whole has reasons to promote WFD. Society's focus will be on the *ecological* results to be obtained. The current type of farming and its demands for water management will lead to a gradual disappearance of the organic (peat) soils, higher costs for water management, and after some generations the end of the typical landscape and nature. Discussion about this has been intensified by the expectation that climate change might make it more difficult to protect the lowest part of the Netherlands from flooding.

In physical planning the main motive for protection of the Green Heart from urbanisation is 'amenities' and manageability of soil and water conditions, not the agricultural production value of the area. In a future Green Heart without farming, soil and water conditions would probably be easier to manage. This implies that in a large 'city park' (some 100,000 hectares), water would be the dominant element, slowing down the process of soil evaporation (oxidation) considerably. The negative side would be the disappearance of the existing cultural landscape - including historic features which are part of national identity - and of peat-meadow bio-diversity.

A rosy scenario characterised by far-reaching wider farm development

At the moment there is a movement towards a system in which governments make contracts with farmers' nature organisations (FNOs) stating environmental targets for areas from 5,000 to 20,000 hectares, followed by contracts between FNOs and individual farmers. However, the resulting financial spillover from 'city to farmers' is relatively small. Generally contracts presume that farmers commit themselves to environmental achievements which are attainable given the existing context for farming, including the regime of water tables. This even applies to the recent area contract between the FNO Waterland and the Water Board.

In a rosy scenario, a contract for many decades is made between 'society' and 'farmers as environmental managers of a spatial system' which enables a much larger spillover and greater concessions from agricultural producers in favour of sustainability of the basic subsystem 'soil + water'. The spatial system named 'peat-meadow district/city's green heart' can be pictured as having four sub-systems:

- soil + water which determines the basic conditions for
- land-based farming, more specifically peat-meadow farming which is the basis for
- typical landscapes and a high bio-diversity being the main 'input' for

- human appreciation of amenities.

From a farmer's perspective, a starting point could be that the current standard for water tables which is imposed without financial compensation (60 centimetres below land surface), already causes income reduction. The next question might be, at what height water tables can be without making dairy farming *technically* impossible. The technical discussion has to answer questions such as:

- at what degree of moisture are higher yielding grass varieties replaced by low yielding grasses? and
- what are the consequences for purchase of concentrated fodder and thus for having to respect the legal limitations set for the farmer in the Mineral Balances Regulation?

The technical discussion might show that dairy farming is technically possible at a certain level between the current level in nature reserves (some 20 centimeter below land surface) and that on most agricultural land (less than 60 centimetres). This would lead to discussion of the financial consequences and bargaining with society - within limits set by the EU - about region-specific cross-compliance payments to farmers. In a rosy scenario the 50 per cent of the national co-financing could be re-allocated to the regional level of the urban area for which the peat-meadow is the 'Green Heart' and this could stimulate a dialogue between city and country about amenities to be maintained or developed.

Some recent institutional innovations in the area are highly relevant for this dialogue function. First, there is growing participation by non-farmer volunteers in farmers' nature organisations. Second, as mentioned before, Waterland is the first area to have a regional contract between the FNO and the Water Board, an institution which is a meeting-point (platform) for farmers and non-farmers in a decision-making context. Third, the FNO Alblasserwaard-Vijfheerenlanden made a preliminary agreement with the provincial organisation for nature and landscape conservation (*Zuidhollands Landschap*) to co-own and manage nature reserves in the peat-meadow district.

Evaluation

Dairy farming, like many other agricultural activities, is firmly integrated in footloose production chains. Re-integration of farming in a local setting would have economic and environmental benefits. This applies especially to physically handicapped agricultural areas like the peat-meadow district in the Green Heart of the Dutch Metropolitan Area (Randstad). This section explores WFD as a learning process for the reintegration of farming in its ecological and social (urban) environment.

The learning process can be examined on four levels:

- the establishment of organisations for exchange between farmers who enlarge their occupation to include 'cultivation' of site qualities, and non-farmers who 'consume' the results;
- an occupational paradigm shift for farmers, especially the conversion from a perception of local specificity as a problem for agriculture to that of specificity as an economic resource;
- acquisition of such new skills for task enlargement, especially for 'cultivation' and marketing of rural amenities and rendering social services to visitors to the countryside;
- effectiveness of WFD as a way of achieving sustainability for this vulnerable spatial system.

Organisations for exchange

The study reveals several types of organisations facilitating information exchange between farmers and society about countryside functions other than agricultural production. They involve a change from solidarity around rural amenities organised on a large geographical scale (top-down EU programs) towards social arrangements on a smaller scale. If this solidarity is organised in an urbanised society on a very small geographical scale, the label 'agriculture in the city's countryside' might be correct. Generally, the organisations are farmers' initiatives (farmers' nature organisations) born out of strategic interaction with governments and environmental stakeholders. They start with ecological widening (nature management) but do not acquire the character of 'agriculture in the city's countryside' until completed with social widening (farmers rendering social services).

Paradigm shift

An occupational paradigm shift takes many years. Such a shift can be seen as mental internalisation of behaviour which started from strategic considerations. Although both study areas are front-runners in WFD, it appears that considerable numbers of farmers are resisting or are complying against their preferences. Many farmers however are enthusiastic about WFD and these are often local leaders.

Generally the acceptance of WFD is rather superficial in the sense that conventional agriculture is still considered to be the core business. Economically strategic considerations remain important (licence to produce) next to intrinsic, expressive or social motivations: nature management is a matter of countryside stewardship (intrinsic), occupational satisfaction (expressive) or social recognition. WFD is sometimes development pluriactivity whose most important function for the farmer is that it generates money for investments in conventional farming. In a few cases development pluriactivity goes the other way around: agriculture is subordinated to the new countryside activity.

However, the most promising type of WFD is a full integration of farming in site management, namely organic farming. Farmers oriented to this option are often considering other types of WFD as well. Organic farming with its emphasis on wholeness has an extra appeal to urban people who are looking for paradise lost in the countryside. The reality is that organic farming in the peat-meadow areas is not 100 per cent harmonious. Organic farming is less compatible with the high water tables which are a pre-condition for sustainable peat soils.

With its emphasis on social services, WFD is mainly supported by farm women. However, they have little room for entrepreneurial initiatives on their own. The division of roles between male and female farmers in the peat-meadow district in general is rather traditional (although not so in Waterland). This is reflected in the limited participation of farm women in the Alblasserwaard farmers' nature organisation.

New skills

Young farmers and farm wives in the Alblasserwaard area were asked their views on the importance and presence (self-perception) of specific occupational capacities.

One cluster of farm families can be labelled 'masculine' and management-oriented. Farms in this cluster quite often have WFD and usually the farmer (seldom the farm woman) is a member of the farmers' nature organisation. Another cluster can be labelled 'feminine' and is oriented towards craftsmanship and social skills. At the moment WFD is less frequent here, compared with the 'masculine' cluster. However, the most extreme case of WFD - dynamic social and ecological widening - can be found here. For far-reaching WFD (agriculture in the city's countryside) this cluster probably provides the best possibilities.

Effectiveness of wider farm development for a vulnerable spatial system

The existing organisations creating a financial spillover from 'the city' to farmers will not solve the income problems of farming in this agriculturally disfavoured area, nor will the vulnerable spatial system be rescued for coming generations. However, there are some institutional innovations which might result in a really sustainable solution. In this case the negative kind of internalising amenities in farm economics could be imposed on the area, because the resulting decreased income would be surpassed by the positive effect of an organisational context which creates a financial spillover from 'the city' to farmers whose environmental performances are above average.

This must be seen against the background of the new situation in which farmers are a small minority in a society interested in the countryside primarily as a location for amenities, and not for agricultural production. Meanwhile the farmers have property rights on most of the land. The case studies show that this leads to much strategic behaviour in bargaining between farmers and 'society' about conditional property rights. Social constructions from the outside world of rural amenities, perceived by farmers as unrealistic, are countered with publicity in which farmers are presented as knowledgeable actors in agricultural production, and above all, in the management of environmental qualities.

Acknowledgement

Many thanks to Anton Jansen who did the interviewing in Alblasserwaard-Vijfheerenlanden. In addition to the 'data-set' he delivered valuable qualitative field impressions. (Qualitative field impressions in Waterland are also based on interviews by the author.) Anton Jansen also commented on the 'final draft'. All remaining errors are solely mine.

9 Reading 'participation in forest management' through 'modern' and 'post-modern' concepts, or: where to start normative debates?

URS GEISER

Introduction

The present paper is concerned with debates over, and practices of, 'participation' in development and natural resource management. Some people see participation as a panacea, or at least an important component to overcome development constraints or achieve sustainable resource use. They search for new approaches, test some, and evaluate outcomes. Many even question their own roles in the development process, and try to apply participatory methods in their professional fields. Others are rather critical about the role of participation. They analyse specific participation contexts and identify shortcomings, hidden agendas, unequal power relations, and unintended consequences - leaving little hope that 'participation works'. While the first group encompasses many development practitioners, the second is usually identified with critical (university) researchers (much in line with what Chambers once called the positive practitioner - negative academic dichotomy; Chambers 1983: 28f). Are these positions irreconcilable?

This chapter approaches this question by discussing 'participation in forest management' using dichotomised research perspectives, formulated in a pronounced, generalised, and partly provocative way, i.e. 'modern' and 'post-modern' (the understanding of these notions is given below). The two perspectives lead to a very different reading of 'participation' and highlight a normative dilemma, i.e. the basic difficulties facing attempts to discuss (and recommend) policy options that should ensure 'sustainable' forest management. This pointed characterisation of the normative difficulties is then used to stimulate a (more differentiated) debate on the 'irreconcilable positions' mentioned above.[1]

The chapter draws on my own insights gained into rural development and forestry in the Indian state of Kerala, in Switzerland, and in the North-West Frontier Province (NWFP) of Pakistan.[2] First, a few hints are given to show that 'participation in forest management' has become mainstream. The third part illustrates what is called here the 'modern' reading of participation, while the fourth part introduces the 'postmodern' one. The normative dilemma is described and discussed in the final part.

Mainstreamed 'participation in forest management'

By the end of the 1990s, 'participation' had become a mainstreamed ingredient in forest management throughout the world. Three examples illustrate this:

In **India**, the National Forest Policy of 1988 explicitly states the need to involve local communities in forest management. Article 2.1. calls for '... creating a massive people's movement with the involvement of women, for achieving these objectives and to minimise pressure on existing forests' (Government of India, 1988). People's participation is further specified as: 'The holders of customary rights and concessions in forest areas should be motivated to identify themselves with the protection and development of forests from which they derive benefits' (Article 4.3.4.2 in GoI 1988).[3] In 1990, the Ministry of Environment and Forests issued a letter to all Indian States giving guidelines for (and thus sanctioning) the involvement of village communities and voluntary agencies in forest related issues (Gupta 1997: 572f). Since then, many efforts have started in the field now prominently known as Joint Forest Management (JFM). JFM is actively implemented in several Indian states (Poffenberger *et al* 1996), while others are in the process of initiating it (e.g. Kerala). In many cases, JFM projects are supported by the World Bank or bilateral agencies.

In **Switzerland**, a new Federal Law on Forests with corresponding ordinances was enacted on 1 January 1993, replacing the earlier Forestry Police Regulation of 1902. The old law emphasised the 'timber function' of forests and prescribed forest planning (Working Plans) for public forests only. The new law gives equal emphasis to timber, protection, and 'welfare' functions (Bundesamt für Umwelt, Wald und Landschaft (BUWAL), 1996c: 11), makes forest development planning compulsory for private and public forests, and (for the first time) makes the active involvement ('Mitwirkung') of the public in these planning activities mandatory (Article 18 Forest Ordinance, November 1992).[4] While the federal law prescribes 'Mitwirkung', the cantons are in charge of operationalising the respective procedures. In the meantime, guidelines were elaborated on how to operationalise participation (BUWAL 1996b; Egli *et al* 1997), and several planning projects were started (BUWAL 1996c).

In **Pakistan**, forests are controlled by the Pakistan Forest Act of 1927, being a 'classical' forest law emphasising forest protection, and limiting community rights (Government of Pakistan/Journalists Resource Centre (GoP/JRC) *et al* 1991: 173). Recent forest policy debates, though, indicate a search for stakeholder participation. The National Forest Policy Statement of

1991 seeks, among other things, to 'promote NGOs and private voluntary organisations to create public awareness for environmental improvement' (Sarhad Provincial Conservation Strategy (SPCS), 1993: 23). The Pakistan National Conservation Strategy (GoP/JRC *et al* 1991) as well as the Sarhad Provincial Conservation Strategy for the North-West Frontier Province (Government of North-West Frontier Province *et al* 1996)[5] call for efforts towards joint management. The forest policy presently practiced in NWFP is characterised by one author as guided by 'concern with, and desire for the protection and perpetuation of forests; promotion of forest tree planting, wherever possible, and to the extent of available resources; desire for extending community participation in forestry, gradually, and wherever appropriate' (SPCS 1993: 24). On the project level under donor support, many efforts towards participation in forest management are going on. A Dutch-assisted social forestry project for example promotes participation regarding afforestation on communal lands and degraded forest tracts (SFPMD 1992), while a Swiss-assisted project even tried to promote people's participation within Reserved Forests (KIDP 1990).

Efforts at mainstreaming 'participation in forest management' at national levels – for which the three above are just examples – are paralleled by efforts at the international level. Examples include International Union for the Conservation of Nature (IUCN 1997) or the forest-related follow-up process of the 1992 United Nations Committee on Environment and Development (UNCED)-conference under the Intergovernmental Forum on Forests (UN *et al* 1999).

The 'modern' reading of participation

Theoretical position

The notion of 'modern' (through which the above examples are to be re-visited) is understood here in close relation to modernity, positivist (and rational) science, and modernisation:

Kumar (1995: 67) describes modernity as a 'designation of all the changes – intellectual, social and political – that brought into being the modern world'. Generally, the 18th century's Enlightment period is seen as the culminating time when the key notions (or 'grand narratives') of modernity started to unfold: history and progress; truth and freedom; reason and (scientific) revolution; and science and industrialism (Kumar 1995: 84). Modernity is conceived – in this reading – as open-ended, involving the notion of 'the continuous creation of new things' (Kumar 19895: 81).

> Western society, as showing the sharpest contrast with earlier or other ... societies, became the emblem of modernity. ... To modernize was to Westernize. Modern society ... was industrial and scientific. Its political form

> was the nation-state, legitimated by some species of popular sovereignty. It gave an unprecedented role to the economy and economic growth. Its working philosophies were rationalism ... and utilitarianism (Kumar 1993: 391f).

Science is a cornerstone of such modernity. Positivism (underlying modern science) claims that nature and society are ordered systems which can be analysed and understood by means of observation and experiment (the main ingredients of rational science) as the basis for scientific progress. 'Modern' science is the key means by which utilitarian knowledge can be gained. Scientists, or scientifically trained experts (see forest experts further below) thus receive a prominent role: '.. those people known as 'scientists' are widely regarded as the purveyors of a superior kind of knowledge which represents the real world with a degree of precision and reliability that makes possible extensive control over its natural processes (Mulkay 1993: 639).

The belief in progressive modernity, supported by and made possible through, positivist-rational science, fostered an understanding of change as a structured process of incremental (and sometimes revolutionary) improvements – the core tenet of modernisation. Modernisation theories (e.g. by Parsons, Hoselitz, Rostow) 'seek to identify in the organization and/or history of industrial countries the social variables and institutional factors whose change was crucial for their process of development in order to facilitate this process in the newly developing countries' (Larrain 1993: 393).

In sum, 'modern' in this paper connotes a *positivistic outlook* especially in relation to *development.* Development is understood as *modernisation*, and is seen as a *shared vision* that brings a *better future compared with the present.* Development can be achieved by incremental improvements, that is solving *existing shortcomings* by bringing *new solutions. Science,* and scientifically trained *experts* are key agents for change. As van der Ploeg *et al* (1994: 2) state for the case of agricultural *development:* '.. the practice of modernization was (and still is) shaped by sets of external interventions, mostly centralized in state-agencies aiming to introduce new organizational models for farming, new interlinkages between farming, markets and market-agencies, new technological innovations meant to replace existing techniques and knowledge, new forms of socialization and techno-economic training, and ... new models for the definition of roles and identities for farmers and their wives'. Here, farming, and farmers, could easily be replaced by forests, and forest users.

In this particular reading of 'modern', 'participation' is one intervention (among others) perceived as promising to achieve improvements towards a specific goal, e.g. a better future for all, the emancipation of humanity, or sustainability [6] – thus, in the following, an interpretation of 'participation in forest management' through such a pointed understanding of 'modern' is offered.

The Indian example. Since 1990, JFM has become popular in a number of Indian states. Though each state is drawing up different rules, JFM generally involves:

> ... forest-dependent villagers agreeing with the Forest Department (possibly through intermediaries in Non-Governmental Organisations) that they will protect areas of degraded forest, normally on Reserved Forest lands. In return for protection activities, the village committees gain legitimate access to various non-timber forest products ... In addition, if commercially-valuable trees are planted or regenerate, the village committees are entitled to shares of the proceeds of the sale of these trees when they are mature enough to be felled (Jeffery 1998: 12).

Kerala state is just about to enter JFM, with the support of the World Bank-aided Kerala Forest Project (World Bank 1998).

A 'modern' reading of participation in this endeavour would be as follows: The project is based on the realisation (by professionals) of earlier shortcomings, for example the (felt) need to change the role of the Kerala Forest Department 'from a predominantly regulatory role to one in which communities and tree farmers are treated as equal partners in the management of forests ..', and the need for 'changes in management systems, that would be based on decentralisation of decision-making and improved planning, monitoring and evaluation, and client-oriented whenever the interests of local communities or tree farmers are at stake' (World Bank 1998: 2). The planned participatory management involving tribal communities and the population along the forest fringe is 'to develop among them a sense of stewardship for the resources. ... Forest Department and [NGOs] will work with Village Forest Committees to develop microplans for rehabilitating forests in their areas, jointly determining which species to plant and how to allocate resources' (World Bank 1997; see also World Bank 1998). In the words of a World Bank staff member:

> Kerala is one of the few states in India where the Forest Department has developed a vision statement in consultation with non-governmental organisations (NGOs), community leaders, and other stakeholders. The vision statement is intended to enable the Forest Department to manage the sustainable development of forests in Kerala in a participative, people-centered way that ensures optimum land use; conserves the fragile ecosystem and protects the natural forests; meets the needs of forest communities and the needs of industries based on forest products (D'Silva 1997: 52).

The Swiss example. Generally, Swiss forest history reports that in earlier times, forests were endangered by peasants in search of new agricultural land and timber for construction and heating. As a consequence, and to prevent further forest destruction, the authorities had to introduce strict forest regulations, to be enforced through a (growing) forestry administration (e.g. Marti 1993: 525). Through these measures, state authorities were able to extend forested areas, protect watersheds, and introduce sustainable forest harvesting (the harvested amount not exceeding the growing stock).

The main argument for the introduction of people's participation in forest management with the new law of 1993 (see above) was that during recent years, societal expectations regarding forests changed. Timber lost its attraction due to low market prices, but demands for nature protection and the use of forests for recreational and sports purposes increased – leading to an increase in conflicts between these interests. Existing procedures within forest management and administration were not in a position to handle such conflicting situations. Therefore, the forest service should know, and incorporate, the expectations of 'the public'. For this – following the 'modern' reading – participation is seen as an adequate method (BUWAL 1996a).

The Pakistan example. As in India or Switzerland, efforts towards 'people's participation in forest management' in Pakistan are justified by acknowledging shortcomings in previous approaches. Increased participation is expected to ensure collaboration of the people, that e.g.:

> ... the Forest Department must accelerate its move towards some form of partnership or arrangement with the local people, based on the recognition that it is in their own interests to protect and nurture the forests. If as a result people withhold herds from grazing or refrain from lopping or cutting, they must be sure they will receive the benefits their abstinence will undoubtly produce (GoP/JRC *et al* 1991: 172).

Increased participation is also expected to help in curbing corruption: 'Securing the collaboration of these stakeholders in the sustainable management of forests can be an effective way to counter the vested interests in the exploitation of forests' (GoNWFP *et al* 1996: 133).

As an example on a project level, the Kalam Integrated Development Project in Upper Swat attempted to develop approaches for people's participation in Reserved Forests. Efforts included public hearings regarding the formulation of Working Plans; the creation of forest committees for afforestation, forest protection, checking timber transports, and controlling grazing; and the training of local people as forest workers, for employment by state authorities as 'petty contractors' in timber harvesting operations (Geiser 1999).

The 'modern' reading

In India, Switzerland and Pakistan, more 'modern' forms of participation are being introduced by forestry officials into forest planning and management, as the above examples indicate. This indeed suggests an 'emerging tendency towards more participatory forms of forest management in modern society'.[7] Scientifically trained experts like foresters or development planners – key people in designing modern forest management – realise that earlier approaches did not lead towards 'a better future for all', or 'sustainability'; thus as a new try (in the sense of an incremental change over past practices) 'the locals' or 'the public' need to be incorporated more into forest planning and management.

A 'modern' reading therefore perceives the present openness of state forest authorities towards people's participation as a logical consequence (and stage) on the way towards better forest management. All the three cases illustrate what is generally seen as central in the contemporary 'stage', that is the changing of relationships between state representatives and 'people': 'The most important aim of [participatory regimes] is to break the mistrust which divides state agencies and user groups, so that local knowledge of the resource and existing social structures can be used to develop more effective strategies for resource use' (Vira 1997: 2). And: 'The new paradigm sees the forest bureaucracy as playing a supportive role, encouraging community initiative and sharing management responsibility' (Vira 1997: 11).

A 'postmodern' reading of participation

Theoretical position

How would these efforts towards introducing 'modern' forms of participative forest management be interpreted through a 'postmodern' theoretical position – if there is something that can be delineated (even in a generalised manner for present purposes) as 'the postmodern position'? Following Kumar (1995), this appears difficult, as postmodern positions are attributed to a wide spectrum of thoughts. To locate this chapter's understanding, first the notions of modernism and postmodernism are touched upon, leading to glimpses at postmodernity debates. Within these, more solid ground is found in poststructuralist notions.

Modernism, postmodernism. Modernity as summarised earlier was already challenged in the 19th and early 20th century by various cultural movements – subsumed under the label of 'modernism'. The core of modernism was its critique of the representation of modernity in arts (realism), and of the negative effects of industrialisation and emerging capitalism. According to Pinkney (1993: 388), modernism consisted of many, often contradictory schools. Some radically broke with established conventions (surrealism; dadaism), while others looked to the past for alternatives (romanticism); still others wanted to utilise the potential of science and technology to create an even better, utopian modernity (Russian Futurism), or searched for a new universal style in architecture (Le Corbusier).

Many of these schools were later challenged as not breaking with, but still being embedded in modernity, i.e. still maintaining modernity's vision and belief of 'humanity' being able to achieve universal goals such as 'a better future'. The attacks by 'postmodern' architecture on the claims of modernism to universalism are well-known. Such universalism is perceived by some postmodern authors as authoritarian because its actual meaning is defined by few. Lyotard (1984: 16) for example denies the possibility of appealing to universal 'truth' to legitimise consensual human interaction. In his reading, consensus is being forced by those in power onto others by suppressing others' way of arguing (language games; see the description of discourse below). Postmodern architecture thus favours diverse positions: pluralism, localism, particularism, and incorporates and accepts 'the present', i.e. mass-cultural, commercial components. But as there were many streams to modernism, so were there now many in 'postmodernism' (Kumar 1995).

Postmodernity. The highly diverse bundle of postmodernism(s) influenced theorists outside the fields of culture and arts to challenge modernity, and especially its 'grand narratives' (history and progress; truth and freedom; reason and [scientific] revolution; and science and industrialism; Kumar 1995: 84; or their 'modern' readings, e.g. 'sustainability') – again leading to an array of positions. Questioning and even denying these metanarratives, some 'postmodern' social scientists see the end of the industrial age (questioning a core metanarrative of modernity, i.e. industrialisation, by referring to phenomena such as flexible specialisation and the increased importance of knowledge), or the 'end of history' (making reference to globalisation and the fall of the Soviet Union in order to destroy the metanarrative of 'history and progress'). Others see an end of modernisation and progress in total, as everything gets encapsulated in global information systems and consumerism, or an end to 'modern' societal values such as 'universal freedom' (due to individualisation and fragmentation) – proclaiming the emergence of a 'postmodern era' of the instantaneous, hyperreal, and cyberspace (for an excellent overview, see Kumar 1995).

Postmodern debates thus span from difficult-to-grasp, eclectic, arbitrary, even cynical descriptions (postmodern as 'post-mortem: obsequies performed over the dead body of modernity, a dissection of the corpse' Kumar 1995: 67) to more accessible critical engagement with issues such as the 'grand narrative' of science as a central cornerstone of modernity. It is here that entry points for the present chapter are found.

(Selected notions from) poststructuralism. Postmodernity/postmodernism is often summarised through Lyotard's (1984: xxiv) statement of the 'incredulity towards metanarratives', and Kumar (1995: 133) adds: 'This is one of the best-known and most generally accepted attributes of post-modern theory. It unifies what would otherwise be a hopelessly diffuse and dispersed series of propositions'. Scepticism towards metanarratives is at the core of poststructuralism too (which some see as part of postmodern debates while others do not; see Kumar 1995: 129). Discourses (Foucault) and deconstruction (Derrida) are

central analytical notions of poststructuralism and represent key tools to analyse 'grand narratives'. Poststructuralists emphasise the importance of power and power relations in discourses:

> Discursive fields overlap, influence, and compete with one another; they appeal to one another's 'truths' for authority and legitimation. These truths are assumed to be outside human intervention, either already known and self-evident or discoverable through scientific inquiry [i.e. metanarratives]. Precisely because they are assigned the status of objective knowledge, they seem to be beyond dispute and thus serve a powerful legitimating function (Scott 1988: 35f).

Modern science is considered by poststructuralists as such a discourse trying for example to devalue 'traditional knowledge' (for a theoretical debate see Lyotard 1984: 7f; for a practitioner's account, Chambers 1983) – which therefore needs to be deconstructed and challenged.

In sum: the notion of 'postmodern' is diffuse, and - by its very nature of negating or questioning grand narratives - escapes being defined ('[contradiction] and circularity, far from being regarded as faults of logic, are in some versions of post-modern theory actually celebrated'; Kumar 1995: 103). Not willing to join too nihilistic strands of postmodernism/postmodernity (see above), there is need for clarification of the term's (still generalised) use in the present paper:

- here, a position is taken that starts from Lyotard's short summary of the 'incredulity towards metanarratives';
- that listens to diverse claims to 'truth', or in the words of Johnston *et al* (1994: 466) 'stressing in ... place [of the grand theory of the modern era] an openness to a range of voices in social enquiry ...';
- that does not consider 'modern' science's claims as more true than other claims; 'truth' is understood here not as objective reality, but as 'constructed' statements;
- that considers – on the methodological plain – narratives as an adequate tool to represent, and make accessible, various discourses;
- – and that finds it very difficult to go beyond narratives and deconstructed discourses, in terms of 'validating' positions, taking (political) sides, and 'recommending'.[8]

'Postmodern' reading of participation. This understanding of 'postmodern' can be translated through the context of rural development to *analyse*[9] questions of 'participation in forest management'. The 'openness to a range of voices' implies that '... issues of policy implementation should not be restricted to the study of ... intervention by governments, development agencies and private institutions [as generally done in the 'modern reading'; UG], since *local groups actively formulate* and pursue their *own* "programmes of development", which may clash with the interests of central authorities.' These (groups of) actors are not 'passive recipients of intervention, but *active*

participants who process information and strategize in their dealings with various local actors as well as with outside institutions and personnel' (Long *et al* 1992: 21, following Giddens' notions of social agency, and of social actors as 'knowledgeable' and 'capable'; emphasis UG).

A 'postmodern' reading of ('modern' forms of) participation in forest management, therefore, describes the *multitude of actors* ('the range of voices') involved, analyses their *discourses* and their 'appeals to truth' (or metanarratives). It also *starts from 'neutral ground'*; all actors using or claiming forests are (analytically) treated equally. Foresters or development experts or peasants are seen as social actors or interest groups, among others, trying to secure their claims on forests backed by their discourses and claims for 'truth'.

The Indian example. Indian forest laws (going back to colonial times) distinguish between Reserved, Protected, and Village Forests. All of these categories are under state control, with village forests allowing communities some use rights.[10] The first efforts toward more community involvement in forest management in the postcolonial era started with the Social Forestry Programme of the late 1970s and '80s. However, these programmes were generally located on farmland or communal woodlots outside state controlled forest lands (Vira 1997: 3f). It was only the 1990 government letter (see section II) that institutionalised collaborative, i.e. 'modern' participative arrangements between state and local communities on state forest land (for the 'modern' reading of the resulting JFM see section III).

But: '.. participation means different things to different people ..' (Vira 1997: 11), or in the words of Jeffery (1998: 13): 'JFM agreements do not arrive in an uncharted sea'. There are already many actors involved in forest use, with their (often varying or even conflicting) claims on forest resources, their discourses and 'appeals to truth', and their histories of interactions and 'participation'.

In the state of Kerala, where JFM is about to be launched, relationships between state authorities and 'the locals' regarding forest use are – in this reading – ambiguous and often conflicting. An analysis of these conflicts along the forest fringe reveals the following rough typology of involved actors (Geiser 1996; Geiser forthcoming):

- The 'accepted' settlers: they arrived in the early 1950s as a result of state- sponsored settlement schemes. The state allocated to them forested land for conversion into agriculture. At that time, settlers and state officials shared the same discourse: to provide food and work to a growing population, and forests were legitimately seen as a land reserve. Early settlers are today considered by the state as accepted partners and are entitled to government support schemes.
- The 'un-accepted' settlers: in the 1960s and especially from the mid-1970s, the state changed its attitude. No forest land was to be given to

settlers any longer, but should be allocated to other users (see below). 'Local people', though, still tried to get land (appealing to their rights to housing and to overcome poverty); but now they were labelled by state agencies as encroachers, and (appealing to ecological statements or 'truths'), seen as a threat to forests. Encroachers are not entitled to government support schemes.

- Political, caste, and religious movements supporting land distribution to the 'rural poor': settlers/encroachers in need of land had to find new alliances (having lost the support of the state) – and often found them with political circles in opposition, 'junior partners' in governing coalitions, or caste and religious organisations. These now supported the demand of the settler/encroachers by labelling them as 'rural poor' and appealing to the need for poverty alleviation.
- The Forest Department (FD) as producer of industrial raw materials: instead of allocating forest land to settlers in the 1960s and '70s, forested land was clear-felled by the FD itself and planted with fast growing Eucalyptus – justified by appeals to scientific plantation forestry, the need for industrial growth (reflecting the 'Zeitgeist' of modernisation theory), and the related opportunity to create urgently required jobs.
- The FD as concerned with conservation: in the 1980s, sections of the FD joined the emerging environmental discourse, and claimed forests to be environmentally important assets that needed strict protection. They appealed to ecology, 'sustainability', and increasingly to biodiversity (each notion with its references to 'modern' science).
- Foreign donors, for example the World Bank (WB): WB projects in Kerala switched discourses (and practices) over the past decades (along changes in modernisation theories). While earlier supporting the FD's Eucalyptus drive (using the same 'truth' as the FD to justify its involvement), it is now inclined to JFM (appealing to other 'truths').
- Besides these, other important actors are state agencies calling upon the FD to produce industrial raw materials; the federal/central state; the tribal peoples; etc.

Most of these actors, and the conflicts between them, are still relevant today, and thus 'chart the sea' onto which JFM arrives. Insights reveal that the Kerala Forest Department seems to be critical towards the JFM discourse and arguments as developed by the WB and by other Indian states. Not to forego other benefits accruing from the WB project – a 'postmodern' reading could suggest – the Kerala FD agreed to a limited 'pilot project' for 'participation'. A Kerala NGO has been entrusted to develop the first models. Discussions indicate that the proposal argues for a soft approach to promoting 'participation' using facilitators from NGOs and the FD, and giving the 'groups of beneficiaries' much room to develop (and implement) their *own* ideas. In addition, it is understood that a formal link to the emerging Panchayat system is proposed[11] especially regarding financial arrangements, monitoring and control – all issues apparently seen differently by the FD (who seem to favour a more controlled approach, including selected NGOs only, and excluding Panchayats).

The Swiss example. In Switzerland, about a quarter of the forests are privately owned (often by farmers or their descendents), while the balance is under 'public' ownership, including forests owned by the federal state (less than 1 per cent), cantons, political communes, by traditional 'local communities' (Burgergemeinden)[12] and by local cooperatives (BFS & BUWAL 1997: 210). Earlier, forests were used by these entities mainly for timber. Trees were cut and either sold on the market, used for private purposes (farmers), or distributed to the entities' members (e.g. Bill 1993). In mountain areas, some forests were, and are, strictly protected, mainly to prevent avalanches.

Today, a multitude of interests and claims on forests are found. A preliminary analysis hints at the following:

- The state[13] owns a few forests only, but its forest service 'supervises' the activities of the direct forest users by means of overall planning and by legally allowing timber cutting (marking of trees). As elsewhere, this is justified by appealing to need for (i.e. the metanarrative of) sustainability, to be ensured through 'modern' scientific forestry.
- The direct forest users: private forest owners often manage forests themselves, while public entities employ professional forest staff. Little research seems to exist regarding the *de facto* influence of the state forest service on forest management, or in other words, the 'room for manoeuvre' (Long) of the direct forest users. One hint is a forester's report that he can not give accurate figures for the volume of timber cut from the forests under his jurisdiction, because permits for cutting are required for marketed timber only, marked trees are often not cut, or forest owners cut trees without a permit (Marti 1993: 526f).
- Environmentalists: nature protection groups call for the increase of 'Waldreservate' (forest reserves in which nature protection has priority; WWF Schweiz 1998). They state that (modern) forest management is an interference in 'natural processes' and claim (on the basis of modern scientific truth) that managed forests have less variety of tree species and less variety in tree age. Therefore, and in order to maintain biodiversity, forest reserves are required as a home ('Heimat') for endangered species (WWF Schweiz 1998).
- Sports groups claiming forests for their uses (cross-country runners, para-gliders, skiers, hunters, horsemen, mountain-bikers; etc.).
- People claiming forests for recreational purposes (often people from adjoining urban areas, or wanderers in remoter forests).

- Federal state providing subsidies: timber prices declined enormously over the last years, and the state increased subsidies to the forest sector from around SFr50 million a year in the 1970 and early 1980s to around SFr200 million per year since 1990 (BFS & BUWAL 1997: 211). State budget constraints, though, call for re-adjustments in forest administration (Linder 1996: 43) and subsidy policies. As one measure, the new 1993 forest law makes 'participation' in forest development planning a pre-condition for forest operators to receive federal state subsidies (Article 39 of the Forest Ordinance 1992).

Conflicts between many of these groups (claiming forests for specific uses, and underpinning these claims with their 'appeals to truth') increased. Forest conservation groups for example accuse the state authorities of delaying the creation of forest reserves; they perceive the decentralised, federal state system as a hindrance ('Hemmschuh') and call for the central state to take action. They also accuse direct forest users, and the forest officials advising them, of planting too much coniferous forest (claiming scientific evidence; WWF Schweiz 1995: 22) which the latter strongly deny (claiming other scientific evidence; Borgula 1996: 303f). People searching for quietness in forests clash with mountain-bikers, and para-gliders come into conflict with bird protectors; skiers (in winter) clash with hunters (because skiers disturb the others' resource of wild animals); etc.

In this 'charted sea', foresters try to implement the prescription of 'Mitwirkung' (participation) in the 1993 forest law. A range of approaches was chosen to involve the public. An analysis of ten projects (BUWAL 1996c) shows that all used the 'Auflageverfahren' (which is compulsory). This method, taken over from regional planning procedures, refers to the placing of the completed draft plan at the commune's office for a specified number of days for public consultation (during this period, citizens can object, and their objections need to be heard, answered, and if felt necessary, considered by the authorities). The 'Auflageverfahren' is a rather 'old' form of participation, and all projects added newer, 'modern' forms such as broader information, working groups, etc.[14]

In all the cases where a broader participation was practiced, the foresters kept overall responsibility for the planning venture, and thus for the definition (and control) of participation, the organisation of the accompanying groups, and the contents of the plans.[15]

A 'postmodern' interpretation of these experiences suggests that the state forest service allowing (or having to allow) for 'modern' forms of participation is a sort of forward strategy. As one author states, the forest service has to re-orient itself in the context of changing frame conditions, but for this depends considerably 'on the increasingly influential public' ('mitbestimmende Oeffentlichkeit'; Linder 1996: 48). This challenges traditional roles. Allowing public participation – but keeping it controlled by the foresters – is seen as a promising strategy.[16] Others within the forest service, though, challenge this forward strategy;[17] one calls the new forest law of 1993 a 'Zeitgeistgesetz', and states that foresters should reassert control over forestry affairs (Weber 1996: 37).[18]

The Pakistan example. Pakistan, as with India, distinguishes between Reserved, Protected, and (forms of) community forests, all being under state control.[19] One form of community involvement is practiced in the Guzara forests of NWFP. Guzara forests are nominally owned by communities, but managed and harvested by the state. Since the early 1980s, local shareholders can organise forest co-operative societies. An assessment, though, states:

> In Guzara forests, local communities with legal ownership have little say over how the produce of the forest will be harvested or otherwise used. They merely receive the sale proceeds. ... Although in theory co-operatives are supposed to represent community interests, the reality in Pakistan is such that co-operatives are quickly usurped by a small number of influential people purporting to represent the community. The 'representative' approach that dominates the organisation of forest co-operatives ensures that representatives have no obligation to the community at large, and that they will plunder the forest in the name of the community they are supposed to represent (GoP/JRC *et al* 1991: 174).

This negative experience with 'community participation' is often quoted by forest officials as argument against the spreading of more 'modern' forms of participation – thus adding to the reinforcement of their traditional claims. But there are other actor interests as well. For illustration, the earlier mentioned Kalam project is an interesting case.[20] The project ended in mid-1998, and its attempts to promote more 'modern' forms of participation within Reserved Forests failed (at least up to now). An analysis shows that in 1972, the state declared the forests of Upper Swat as state forests and categorised them as Reserved in 1975. Since then, the following scenario emerged:

- The Forest Department (FD) is *de jure* in charge of the management of these forests. For this purpose (as in India or Switzerland), it produces – following 'scientific forestry' – Working Plans, marks the trees for felling, and polices the forests. To justify state control over forests, the FD generally blames 'the locals' for over-utilising and destroying forests, and the need for scientific forestry.
- Timber harvesting is mandated to the Forest Development Corporation (FDC) which in turn subcontracts harvesting lots to private entrepreneurs (contractors). The FDC justifies its operations with the need to separate forest management from timber operations, and the need to supply timber to the Pakistan market.
- The contractors have to submit tenders when FDC announces forest compartments for harvesting. FDC is obliged to choose the cheapest bidder,

and contractors justify their often poor investments in harvesting by appeals to 'economics', i.e. financial constraints arising out of these existing procedures.

- The local tribal groups, on the other hand, continue to perceive the forests as their property appealing to 'their truth', i.e. traditional use rights (or 'indigenous', with the whole connotation to indigenous science) as specified among the various groups living in the area. Their resistance to the state taking over the forests has made land settlement, including the delineation of state forests, impossible up to now.

The (temporary) arrangement between tribal groups and the state authorities is that the proceeds from timber sales are shared among the locals, FDC, and the general state exchequer. However, there is a lot of mistrust and conflict among these actors (note that timber prices are very high in Pakistan). FD and 'the locals' blame the harvesting contractors for overexploiting the allocated lots; FDC blames FD for harassment through delayed marking or hindering timber transport; the locals blame FD, FDC, and the private contractors for forming a nexus of exploitation, resulting in damage to the forest and lower royalties to the locals; FD blames the locals for selling their royalty rights to timber contractors, etc. (see also Knudsen 1999).

Within this setting, it seems rather obvious that the individual actors' 'appeals to truth' are simply seen by the other actors as strategic means. Though talking of the need for scientific forestry in view of sustainability (especially among higher levels of the forest administration), FD for example is perceived by many as primarily trying to grasp more from the high valued forest resources. To introduce more 'modern' forms of participation within such an arena becomes difficult – as the example of the Kalam project demonstrates. Though people were allowed to develop and test models within the setting of the project, their proposals were not taken over after project completion.[21]

The 'postmodern' reading

The above examples indicate the meaning of a 'postmodern' (as defined here) view of 'people's participation in forest management'. This view sees 'people' as forever having participated in forest use – and not just as a result, or an innovation, of the 'modern' developments in forestry: farmers cutting trees to generate farmland or to produce timber for self consumption or markets; cooperatives supplying forest products to their members; state forest officials trying to implement their agenda, etc. And all of these actors justify their practices by making reference to specific discourses and 'truths' built up over time.

Participation, through a 'postmodern' reading, appears as a contested field, with multiple alliances and contingent consequences. Participation is also a relative notion, depending on specific actors. The 'modern' understanding of participation can for example be attached to the forest officials' discourse. Within the state forest sector, 'participation' became a prominent slogan, a part of the discourse – and a strategy – over the last years. State foresters use this strategy as a resource to support the claims of the forester profession – challenged by (not so) different reasons in different localities. But to avoid letting 'participation' slip out of their hands, the creation of the space for participation is controlled by the forest authorities, justified by making reference to 'scientifically' underpinned issues such as sustainability. Other actors, though, seem to be sceptical of this practice of participation because they see its political (and actor-related) nature. They often challenge this understanding, question the validity of the underpinning 'scientific truth' – and learn to play with the new strategy adopted by the state.

The normative dilemma

The dilemma

Analysing – through a notion of 'postmodern' – forest management as a contested domain with various actors claiming to be 'right', leads to a dilemma on the normative (so-what) side. The dilemma arises because on the one hand, many actors (among them scientifically trained forest officials) emerge as active participants in forest management. There are no 'key agents for change' (as in the modern reading), but all the actors are 'knowledgeable and capable' (Giddens) in their fields. On the other hand, the 'postmodern'/poststructuralist reading suggests that each actor uses some sort of discourse to justify practices, appealing to some sort of truth. In a modern reading, objectivity, or universal truth exists (i.e. the 'modern' metanarratives) – and being objective, this helps to gauge and identify the actors who appeal to the 'correct truth'. An 'objective' understanding of sustainability, informed by scientific insights (especially into forest ecology), acts as a guideline and orientation for forest management. However, in the 'postmodern' view (and it is in these normative dimensions where a 'postmodern' position differs most from a 'modern' one) 'truths are statements within socially produced discourses rather than objective "facts" about reality ...' (Peet and Watts 1996: 13). The notion of 'postmodern' either questions, or outrightly dismisses (as social constructs) claims at overarching 'truths'. Which 'truth', then, is to guide practice?

In other words: There is no doubt that 'scientific' justifications (or justifications based on other general 'values' of humanity) can be found to support Keralese farmers' claims to convert forest land to agricultural uses (though ecology-oriented Forest Department staff would find counter-evidence); or to support a Swiss nature-protection NGO calling (based on scientific arguments of a 'natural forest') for the cutting of trees in a forest to create an 'enabling environment' for a rare variety of birds (though forest officials could claim scientific arguments to challenge this case by defining 'natural forest' differently); or to justify a Pakistan contractor negotiating with local royalty shareholders to cut (more) trees for one reason or another.

The normative dilemma for 'participation in forest management' therefore reads: Who can participate in forest management and who not? Whose claims on the forest are to inform planning (and subsequent forest management practice)? On whose 'objective' assessment of consequences is forest management to be oriented? Whose claim for truth is 'right'?

Following 'radical' postmodernism/postmodernity, there is no such truth; there are no 'grand narratives' one can appeal to for help. There is actually no solid ground. All is possible, all is acceptable.

Searching for legitimacy 'after the metanarratives'

The 'modern' as well as the 'postmodern' reading of participation is based in this chapter on exaggerated formulations of the respective notions, and the resulting dilemma may be too pointed. Still, the dilemma questioning the justifications for forest management needs discussion – because: who defines 'improvement', and whose definition is to guide action? In the words of Lyotard (1984: xxv): 'Where, after the metanarratives, can legitimacy reside?'.

The easiest way out of this normative dilemma is – of course – to deny the 'postmodern' reading. Habermas (*Theory of Communicative Action* 1981) positions himself as an anti-postmodernist. He *is* concerned with the short-comings of modernity, but perceives modernity as incomplete. Communicative rationality and competence[22] (Habermas' metanarratives, being central to human interaction and societal reproduction at large) have deteriorated in modern societies, and therefore, according to Habermas, the suppressed potential of communicative competence needs to be re-emancipated. He analysis communication as action guided by universal rules ('Geltungsansprüche'), namely the search for 'Wahrheit' (truth), 'Richtigkeit' (correctness), 'Wahrhaftigkeit' (veracity) and 'Verständlichkeit' (intelligibility). Communicative interaction is not strategic or success-oriented (geared by 'instrumental rationality', a metanarrative of modernity criticised by Habermas), but geared (by communicative rationality) towards common understanding.[23]

Being directly concerned with questions of participation, and carrying a positivist outlook, Habermas' theory has become popular among many practitioners of participation. Practitioners often tend to see communicative competence as possessed more by the 'marginal', the 'tribals' and 'the locals', and more by women than men.

But as one author states, as soon as the underlying assumption/ metanarrative (actors are universally equipped with communicative competence) is withdrawn, then the entire Theory of Communicative Action collapses.[24] In the words of Kumar (1995: 136), Habermas

> looks to the achievement of a rational 'consensus' through a dialogue between free and equal actors. Such a hope still rests on the Enlightment belief in humanity as "a collective (universal) subject", which attempts to achieve its "common emancipation" through finding a structure of agreed or general rules governing all forms of interaction. But there are no such universal rules of the game ... and no prospect therefore of consensus.

Having been 'intoxicated' by the insights gained into the existence of various actors, 'range of voices', actor interests, justifications of these interests, and being critical of grand narratives, Habermas' 'communicative competence' is no option for a point of reference to validate discourses. What might be needed instead of a (normative) theory of communication and consensus (being dominated by a few) is, to follow the words of Lyotard (1984: 16), a 'theory of games which accepts agonistics as a founding principle', and thus a theory to handle dissensus – not in a nihilistic sense, but to advocate plurality on all levels.

Lyotard's more normative outlook, therefore, searches an alternative to consensus enforced by metanarratives:

> Consensus has become an outmoded and suspect value. But justice as a value is neither outmoded nor suspect. We must thus arrive at an idea and practice of justice that is not linked to that of consensus (Lyotard 1984: 66).

To operationalise this vision, he proposes two (initial) methodological steps:

> A recognition of the heteromorphous nature of language games is a first step in that direction. This obviously implies a renunciation of terror, which assumes that they are isomorphic and tries to make them so. ... The second step is the principle that any consensus on the rules defining a game and the 'moves' playable within it must be local, in other words, agreed on by its present players and subject to eventual cancellation.

For Lyotard, a 'postmodern science' would have to search not for universal laws – homological metanarratives (to be forced upon all humanity) – but for differentiated, differentiating 'para-logical'[25] and 'little narratives'[26] a science

> ... in which a statement becomes relevant if it generates ideas', that is, if it generates other statements and other game rules. Science possesses no general metalanguage in which all other languages can be transcribed and evaluated. This is what prevents its identification with the system and, all things considered, with terror. ... The function of differential or imaginative or paralogical activity ... is to point out these metaprescriptives (science's 'presuppositions') ... and to petition the players to accept different ones. The only legitimation that can make this kind of request admissible is that it will generate ideas, in other words, new statements.

Towards a (personal) way out of the normative dilemma

Two fields of tension need to match to sketch a possible way out of the normative dilemma. One is personal: there is a deep resistance (somewhere in my mind) to radical postmodernists' 'nihilistic rejection of all forms of political belief and commitment' (Kumar 1995: 180) – but there is also a strong (personal) uneasiness *vis-à-vis* much discourse and practice of 'development' in the name of modernisation (such as partly shown in the modern reading; section III). The other field of tension has to do with the wide spectrum of postmodern thinking. Poststructuralist notions for example are powerful tools to critically discuss certain practices of development, or forms of participation. Radical postmodernism, though, negates all 'development'.

The two fields of tension seem to meet at a point that suggests (i) we should be critical *vis-à-vis* grand narratives, but (ii) not give up on 'political belief and commitment'. Watts and Peet (1996: 261) call for something like such a middle path. They start from the statement that 'branches of poststructural [they seem to refer more to other strands of postmodern debates; UG] and ecological thought which abandon reason and science as guides to human action ... run the grave risk of an idealism which throws the baby out with the bath water'. They then call for a re-construction of some sort of reason, or logic, a process they call *reasoning* (as opposed to the notion of *rationality* used in the 'modern' discourse, being linked to metanarratives): 'If we call "reasoning" the contested process of constructing logic - in contradiction to "rationality" as the discovery of a logic latent in things - then reasoning as an active process means contemplating the consequences of practice *a priori*; "reasonable behavior" with regard to nature involves practices disciplined by prior knowledge of their effects' (Watts and Peet 1996: 261). And to this could be added the position of Lyotard that 'any consensus on the rules defining a game and the "moves" playable within it must be local, in other words, agreed on by its present players and subject to eventual cancellation'.

Still rather general and abstract thoughts; and what could they mean for the concrete issues in forestry? A preliminary attempt at translating these ideas to the case of 'participation in forest management' leads to the following argument to escape the normative dilemma:

As a starting point for the construction of a critical, personal, normative position, participation is understood here as the *purposive interaction of specific social actors with other social actors with a view to achieving specific outcomes. Recommendations for action need to be negotiated among those involved through a process of reasoning (Watts/Peet), and based on the readiness that 'agreements are subject to cancellation' (Lyotard).* In other words - there is no general 'guideline' to justify 'participation' leading in one specific direction of forest management; instead, reasons and rules for participation are under constant negotiation – applied 'paralogics' (or avoiding one appeal to truth becoming dominant). In more programmatic terms:

- Analysing 'participation in forest management' starts from the realisation that many actors are involved.
- Many of these actors will have their own reasons for becoming involved in a specific participation venture.
- Forest or development professionals often rationalise their involvement in 'participation' as a logical further development of their profession's (and the 'science' backing the profession) skills and understandings.
- These forest professionals' statements need to be analysed as a discourse, and a practice, and need to be deconstructed, to show the nature of their arguments, the power at their disposition, the means by which they try to 'enforce' their point of view, and 'truth'.
- In the same way, the discourses and practices of other actors involved in the specific participation arena need to be analysed and deconstructed critically.
- Through these analyses, the specific arena becomes more transparent. And it can support all involved actors to realise their 'room for manoeuvre'.
- The negotiation process itself among the involved actors is to be based on the listening to the others' reasons as good, as transparent, and as 'fair' as possible. But the process is based on the acceptance of agonistics and conflict.
- Conflict solution is not to be based on appeals to science, or 'humanity', or 'the good in people' as understood in the 'modern' sense with its homologics, but – based on paralogics – on the temporary acceptance of one or the other (or a blend of various) actor-related positions and ways of reasoning. Based on this, temporary guidelines for forest management emerge.
- Having guided practice along the agreed position for some time, an open re-negotiation of the 'rules of the game' is called for.

But who is the acting subject that 'analyses discourses', 'makes the arena transparent', 'helps to realise room for manoeuvre', etc.? Not (necessarily) scientists' as they are actors (with their vested arguments) among many (except – maybe – that they are open to critical self-assessment). Looking at today's 'reality' in Kerala, Switzerland and North-West Pakistan, many societal entities are present, active, and influential – also regarding the question of 'forest management': various sections of the state; non-governmental organisations, trade unions (e.g. of the forestry staff); etc. And it might not be too much of a statement that these entities are 'today' more outspoken, more known, more prominent, to be taken more seriously by other entities than say

30 years ago – indeed a phenomenon of 'late modernity'. All these entities are *de facto* already active participants in the reasoning for forest management.

So far this is a preliminary proposition for a possible approach to 'participation in forest management' – an understanding, though, that needs empirical analysis to assess possible contradictions, and thus to help refine the proposition – but not as a general recommendation (or a new grand narrative), but as a 'small narrative' which might apply in one or another arena.

Notes

1 This methodology is inspired by (my reading of) Lyotard's (1984: 7) statement (regarding another generalisation he makes) that 'this scenario ... makes no claims of being original, or even true. What is required of a working hypothesis is a fine capacity for discrimination'. Whether the discrimination allowed by my hypothesis is 'fine' needs discussion.

2 Insights in Kerala and NWFP are based on my own case studies; in Switzerland on an involvement in forest planning - case studies here are to follow.

3 Article 2.2 gives the principle aim of the Forest Policy as: '.. to ensure environmental stability and maintenance of ecological balance including atmospheric equilibrium which are vital for sustenance of all life forms, human, animal and plant. The derivation of direct economic benefit must be subordinated to this principal aim' (GoI 1988). Compare this with the new Swiss Forest Law of 1993, where the economic function of forests is subordinated, or at least put at the same level, as other forest functions.

4 Swiss text of Art.18 Abs. 3 WaV reads: 'Die Kantone sorgen bei Planungen von überbetrieblicher Bedeutung dafür, dass die Bevölkerung: a. über deren Ziele und Ablauf unterrichtet wird; b. dabei in geeigneter Weise mitwirken kann; c. diese einsehen kann'.

5 A major portion of Pakistan's forests is located in the NWFP.

6 Lyotard (1984: xxiii) summarises modern science as 'any science that legitimates itself with reference to a metadiscourse ... making an explicit appeal to some grand narrative, such as the dialectics of Spirit, ... the emancipation of the rational or working subject, or the creation of wealth'.

7 Quoted from the announcement for Working Group 9 of the XVIII ESRS conference.

8 Beck (1991) introduced the notion of 'reflexive modernity' to describe a similar context of critical debate of modernity. He uses the new risks that emerged with modernity (especially ecological), and the related broad response by various sections of society to define a new situation in which established norms are being questioned: 'Dies bedeutet nicht das Ende jeder Gemeinsamkeit ..., aber das Ende jeder vorgegebenen, abrufbaren, berechenbaren Gemeinsamkeit und der Beginn einer anderen Art von Gemeinsamkeit, die nicht mehr von oben herab regiert und nach dem funktionalistischen Trichtermodell der Sozialisation in die Menschen hineingerieselt werden kann, sondern von unten herauf abgewartet, gewonnen, erstritten, erfunden, ausgehandelt werden muss' (Beck 1991: 186f). How far Beck questions underlying metanarratives, and thus whether 'reflexive modernity' is similar or different to certain streams of postmodernism/postmodernity, I need to study.

9 Here, emphasis is given to the analysis of participation contexts through postmodern concepts. The question of 'so what', i.e. normative issues, – and what 'postmodern' concepts have to say on this point – are dealt with later.

10 In India, forestry is under the central, or federal government, which controls the forest-related activities of the individual states.

11 Panchayat: local government at village level.

12 The Swiss notion of 'Burgergemeind' is translated in the English version of the Swiss government's Environmental Report as 'local communities' (BFS & BUWAL 1997: 210). 'Burgergemeinden', though, are not general representatives of the 'local people', but traditional associations of those residents whose family origins are from the locality. As such, they represent a small part of the 'local community' only, and often include the more influential class.

13 In Switzerland, forests are a subject at the level of cantons, with the federal state giving general guidelines.

14 These more 'modern' forms of participation followed the 'representative model' and include:
Information for the public through the press and/or information events: this approach facilitated a speedy planning process, but it is questioned whether this really helps to fulfil the objective of clarifying forest use conflicts because of the lack of direct involvement of non-forest interests (BUWAL 1996c: 43).
Formation of an internal working group to accompany the planning process: here, generally forest service representatives and forest owners are included. It is reported that this facilitated the planning process, but that difficulties arose due to the non-involvement of other interested groups (BUWAL 1996c: 31).
Formation of a more open working group to accompany the planning process: in addition to forest service and forest owners, other representatives of interest groups are allowed in, e.g. from nature protection, sports clubs, hunters, tourist associations, agriculture, etc. It is reported that this broader approach created 'less problems than expected', and helped to foster a climate of dialogue. It however needed more time than a planning done by foresters alone (this conflicts with the expectation upon the more modern forms of participation to help reduce time-consuming conflicts).

15 This contrasts with a general recommendation given in the manual for the new forest planning (BUWAL 1996b) which recommends that the groups are moderated by 'neutral' persons (BUWAL 1996: 80). Swiss texts read: 'Der Kreisoberförster hatte eine zentrale Rolle im ganzen Planungsprozess. Die wichtigsten Aufgaben können mit Informations-, Motivations- und Koordinationstätigkeit umschrieben werden' (BUWAL 1996c: 35). And: 'Der Kreisoberförster hat als gleichzeitiger Projektleiter und -bearbeiter inkl. Information alle Projektfunktionen inne' (BUWAL 1996c: 24).

16 Some forest circles even call for the foresters to search for new opportunities outside the traditional realm of forestry, e.g. to become custodians of the rural areas (Siegenthaler 1998).

17 In Switzerland, a discussion seems to exist within forest service circles whether or not to separate the forest policing and protection function from forest advisory functions (Linder 1996: 46). It is interesting to note that a similar debate (with related practices) is ongoing in the Northwest Frontier Province of Pakistan.

[18] The Swiss text reads: 'Wer zu spät kommt, den bestraft das Leben – oder: Wenn nicht wir Forstleute das "Waldblatt" wieder fester in die Hand nehmen, dann wird es uns aus der Hand genommen' (Weber 1996: 39).

[19] In Pakistan, forestry is a provincial subject, with the central/federal state providing general guidelines.

[20] Where not mentioned differently, this account is based on own insights gained since 1985.

[21] The official argument given is the timber harvesting ban announced by the central government in September 1992, following heavy floods. However, the proposed petty contract system could still have been absorbed into the regular procedures since, which did not happen.

[22] 'Dieser Begriff kommunikativer Rationalität führt Konnotationen mit sich, die letztlich zurückgehen auf die zentrale Erfahrung der zwanglos einigenden, konsensstiftenden Kraft argumentativer Rede, in der verschiedene Teilnehmer ihre zunächst nur subjektiven Auffassungen überwinden und sich dank der Gemeinsamkeit vernünftig motivierter Ueberzeugungen gleichzeitig der Einheit der objektiven Welt und der Intersubjektivität ihres Lebenszusammenhangs vergewissern' (Habermas 1981: 28).

[23] 'Im kommunikativen Handeln sind die Beteiligten nicht primär am eigenen Erfolg orientiert; sie verfolgen ihre individuellen Ziele unter der Bedingung, dass sie ihre Handlungspläne auf der Grundlage gemeinsamer Situationsdefinitionen aufeinander abstimmen können' (Habermas 1981: 385).

[24] 'Wenn es nicht gelingen sollte, überzeugend darzutun, dass in der Tat in jedem kommunikativen Akt immer schon die Geltungsansprüche Verständlichkeit, Wahrheit, Wahrhaftigkeit und Richtigkeit konstitutiv erhoben werden, dass wir also qua unserer Ausstattung mit kommunikativer Kompetenz eine Potenz in uns tragen, die Habermas "kommunikative Rationalität" nennt, dann bricht das ganze Gebäude zusammen' (Gripp, quoted in Treibel 1994: 161).

[25] 'Postmodern science – by concerning itself with such things as undecidables, the limits of precise control, conflicts characterized by complete information, ... catastrophes, and pragmatic paradoxes – is theorising its own evolution as discontinuous, catastrophic, nonrectifiable, and paradoxical. It is changing the meaning of the word knowledge, while expressing how such a change can take place. It is producing not the known, but the unknown. And it suggests a model of legitimation that has nothing to do with maximised performance, but has as its basis difference understood as paralogy. .... [and quoting Medawar] ... "having ideas is the scientist's highest accomplishment," ... that there is no "scientific method," .... and that a scientist is before anything else a person who "tells stories." The only difference is that he is duty bound to verify them' (Lyotard 1984: 60).

[26] 'We no longer have recourse to the grand narrative – we can resort neither to the dialectic of Spirit nor even the emancipation of humanity as a validation for postmodern scientific discourse. But ... the little narrative .... remains the quintessential form of imaginative invention, most particularly in science' (Lyotard 1984: 60).

Part IV

Rural development, policy and civil society

10 Becoming a part of the union – changing rurality in East Germany

ROSEMARIE SIEBERT and LUTZ LASCHEWSKI

Introduction

The ideological dust and early enthusiasm, which arose when the iron curtain fell, has settled. Nevertheless, despite all difficulties and failures during the last decade, and a greater realism about the evaluation of and limitations on institutional reforms, the dominating perspective is still that Central and Eastern European countries have to implement institutional reforms towards market economies.

Within the general debate the dominating rural issue has been agricultural reform, whereas comparatively little attention has been paid to rural society as a whole. This is surprising insofar as the concept of a unity of economic and social policy has been at the core of socialism. The socialist agricultural firm has been much more than just an agricultural production unit. Therefore, the dominant research focus, which concentrates on looking for the new peasantry and new forms of agriculture, underestimates the contributions of former co-operatives and state-owned farms for the provision of social security, social integration, education, infrastructure services and the structuring of power relations.

However, there are several general experiences from the agricultural reform debate. Privatisation has been contested, is occasionally incomplete and usually accompanied by neither a rational loan system, an efficient development of infrastructure nor an establishment of organisations of political representation (Granberg and Kovách 1998). Besides the meagre or even adverse effects of privatisation on agricultural productivity, negative side effects, such as unemployment, poverty, emigration and social disintegration come more and more to the fore (Granberg and Kovách 1998; Barbic 1994; Keliyan 1994). Finally, other issues, such as the impacts of privatisation on environmental sustainability, have yet to be integrated into the debate (Hagedorn *et al* 1999). The example of the environment demonstrates that in many cases institutional models have been transferred to former socialist countries which are already under pressures for reform in the West itself. Institutional models are therefore oriented towards positive experiences in the past rather than future problems.

Overall there is an underlying assumption that if institutional reforms go far enough, they will eventually create the positive results which are intended. In this sense institutions provide a framework of regulative norms to which individuals will adapt. It is meanwhile widely acknowledged that due to socialisation under socialism actors do not adapt to new institutional settings as quickly as institutions are changed (Wollmann 1996). At the same time it is often overlooked that cultural conditions can not only be traced back to the socialistic past, but are also rooted in longer national or regional traditions, and in given geographical conditions. Moreover we find a considerable amount of resistance against changes in particular by groups who have something at stake. In practice this does have several effects. On one hand the implementation of institutional reform is usually not a straightforward, but a contested and contingent process. On the other hand the same institutions may not have similar effects under different conditions (Miegel 1994).

In this context East Germany is an interesting case to study. In contrast to other Central and Eastern European Countries, institutional reform in East Germany has been more complete, due to German unification. West Germany provided an existing institutional framework, which simply had to be transferred. This is an 'ideal' situation, if transformation from a planned to a market economy is the political objective. In this paper we analyse the effects of unification on rural restructuring. Starting from the situation before unification we describe major structural and institutional changes, and finally draw some conclusions about the limitations of institutional transfers to Central and Eastern Europe, for instance within the context of the enlargement of the European Union.

Rural development and rural life in the GDR

Rural restructuring in East Germany after unification can only be understood if we understand rural development under socialism. However, we also find considerable regional differences between the very remote agrarian north, and the more urbanised and industrialised south (Kind 1997). After the World War II the polarisation between these regions deepened (Figure 10.1).

In comparison with West Germany, the East German *Länder* are more 'rural' in terms of population density (Figure 10.2). The repeatedly proclaimed objective of regional policies in the GDR was the upgrading of rural areas and therefore the adjustment of working and living conditions between urban and rural areas (Schaefer *et al* 1992). Bringing rural-agrarian areas with lower economic and social standards up to the higher standards of urban-industrial regions was seen as the main task of regional policy (Kind 1997). The way to achieve this objective was through industrialisation.

Figure 10.1 German *Länder* and their population density

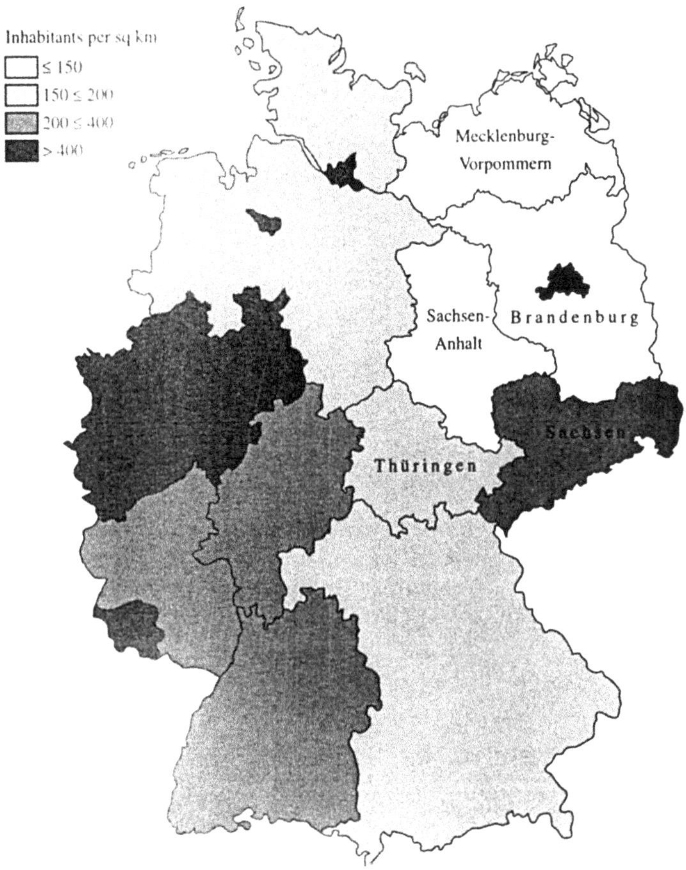

Source: Berichte des BBR 1998.

Figure 10.2 Population density in East Germany

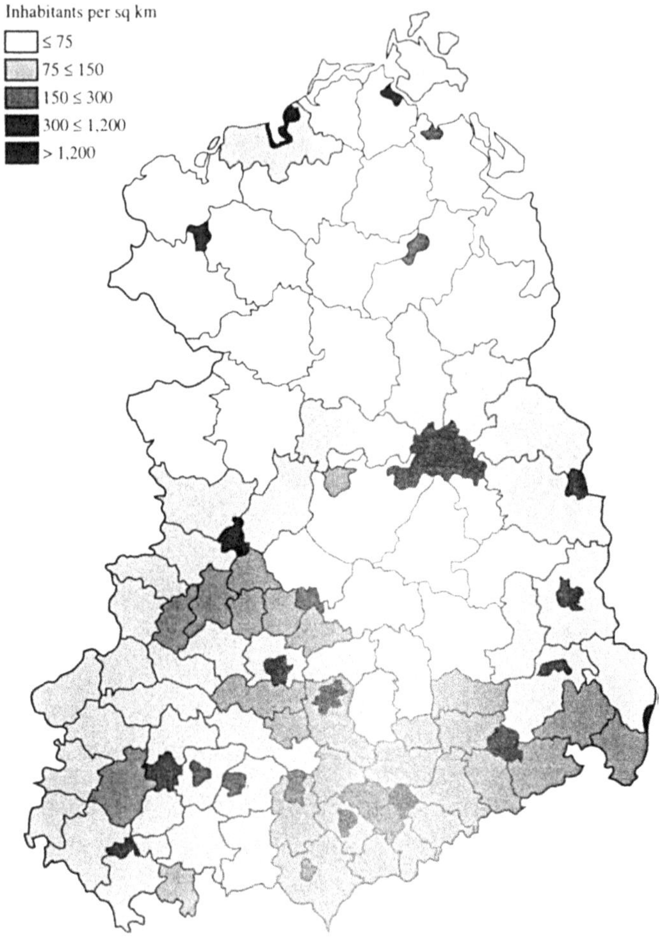

Source: Berichte des BBR 1998.

Industrialisation in rural areas included two strategies – the establishment of industrial production, and the industrialisation of agriculture. In the remote areas of the north, large-scale industries (Eisenhüttenkombinat Ost, Petrolchemisches Kombinat Schedt, expansion of coastal towns etc.) were established. The development of further smaller industrial estates followed, which were intended to use an existing labour surplus. The addition of firms dedicated to the use of female labour may also be understood from this perspective. Between 1950 and 1970 large numbers of new industrial estates were established, especially in the north. Some new industrial towns were built, but mainly it was the district towns (which were also the administrative centres of the regional districts) which were transformed into industrial centres. Many of these newly established industries were not able to survive after unification. Yet, because of the relative success of such industrial projects in the GDR, the policy had been pursued right up to unification.

Despite the establishment of industries, which were concentrated on a few central places, agriculture remained the most important economic sector, particularly in the north. Socialism in the GDR established a system of large-firm paternalism. In rural areas large firms were predominantly agricultural co-operatives (*Landwirtschaftliche Produktionsgenossenschaften* - LPG) and state farms (*Volkseigene Güter* - VEG). These firms did not just provide employment, but also wide range of social services from childcare and cultural events to organised holidays.

Within the north-eastern districts, which now form the *Land* Mecklenburg-Vorpommern, the share of agricultural employment in 1989 was about 20 per cent. Collectivisation was meant to introduce industrial production methods into agriculture. The farm buildings of the agricultural estates were grouped outside the villages. Finally animal and plant production were separated into specialised units. This led to a spatial segregation of village housing and working, and animal and plant production estates.

As previously mentioned, firms were not only important in providing employment, but also in their provision of a wide range of social and cultural services. The co-operatives and the state firms disposed of funds for cultural and social use, financed social services, and provided services in kind. Collective farms like other industrial firms were responsible for childcare facilities and canteens, maintained holiday camps and hostels. They initiated and/or supported youth clubs, riding schools, choirs and music groups. In many cases they built village halls (*Kulturhäuser*) and pubs, and even established village shops (Rodewald 1994). However, most village shops were established by consumer cooperatives. Consumer Co-operatives were obliged to maintain shops even in the most remote and tiniest villages (Laschewski 1998c). Therefore, they were the complementary organisation to agricultural co-operatives in providing services in rural areas. The local '*Konsum*' (or Co-op shop) was in many cases also a centre of social life in rural areas.

Collective farms were also involved in building the local infrastructure, constructing both housing and roads (Zierold 1997). House-building was necessary to attract labourers to work in agriculture. Agriculture had suffered a continuous shortage of workers since the 1970s. Road-building was important because of the industrialisation of agricultural production, which went along with consolidation of farmland.

But due to the concentration of agricultural estates, a new hierarchy emerged between the villages. Agricultural co-operatives and state owned farms were combined in so-called '*Kooperationen*'. Such '*Kooperationen*' included one specialised plant production unit, and normally two to three farms specialising in animal production. On average they covered a territory of approximately 5,000 ha., which usually included between 8 and 12 villages. Those villages in which the administrative centres of the collective farms were located profited very handsomely from their social and cultural activities, while smaller and medium-sized villages were predominantly neglected (Parade 1991).

While the agricultural co-operatives provided a lot of services and employment, municipalities were comparatively weak. In many communities the chairman of the agricultural co-operative had more influence, because of its economic power, than the local mayor (Zierold 1997). But the vitality of the municipalities also suffered due to the dominant role of the Socialist Unity Party, and the centralised political system (Humm 1999). An important role of the municipalities was mobilisation of the population for voluntary actions and political mass events, such as the May Day celebrations.

Since the 1950s mobilisations of labour were initiated regularly to improve the local infrastructure beyond the frame of the '*Nationalen Aufbauwerk*' (NAW) such as building and maintaining the village shop, and generally to construct local social buildings, organise local events, etc. Participation in such initiatives should also demonstrate the will of the population to contribute to the erection of socialism. The dominant idea of those mobilisations was not to ensure participation, but to use the people for the realisation of political plans. The experience showed that the people responded to these initiatives differently. If they met recognised needs of the population, the response was very strong, and many citizens spent long voluntary and unpaid working hours in such projects. But many initiatives were boycotted when people felt that they were simply being used for political goals which did not take their needs into account (Humm 1999).

Similarly to the NAW, the May Day celebrations tended to become less and less political. Apart from citizens who attended the celebrations to avoid inconveniences and social pressure, those events were largely combined with sport and dancing to encourage people to attend.

Yet the subsumption of public life under political purposes encouraged tendencies towards a withdrawal of social life into privacy. The equivalent of the townspeople's '*Datsche im Grünen*' (allotment gardens) was individual farm household production (Zierold 1997). Co-operative members were entitled to use 0.25 ha. per person up to 0.5 ha. per family of agricultural land for individual purposes. Although individual production was an important source of additional income and supply of certain products, it also determined family relations and neighbourhood co-operation. Finally, private networking in procurement systems was also very important as a means of coping with the permanent shortage of everything. This extensive informal exchange economy produced a second reality beyond the official planned economy.

Social processes in rural areas after unification

In the previous section we argued that rural life under socialism was shaped by a paternalistic industrial system and public mobilisation through politics, but also by a second layer of informal networks and privacy. Under such circumstances, local identity was closely linked either to the firm (as a form of local corporate identity) and/or to the informal social networks. In relation to the transformation process since unification, we now try to show that the effects on rural areas were twofold. On one hand, the destruction of the socialistic order of rural life was fundamental and total. On the other, certain patterns have been reproduced, and certain trends were enforced.

(Un-)employment and migration

Integration into the European Market was accompanied by an industrial and agricultural decline, which wiped out most of the industrial firms within a few years or forced them to reduce their activities to a much smaller scale. Although the re-transformation of agriculture into a family farm sector remained incomplete the effects were significant (Laschewski 1998a). Within two years employment in agriculture had been cut down to little more than 20 per cent of the starting level (BMELF 2000). This process went along with extremely sudden, and far-reaching, employment reductions in the industrial centres. In particular, those firms which had been established after the World War II in many cases proved to be uncompetitive and are now (as far as they still exist) dependent on subsidies. Today unemployment is the central problem in most villages (Figure 10.3).

The real level of unemployment is not reflected at all by the official unemployment rate of about 20 per cent during recent years. Without a wide range of social and work-providing measures the extent of unemployment would be much larger. To be in employment or to be unemployed is the major dividing line between winners and losers in the villages. In the beginning, job opportunities were segregated by individual position in internal firm hierarchies (Siebert and Zierold 1997). High levels of unemployment among women are in fact the result of the gendered vertical and horizontal divisions of labour in the GDR (Fink *et al* 1994).

Figure 10.3: Regional distribution of unemployment in East Germany in 1998

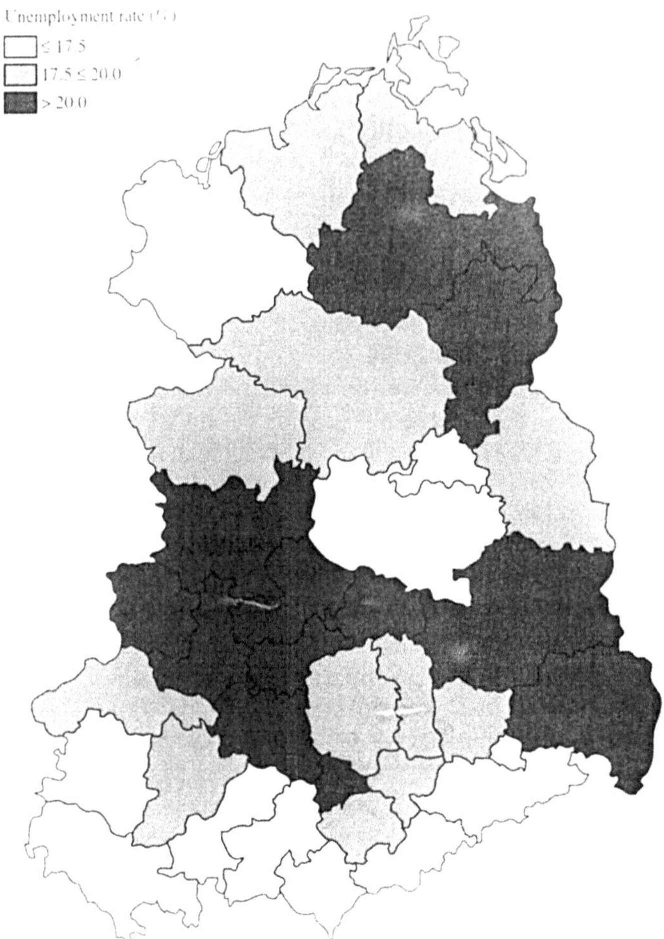

Source: Bundesanstalt für Arbeit 1999.

A second response to the crisis was emigration. Between 1989 and 1993 net migration from East to West Germany was about 1 million people (Figure 10.4).

Since then net migration is continuing, but on a much lower level. Due to the fact that it was predominantly young people who migrated, the effect of unification on the demographic structure in East Germany is considered to be even more significant than the World War II (Bertram 1996). This situation became even worse due to the fact that the birth rate has sunk to a historical low. In many rural areas the rising proportion of old people is going to become the most important social problem. Since 1993 a significant intra-regional migration/sub-urbanisation has also been observed. The population has moved from rural areas and from the inner city into the urban fringe.

The population decline in rural areas is by no means a new trend (Becker 1997). In fact current developments can be seen as a continuation of migration trends which already existed in the GDR on a lower (but still comparatively high) level. For example, studies showed that some regions in the north lost approximately 20 per cent of their population due to emigration between 1971 and 1981 (Grundmann 1997). People left the villages to move to the industrial centres. Emigration existed despite or even because of the industrialisation of rural areas (Grundmann and Schmidt 1988).

Institutional changes and regional planning

Since unification we may distinguish different periods of regional planning. Between 1990 and 1993 almost no regional planning activities were achieved. During this period politicians and the administration had to cope with the most dramatic impacts of the Economic, Monetary and Social Union which was implemented in the summer of 1990. Winkel (1996: 386) describes the planning problems in those years as follows:

> On the local level total over-estimation of the new local autonomy and therefore also the planning autonomy was united with a lack of experience of the large number of 'chance politicians'. This went along with uncertainties about the planning procedure and a lack of personnel in administration, and orientation at the economic structure to the economic development strategy. Finally, we should also mention the negative impact of some dubious planning offices and consultants from West Germany (translated from German).

At the same time the federal system had to be established; therefore five new *Länder* have been built, merging administrative districts of the GDR. Only East Berlin merged with West Berlin, and because West Berlin already had the status of a federal state, the functions of the state were merely extended to the East. Although the idea was to build the *Länder* in the way that they

Figure 10.4: East Germany – change of the population between 1989 and 1992

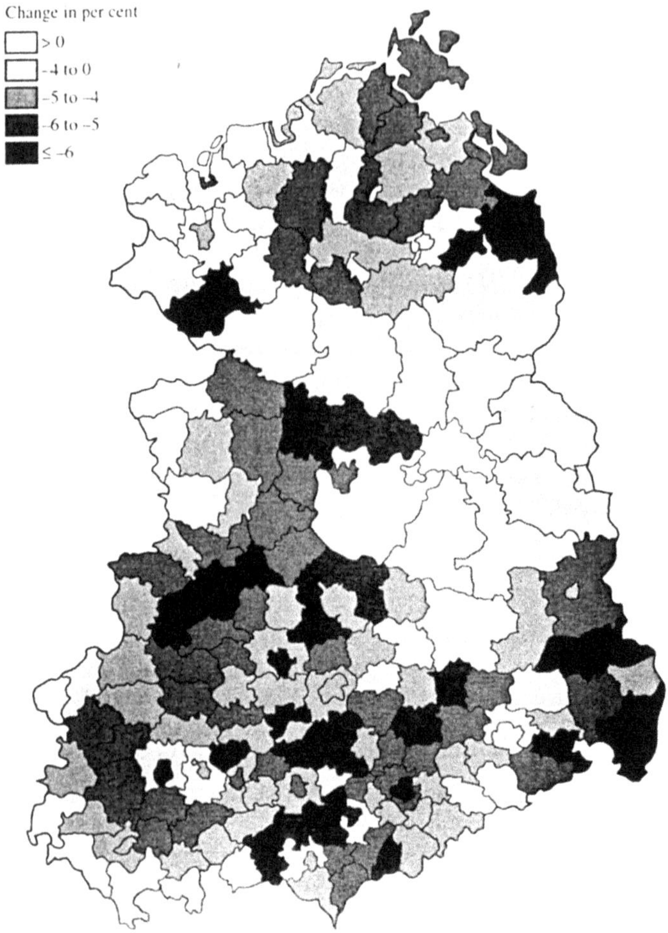

Source: BfLR 1995.

existed after the War and prior to 1952, the outcome was slightly different because the former GDR districts did not represent the historical regional states. Therefore, regional borders changed considerably in the following years, because municipalities (*Gemeinden*), and even *Kreise*[1] located at the border could vote to become a part of a different *Land*. There were also decisions by East German communities to become a part of a west German *Land* and sometimes it even went the other way round. Because of the nature of borders, those discussions were overwhelmingly rural. Although the political legitimation for these decisions was historical or cultural, the discussion about borders was also grounded in very practical concerns (expectations of higher funds, administrative issues etc.).

Beside the basic problems of transforming merged districts into federal states, each with its own government and parliament, and the right and duty of regional planning, a further important issue was the process of replacing the old élite. In contrast to other post-socialist countries the replacement of the former élite has been far-reaching. Former bloc party members (mainly the CDU), who were not members of the SED (the socialist party), a minority of activists of the democratic movement, and a large number of west German bureaucrats and second-row politicians formed (and still form) the new regional élite. The new *Länder* also had administrative support from West German *Länder*. Because of that it was not only politicians who were imported, but even civil servants on every level of the administration.

In practice this has led to a situation where the East German Länder simply copied legislation, political programmes and initiatives from their West German partner *Land*. During the first years after unification only a low level of political planning was achieved. Regional development therefore proceeded almost unplanned. In an unspecific way, state funds supported improvement of (the existing) infrastructure, and social security to cope with the exploding numbers of unemployed. The dominant development pattern was the rapid growth of settlements in the urban fringe and along the main traffic axes. On the other hand, cities suffered from difficulties in restructuring their city centres while most rural areas suffered from remoteness (Winkel 1996). The impact of these early developments is so significant that it is likely that many towns, sometimes even big cities, will not be able to revitalise their centres.

In the second period the institutional context for regional development changed considerably. Since 1994 East Germany has been integrated into the Common European Regional Policy and has been designated an Objective 1 region. Therefore, it is entitled to receive EU funds for Rural Development and has also been fully integrated in the LEADER II framework (Akademie für Raumforschung und Landesplanung 1994).

Moreover, the early period of establishing political institutions ended with a slow re-emergence of regional planning. Saxony was the first federal state to finish its regional plans, which became the legal basis for regional development[2] for its whole territory in 1996. Others followed rapidly, Saxony-Anhalt being the last to complete this legal procedure in 1999.

A commune reform forced local authorities either to merge or to build common administrations, and reduced the number of *Kreise* significantly from 191 to 87. The intention of the reform was to reduce public spending and to establish more powerful local authorities. It has also been expected to build up local political structures similar to those in West Germany. But it faces strong countervailing powers for two reasons. On one hand the dynamics of transition empowered local communities, and especially local mayors, to implement many projects comparatively unhindered by regional planning and supplemented by generous financial support. Despite numerous problems, local communities were able to maintain many services, while the economy as a whole declined dramatically. Now, because of the end of financial support through the expiry of special unification schemes, in conjunction with the reform of local governments, municipalities are contributing to unemployment. This is taking place in a situation when unemployment levels were already very high, and services were reduced at a time when many other local actors had already abandoned their activities.

Local communities and rural development

The central objective of institutional reform was the functional differentiation of economic, policy and voluntary sector institutions. The unanimous all-embracing role of collective or state-owned farms and firms in rural life had to be transformed into three sets of institutions: firms, municipalities and the voluntary sector. This process has been dramatically affected by economic crisis, emigration, and a time lag of several years just to put the new institutions in place. Although the new institutional order does in principle guarantee a de-centralised, polycentric power structure at the local level, dependency on external factors (markets, planning systems and external funds) places strict limitations on the action space for local actors. For instance, due to the degree of unemployment the (national) social welfare system is much more important to the everyday life of large parts of the population than local policy. Case studies indicate that there is a linkage between the perceptions of local policy (Bußmann 1998) and voluntary activities (Hainz 1999) in East German villages and the local economic conditions (employment opportunities). However, this does not appear to be the only important factor. Beyond that it is the geographical setting (accessibility), the existence of a peasant tradition, the history of industrialisation, and specific cultural conditions, which also influence the way in which communities deal with the transformation process.

In East Germany we find a North-South divide. The fundamental difference is the historical farm structure, which also influenced the industrial history. While most northern regions were dominated by large farm estates, the southern parts have a predominantly peasant tradition (Laschewski 1998a, Bruckmeier and Grund 1994). It has been argued that peasants, despite all the political pressure, remained a distinctive and influential group within rural communities, and have regained a higher status due to unification (Brauer, *et al* 1996, Laschewski 1998b). The distribution of land property rights also had an impact on the agricultural restructuring process and farm models (Laschewski 1998a, 2000) besides community life and migration trends. The earlier observation that unification enforced some already existing trends in the north-east is particularly true for regions which were historically dominated by large (feudal) farm estates and which were the object of the Soviet land reform in 1945. Such areas are usually less populated, which is linked to late industrialisation and a comparatively high degree of post-war settlement with refugees.

In a historically Protestant, and nowadays mostly atheistic East Germany, Catholicism is also a culturally distinctive feature. Small Catholic territories managed to keep a certain degree of distinctiveness and persistence even under the politically oppressive system of the GDR and earlier. A particular case is the Catholic Sorb minority in Saxony (Zierold 1995).

The historical and cultural context is also important for understanding changing power relations in local communities. Although there has been a power shift from the collective farms towards the municipalities, the position of the remaining co-operatives - in particular in remoter areas - remains strong because they are very often the only employers (Siebert and Zierold 1996). Municipalities are now as weak as before, although for a short period after unification the situation seemed to be otherwise. But significant budget constraints, the centralisation of public services because of local government reform, and the rising cost of social welfare due to long-term unemployment contribute to the continuance of a trend which had already started before unification - the loss of functions in smaller villages due to their concentration in larger communities. In many regions, and particularly in smaller villages, the situation may be described as a devastation of social infrastructure, and an impoverishment of local communities. Many local initiatives depend on providing work, through measures financed by the labour exchange. In many communities more people are employed as a result of such measures than by local businesses (Rodewald and Siebert 1995).

However, in the early years of unification, due to the lack of working political structures, the potential of local policy was very much dependent on individuals, in particular the local mayor. The sudden increase in responsibilities, the necessity to make many quick, but nevertheless important, decisions as well as the replacement of local leaders by newcomers in many cases asked too much from people in charge. Due to the selectivity of emigration processes, regions with a long-term emigration history also limited the personal alternatives for leadership (Meyer-Engelke *et al* 1998).

Economic restructuring also meant that all sorts of social and cultural services could no longer be provided, and municipalities were unable to maintain most such services. The existence of countless small village shops proved to be one millstone (among others) around the neck of the consumer co-operatives which finally led to a dramatic decline in the consumer co-operative sector, and the end of most village shops (Laschewski 1998c).

Although there are increasing social differences, income distribution appears to be comparatively homogenous. This is a result of the social welfare system, from which pensioners in particular benefited very much. Nevertheless, there is a new and clear dividing line between the employed and the unemployed (Siebert and Zierold 1996). Among the winners in the agricultural restructuring process in East Germany are also many of the newly founded large agricultural enterprises, partly constituted by former local cadres, and partly by newcomers from West Germany and the Netherlands. These farms are highly profitable because they are favoured by the current subsidy structure and are cost-efficient because they use a minimum of (expensive) labour. On the other hand, the neglected majority of farms is owned by part-time farmers and smallholders. There are visible consequences of these policies in many villages. The restored farm buildings are very likely to belong to a new farm enterprise; the agricultural co-operatives do have new machines, but have (too many) old buildings, and small and part-time farmers must work in old buildings with old machines (Schmidt 1996).

High unemployment with its adverse effects on people, the high proportions of elderly people - in particular in more remote areas - and the loss of almost all public places, also enforce once again a withdrawal into privacy, especially in the isolated areas of the north-east (Hainz 1999). But this time it is not an active, but in many cases a passive, frustrated privacy. Furthermore, it is no surprise that despite a number of positive cases, development projects very often suffer from apathy and weak local identities. It is among the ironies of the current situation that local initiatives usually draw much more upon those who are already employed, and therefore are in many cases short of time.

Informal exchange networks surrounding household production also became useless overnight. Products were suddenly available and cheap. Migration and the sudden necessity to commute long distances (for example to West Germany during the week) also weakened informal networks. Some studies indicate that disintegration in local communities goes along with a commodification of social relations. 'Unification' solidarity in the early period of the unification has been replaced by economic individualism (e.g. Zierold *et al* 1998). On the other hand, informal networks also suffered due to the reshuffling of social hierarchies, and with the loss of their economic functions informal networks seemed to have disappeared in the earlier period. Now, after the new social hierarchies have been established and due to the ongoing economic crisis, there are new signs of mutual self-help and informal co-operation.

Voluntary association as a mode of social integration is under-developed in East Germany in general (Seibel 1997) and in rural areas in particular (Hainz 1999). There is a general weakness in the capacity for self-organisation of the voluntary sector. State and municipalities are still dominant due to financial and infrastructure support and because the *Länder*, or local authorities, run organisations such as 'Quasi-non-Governmental-Organisations' ('quangos'), which are usually seen as part of the voluntary sector (Seibel 1997). There is also the continuing expectation that many services should be provided by the state. Therefore, the high dependency of voluntary associations on public funding is not necessarily perceived as a problem (Seibel 1997: 485). It is therefore also not surprising that rural development initiatives, which are based on the idea of an existing 'active society', such as LEADER, lack active local initiatives and function as rather conventional additional rural development programmes (Bruckmeier 2000). However, although there is a lack of functioning voluntary institutions as part of a civil society, Seibel (1997, 490/491) argues that the lack of change in the voluntary sector provided milieu continuity for the population, and a refuge for the former functional elite, and therefore contributed positively to political stability.

Summary and conclusions

We addressed the issue of rural development in East Germany as a specific case of post-socialist transformation. Many trends in East Germany may be in similarly observed in other Central and Eastern European countries with a socialist past. However, due to German unification, and the integration of East Germany into the European Union, institutional changes have been far more radical. East Germany has been thrown into the cold water of the market economy, although supplemented with huge social benefits. Legal institutions and administrative structures have been implemented which stemmed from a completely different social context. Despite a significant learning and adaptation process among the population as well as the administration, there is a tentative relationship between social structures and the action orientation of the population to the institutional framework.

The lesson therefore is that institutional reform cannot be achieved by a simple institutional transfer from the West to the East. Further development also depends on structural and cultural realities, which to a large extent are rooted in the socialist past, and persist after the institutional transformation in East Germany (Eisen and Kaase 1996). Similarly, Swain argued in relation to the transformation in Central and Eastern European countries 'that current developments in the post-socialist countryside cannot be understood in isolation from the experience of the socialist past and the constraints that it imposed' (Swain, 1998). Political reform and social change happen at different speeds. As a consequence a gap remains between the institutional framework and the socio-cultural context. Because of that, a similar impetus can bring about different responses in East and West.

The persistence of socio-cultural structures is also part of the observation that not all current changes can be traced back to institutional reform. Although unification brought about radical changes in rural areas, the origins of many problems and the trajectories of many trends lie in earlier developments. However, institutional reform has usually increased such problems. In principle, many of the problems which existed under socialism are nothing more or less than difficulties related to modernisation processes in general. Such 'modernisation' tendencies are reflected in the perception of qualitative characteristics of social relationships, such as commodification, bureaucratisation, and individuation. The hopes of 'modernisation' losers from unification have therefore been disappointed. Direct political patronage has been replaced by anonymous market forces, which very often work in the same direction.

In rural areas, tensions between institutions and local actors and structures affect participation patterns. Indigenous initiatives often lack support, while 'participatory' approaches become tools of state policy. Intermediary institutions, as well as the third sector, are financially dependant on state subsidies. Instead of a development of self-governed institutions, rather we find a process of 'quangoisation' of the third sector. However, this process also reflects behavioural patterns and social expectations in East Germany.

The transfer of existing institutions is not only connected with the question of adaptation, but also with the consequence that all known problems are transferred as well. It appears that established institutions have been transferred, and subsequently the East follows the West in the institutional reform process, after a certain time lag.

Several conclusions can be drawn form these observations. Due to the importance of social integration, historical experiences and socio-cultural conditions are fundamental and have to be taken into account. Therefore, transferred institutions may function differently in transformation countries than they do in the country where they evolved. In the context of rural development we see in particular difficulties with the transfer of 'participatory' approaches. Experiences with East Germany also demonstrate that financial support is limited in its effects. It may be useful to improve infrastructure, but it can hardly compensate for radical changes, or even generate a new basis for self-reliant development. The financial expectations of the new EU member states to benefit from rural development funds and agricultural subsidies may be self-deceiving. Therefore we have to question how easily current popular EU development approaches can be transferred to Eastern Europe, in particular to the candidates for the EU enlargement. It appears that there have to be new models, which take the different history and social structures in Rural Central and Eastern Europe into account. Finally, further research is needed to understand the local lifeworlds, and the effects and transformation of policy reforms at the local level.

Notes

1 The *Kreis* is an administrative unit between municipalities (*Gemeinde*) and the federal state (*Land*). It functions as an association of municipalities as well as administrative level for the *Länder*.

[2] In the German Regional Planning System different laws are necessary. Here we refer to the '*Landesentwicklungsplan*', a planning framework which is binding upon decisions at the local level.

11 Europeanising rural Hungary - rural policy networks and policy representations of the countryside in Hungary in the 1990s

ANDRÁS CSITE

Introduction

In recent writings on Europeanisation *policy network* has become a fashionable and in turn widely discussed term (Börzel 1997; Cole and John 1995; Coleman 1993; Conzelmann 1998; Grundmann 1997; Kassim 1994; Knill and Lehmkuhl 1999; Marks and McAdam 1996; Peterson 1997; Peterson and Bomberg 1999; Smith 1992; Van den Bulck 1992). One of the core issues in relation to a policy network is whether it is only a research tool, a mere descriptive term, or a central institution of a new policy-making system having been brought about by the progress of European integration. Exponents of the latter view argue that there is a multi-level system of governance coming into existence in the European Union. Emerging 'European governance proceeds through negotiations in policy networks linking public and private actors of different levels and dimensions of government' (Börzel 1997: 9). A multi-level system of governance (MLSG), in which negotiation is the core social action of policy agents, like *hierarchy* and *market*, is an ideal type of policy-making. The reason European governance is approaching the ideal type of multi-level governance is its effectiveness. In late modernity's dynamic, complex and diversified policy environment public actors, both at the European and the national levels, are increasingly dependent upon the resources of other public as well as private actors. In this policy environment hierarchical co-ordination has become inefficient, while the scope of non-hierarchical co-ordination through de-regulation and privatisation is limited due to market failures.

The evolution of a European governance system may have diverse impacts upon national policy actors and institutions. Knill and Lehmkuhl (1999) distinguished three mechanisms of Europeanisation when considering domestic institutional change caused by European policies:

- *positive integration*: EU policy 'positively' prescribes institutional models to which domestic arrangements have to be adjusted;
- *negative integration*: European legislation affects domestic arrangements by altering the domestic rules of the game. In this case European influence is confined to altering domestic opportunity structures, and hence the distribution of power and resources between domestic actors;
- *framing integration*: in its weakest form European policy neither prescribes an institutional arrangement to be applied by national actors, nor modifies the institutional context for strategic action, but affects domestic arrangements by altering the beliefs and expectations of domestic actors.

Different policy environments, varying in the extent of institutionalisation of a policy domain, the nature and influence-potential of domestic and European policy networks, and national policy-making styles etc., cause variation in the functioning of Europeanisation mechanisms. Policy domains vary in the level of institutionalisation, existing resources, organisational structure, core issues and types of policy networks.

According to Rhodes (1990, 1992) a policy network is a group of organisations linked together by a mutual dependency on resources. A policy network is isolated from other networks along the break lines in the structure of the resource-dependencies. Rhodes proposes to make a distinction between policy networks on two considerations: first, on the basis of the degree of integration of the network, and secondly, on the basis of which interest groups belong to the network and how they share their resources. In his analysis of the relationship between the British central and local governments, Rhodes distinguished 5 types of policy networks:

- *policy communities* characterised by a high degree of integration;
- *professional networks* dominated by experts;
- *inter-governmental networks* based on the organisation of interest representation of the local governments;
- *producers' networks* separated from one another on the basis of economic interests; and
- *issue-networks*, weakly integrated and fraught with conflicts.

From the chapter's point of view there are three relevant policy domains: agricultural policy, rural development policy and regional development policy, which are for the sake of simplicity concisely called rural policies. The boundaries of these policy domains are blurred and in many fields overlap each other. These policy domains may vary in countryside-definitions, in the progress, and in turn, the impact of Europeanisation. Studies on policy-networks in farming have come to the conclusion that agricultural policy-making is determined by closely knit policy communities. Peterson claimed that:

> most EU agricultural policy communities are relatively cohesive. At their core is a solid alliance between the Committee of Professional Agricultural Organisations (COPA), one of the oldest and best-funded of all trans-national interest groups, the Special Committee of Agriculture, management communities for specific products, and DG VI of the Commission. The mutual dependence of these 'core' actors is clear and powerful, particularly given new threats to the privileged place of the CAP posed by the prospects of eastern EU enlargement. Actors in CAP networks have powerful motives to speak with a single voice on most issues and 'deliver' on the terms of bargains struck (1997: 2).

Adshead (1996) also employed a policy network approach to her analysis of Irish agricultural networks. She concluded that the closed agri-policy community of Ireland did not merely endure EU accession but also turned it to its advantage:

- Despite the declared transparency and openness of the European Union, only nationally recognised interest representations are entitled to take part in EU policy-making.
- Organisations that do not have a Brussels representation are in a weaker position at negotiations.
- Reform of EU decision making-structures changed the relationship between national governments and national interest representations: at the EU level the Minister for Agriculture functions as a representative of national agricultural interests.
- The market positions of Irish farmers have looked up since the accession.

Elsewhere, Csite and Granberg (2000) have argued that Europeanisation in Finland had a threefold impact upon rural policy-making. The Finnish agri-policy network, that we conceptualised as a closely knit community of party politicians, civil servants, experts, and farmer unions, has maintained its grasp on the policy process, while adopting new environmental and support measures that were in line with the EU regulations. The regional development policy network has increased its influence in terms of financial means to be used for regional development measures. Finnish regional developers, earlier a part of the cross-Nordic network of regional experts, built up a network of regional institutions, and by making use of the chances offered by the country's EU accession have made a decisive step in the institutionalisation of regional development. Rural development policy is rather a new phenomenon in Finland, and until the country's entry to the EU the policy network of rural developers had been a loosely integrated issue network of professional organisations, a country-wide organisation of local development groups and a few civil servants in the Prime Minister's Office. From 1995 on EU's Objective 5b, 6 and LEADER II programmes have given a new financial and institutional impetus to rural development, and the policy network started transforming itself into an institutionalised policy domain with a widening circle of local and regional politicians, experts, civil servants, and local developers.

In a paper on East Anglia's Objective 5b project, Ward and Woodward (1998) suggest that European policy is becoming increasingly important for rural development actors not only in financial terms, but also in terms of institutional practices and procedures. Their level of analysis is the region, and they interpret Europeanisation as a learning process of the 'European game' by rural development policy communities in regions, i.e. how to create new representations of a region to make it eligible for EU funding, and how to set up formal and informal links to EU decision-making officials, both in Brussels and in the respective national ministries.

In a paper on the Europeanisation of regional development policies, Conzelmann (1998) argued that the domestic institutional context of policy-making in individual member states has to be taken into account when discussing the question of policy changes under the influence of EU factors. This context may matter in different ways:

- challenges posed by the presence of EU regional policy are likely to be different according to the respective country specific variables
- existing institutional and cognitive structures condition how supranational factors come to be felt in domestic politics
- national policy styles also influence whether the process of policy change is more likely to advance through unilateral imposition, processes of consensus-oriented problem-solving, or pluralistic political conflict.

Grabher and Stark (1997), one of the first users of a network approach to post-communist transformation, state that a significant part of the communist networks had been eliminated in the transformation era, but there are also new networks coming into existence, often on the ruins of communist networks.

The first question we address in this chapter is the type of policy networks functioning in the Hungarian rural policy arena. Hierarchical co-ordination, the key feature of communist policy-making, theoretically conditions the existence of closed policy communities. But what happens to them after the elimination of the one-party system? Post-communist political transformation, the de-regulation of political activities, theoretically creates a policy environment wherein potential policy actors can build new policy networks, i.e. employ collective strategies that are expected to be advantageous to all the parties involved. And what is the impact of Europeanisation on the institutional structure of rural policies? How and to what extent do mechanisms of Europeanisation (positive, negative, and framing integration) cause changes in the internal structure and external relationships of rural networks? These are the issues on which we attempt to focus in this chapter, hoping that the reader will be provided with a few new insights into the rural policies of a country which is not a member state of the European Union, but where Europeanisation is the main driving force for the transformation of domestic policy structures.

The chapter's findings are based on extensive interviewing of core executives of Hungarian rural policies, as well as of regional and rural developers. Between 1997 and 1999 I carried out interviews with civil servants from the Ministry of Agriculture and Regional Development and with activists from the Hungarian Rural Parliament; I also analysed rural-related articles in the major daily 'Népszabadság', and bulletins from the Ministry of Agriculture and the Hungarian Chamber for Agriculture.

Elsewhere, I have published papers on the structure and countryside representation of policy networks acting in the rural arena (Csite 1997, 1998, 1999). This chapter is an attempt to summarise my earlier findings and draw a comparative conclusion on the evolution and linkages of the actual networks. In earlier papers I have identified the following policy networks that exist in the rural policy arena: the 'big business' network, the 'family farm' network, the 'regional developers' network, and the 'rural developers' network. Their evolution and structural composition was investigated through the following policy events:

- 1992/3: passage of bills on the transformation of state and collective farms, and on the restoration of family farms
- 1996: passage of the bill on regional development
- 1997: farmers' demonstration
- 1998: establishment of the Hungarian Rural Parliament and of the Department for Rural Development in the Ministry of Agriculture and Regional Development
- 1999: completion of the first version of Hungary's SAPARD programme.

In this chapter I will proceed through the following steps: first I briefly sketch the rural legacies of communism, then I focus upon the early period of post-communist transformation, which will be called 'the fragmentation period'. The main feature of this era was dissolution and in many cases forceful elimination of communist policy networks. This was the time when the major organisations of a new policy network of family farms was created, and in legislation the basic institutions of a family farm-based agriculture were set up. The period's other adventure of major importance to our argument was the reconstitution of democratic local governance in rural settlements.

Later, I outline the structural characteristics of rural policy networks, first the agricultural ones, then the regional development and finally the rural development network. 'Europeanisation', although being a permanent element of the Hungarian policy-process of the 1990s, became the ultimate factor in rural policy changes in 1996. This was the year when the law on regional development - the first in the country's history - was passed, when farming subsidy schemes were reshaped into a EU-conformist framework, and when the cabinet, pleading pressure from Brussels, did away with small farms' exemptions from taxes. Finally, I will argue that in the second half of the 1990s the main mechanism of Europeanisation in Hungarian rural policy-making has been framing integration, i.e. lacking legal means (positive or negative legislation), the European Union impacts upon the beliefs and expectations of domestic policy actors. However, this is not inconsiderable given the pace and extent of policy changes taking place in the rural policy arena.

I feel it is in order here to make some remarks on my approach to the relationship between policy networks and the narratives or discursive representations constructed by policy actors. One way to understand this relationship is to focus on actors' resources and assume that policy actors have clearly defined interests and the reason they co-operate is to create extra benefits from networking. Narratives developed by these co-operating actors are a sort of 'weapon' in the hands of actors prompted by common self-interests. Another approach stresses that networks evolve from discourses: existing narratives thought to be plausible or useful by unassociated actors trigger off co-operation between them (Frows 1998, Hajer 1995). My proposal to bridge the abyss between these approaches is to focus on the extent of *institutionalisation* of the actual policy domain: if the given policy is highly institutionalised, there is a widely established structure of policy measures and organisations, interests at stake are deeply embedded, and narratives play a comparatively less important role in the policy process. In other policy domains, whose institutional structure is less developed or practically non-existent and policy actors pursue official recognition and institutionalisation, the construction of common understandings is an essential pre-condition for effectual co-operation.

Rural legacies of communism

In Hungary the collectivisation of agriculture was completed by the end of 1961. Between 1958 and 1961, the area of land under cultivation by the agricultural co-operatives expanded from 797,000 hectares to 5,313,000 hectares. During the same period, the number of private plots of greater than 0.5 hectares decreased from 1,366,000 to a mere 160,000. The proportion employed in agriculture fell from 37 per cent in 1961 to 27 per cent in 1965. During this same period, more than 450,000 people left agriculture (Donáth 1977, Orbán 1972). Agricultural collectivisation put an end to the liquidation of private capital ownership (Balogh and Pölöskei 1979; Csizmadia 1962). With the completion of nationalisation, any obstacle to the communist implementation of industrialisation was thus removed. During the 1960s, favourable market conditions increased industrialisation. The economic growth of the Soviet Union and the 'Capitalist World' produced larger markets for Hungarian products and hence brought about the expansion of the Hungarian economy as well.

The launch of economic growth also meant that, in principle, enough income originated from the national economy to enable industrialisation of agriculture to commence. The leaders of the Hungarian communist party did not wish to repeat the same mistakes as those made in the 1950s, thus profit was not taken out of agriculture. The successful collectivisation strengthened the positions of a 'Peasant Fraction' led by Lajos Fehér in the leadership of the party, and soon it gained a large autonomy in the management of agricultural policy. By the end of the 1960s it developed into a closed policy community, integrating communist 'peasant' politicians (like Fehér), a network of agricultural universities and research stations, collective and state farm leaders (who were in increasing number experts trained at agricultural universities) and their national interest representation, the Agricultural Cooperative National Council (TOT) (Varga-Suplic 1988), which continued to exist until the change of regime in 1989. During the twenty years of its existence, the TOT became one of the strongest lobbying organisations in communist Hungary. It had a say in decision-making which affected both agriculture and the villages – it catalysed and represented the co-operatives' reformist endeavours and demonstrated that agricultural intellectuals were dominant in rural Hungary (Erdei and Fazekas 1970; Erdei 1977, Ivanics 1982; Juhász 1983; Pünkösti 1986a, 1986b, 1988).

One of the major successes of the agricultural policy community was the reform of collective farms and agriculture in 1967/68 (Harcsa *et al* 1995; Fehér 1970). The argument behind the reform was that obstacles to the expansion of agriculture came from the fact that neither the leaders nor the workers received a large enough share of the rewards generated by their labours. The solution was that the workers would receive a financial share of the co-operatives' rewards and small-scale production of farm members was integrated into large-scale production by various forms of putting-out, production contracts, etc. The effects of the reform meant that the co-operatives became 'legitimate' and preferred as state-controlled industries and, in addition to this, they also received supplementary support from the state. The result of this was that it accelerated industrialisation and mechanisation of agriculture, increased the use of chemicals and speeded up co-operation between the production and processing industries (Romány 1983, 1996). The most significant part of these developments was 'The Hungarian Agricultural Miracle': integrated, industry-like production systems in the corn, wheat and poultry sectors (Kovách 1988).

Besides collectivisation, another policy to transform rural society was the introduction of a strict settlement hierarchy, whose framework steered rural modernisation (Juhász 1988, Vági 1982). The development of the National Settlement Network Conception (OTK) of 1971 defined the settlements' main functions, and thus how infrastructural and industrial development resources were to be distributed. The main settlements were the receivers of these funds, whereas those not considered significant enough received little or nothing. In the small village districts, the local councils were united, and five or six settlements came under the authority of one council. Each council district had one school, one kindergarten, one doctor and one vet etc. (Navracsics 1997). 1,998 settlements (65 per cent of the total) were defined as 'other', during the period from 1970 to 1990, and received almost no support from the development resources. Their populations decreased and the young moved away to larger settlements (Szelényi 1990, Vági 1982, 1991).

Post-communist fragmentation

From the mid-1980s, rural societal issues, chiefly the settlement hierarchy and collectivised agriculture, had been re-problematised by diverse political groupings emerging in the changing political climate (Vági 1991; Swain 1993; Csite 1997; Harcsa *et al* 1995). New parties coming into existence on the eve of the communist climax stressed the necessity of reshaping rural economic and political structures.

One of the first actions of the post-communist government, consisting of the national conservative parties, was the abolition of local councils and the re-creation of local self-governance. In the reorganisation of the administrative system and the re-direction of development funding to the countryside, compensation was a predominant factor. Significant infrastructural developments were started in those settlements which had not received financial support under the previous system (e.g. gas supplies, sewerage and telephone systems were installed or modified.) From 1992, 'disadvantaged' villages were in receipt of extra support from the state.

It is of some importance to mention here that the reason behind the disadvantageous state of the villages was due to the communist settlement hierarchy. Because of this, the framework of the compensation was set out in a similar way to that of the hierarchy: the imbalances caused by socialist industrialisation were thus reduced. Agricultural reorganisation and the evolution of local self-governance shifted the balance of local power: Previously, the development of the settlements had been decided by the co-operatives, whereas now this was controlled by the local self-governments (Bánlaky 1996).

Disintegration of the agricultural co-operatives was accelerated from 1992-93 (1992 being the year when, by law, the co-operatives had to be dissolved). Between 1990 and 1995, two thirds of the co-operatives' lands were handed over to the economic units which were founded in their place.

Table 11.1 The distribution of land between economic sectors, 1990-1997

	1990	1991	1992	1993	1994	1995	1996	1997
Business Organisation	816	959	981	1235	1163	1086	1090	967
Co-operative	4378	4058	3577	3267	2305	1886	1748	1611
Subsistence Farm, other	1277	1441	1576	1626	2652	3205	3345	3615
Total	6473	6459	6135	6129	6121	6179	6184	6194

Source: Kovács 1998: 187

Disintegration of the co-operatives led to a dramatic decrease in the size of farms. Small businesses, which employed less than ten people (family farms, limited partnerships, etc.) became the defining agricultural farm type (Kovács 1996, 1998). In addition to this there was a dramatic fall in the number of people employed in agriculture. At the beginning of 1990s, nearly one million people worked in agriculture; today this number is a mere 350,000.

Rapid reshaping of rural, social and economic structures took place without significant collective resistance by disadvantaged rural actors. One cause of resignation is the aggressive witch-hunting against 'green barons', i.e. collective farm leaders and agronomists (Swain 1993), who, having become scapegoats for forced collectivisation, lost their ability to oppose collectively the way of de-collectivisation. The post-communist government also eliminated the domination of the closed policy community of collectivist agriculture in rural issues. However, the politically motivated campaign against large-scale collectivist farming did not result in the complete disintegration of the co-operative policy network, but yet in its transformation.

Table 11.2 The number of corporate bodies in agriculture, 1990-1996

Number of employees	1990	1991	1992	1993	1994	1995	1996
Under 11	-	-	-	1669	2420	3834	7256
11-20	413	1031	1471	991	1012	624	605
21-50	106	297	437	730	761	799	743
51-300	860	1060	1200	1188	1209	1061	963
Above 300	722	666	378	129	126	95	72
Total	2101	3054	3486	4707	5528	6413	9639

Source: Kovács 1998: 189

Agri-policy networks in conflict

At the end of 1996, Parliament passed amendments to the bills on family farms' taxation, social insurance and economic registration. As a consequence of the new regulations the privileges of small-scale agricultural producers, which they had enjoyed since the second collectivisation of the early 1960s, were annulled (Szabó 1998).

In January 1997, wine-growers of Kiskõrös (one of the major viticultural areas) issued an appeal in which they asked for withdrawal of the new regulations and the re-establishment of the 1996 conditions. Their appeal got support from the Hungarian Union of Smallholders' Clubs (MAGOSZ) and the Peasant Union (MPSZ), as well as from local agrarian clubs and party organisations. 'The possibility of integrating fragmented small-producers scattered in diverse production sectors came about along the common interest against the manifest adversary, the cabinet and the financial lobby' (Szabó 1998: 203-204). Figure 11.1 below exhibits the number of participants in the demonstrations.

In the last week of February 1997 the cabinet attempted to pacify protesters by initiating negotiations with representatives of the Kiskõrös producers (by then they had set up their cover organisation, the Interest Representation Organisation of Agricultural Producers, METÉSZ), and by convening the Agricultural Interest Reconciliation Council (FÉT) and launching negotiations on the National Agricultural Programme. After two days of heavy debates, representatives of the government and the Kiskőrös producers adopted the draft of a future compromise. Although protesters refused to accept this compromise, the cabinet held it to be obligatory and was not willing to make further concessions. Prime Minister Gyula Horn invited the leaders of the protesters to a meeting on March 4, but after unsuccessful Interest Reconciliation Council negotiations they remained aloof from further co-operation. Thereafter the government refused further negotiations; however, it introduced amendments to the taxation and social insurance bills which were passed by the Parliament on March 25. After March 11 no further demonstrations of great importance took place, and the government, opposition parties, interest representations and ministries attempted to come to an agreement within the framework of NAP negotiations.

Figure 11.1: Number of participants in the demonstrations

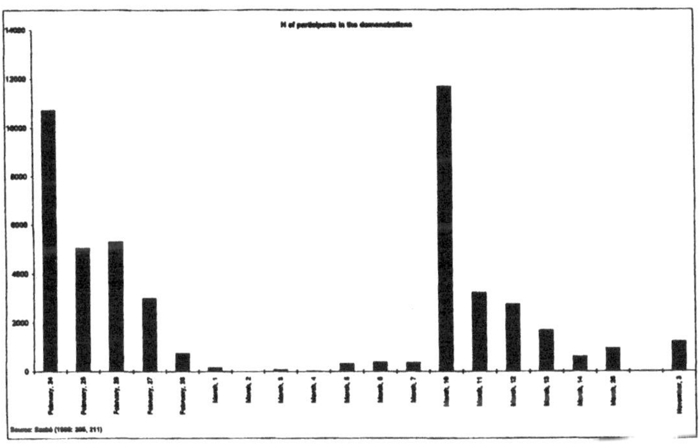

In 1994 the incoming socialist-liberal coalition had considered the National Agricultural Programme (NAP) as a necessary package of measures to settle agrarian issues for the period before the country joined the EU. The function of NAP was to transform Hungarian agriculture into one compatible with EU structures and to create stability within the sector. Although before the 1997 farmers' demonstrations the Ministry of Agriculture had prepared drafts, the February protests speeded up the process and the cabinet came forward with its first official proposal in early March. This was the topic of a meeting held in Gödöllõ on March 13 when all the major political parties, agricultural interest groups, ministry officials and researchers came together. In April the Smallholders' Party (FKGP) initiated an Opposition Agricultural Round Table (OART). They invited opposition parties and major interest representations to work out a stand on the government's NAP proposal.

On April 25 the government held a two-day National Agricultural Round Table where participants decided that the first draft of NAP be shortened and the new version be worked out in a more detailed way by the summer.

On September 26-27 participants at a second Gödöllő meeting (among others the president of the republic, ministers and leaders of interest representations) passed the final version of NAP. Henceforth, NAP was removed from the political agenda, although only some parts of minor importance have been fulfilled (e.g. the establishment of a Credit Bank).

On June 26 1997, the cabinet decided upon a proposal to amend the law on land ownership which would have made it possible for legal entities to buy land in Hungary. This would have caused a basic change in the structure of land ownership, for according to the regulations acquisition of land by non-natural entities had been forbidden since 1994. In August four opposition parties (MDF, Fidesz, FKGP, KDNP) and three interest representation bodies (MAGOSZ, MPSZ, Metész) initiated a referendum against the acquisition of land by foreigners. According to them the amendment would have made possible the acquisition of land by foreigners - this was later also admitted by governing parties, too. The cabinet argued that Hungarian agriculture badly needed foreign capital investment which is only possible, however, if foreigners can also acquire land. Opponents pointed out that the real reason foreigners wanted to buy land was financial speculation due to differences between land prices in Hungary and the EU. Foreign land ownership would not mitigate the hardships of Hungarian producers, but cause more severe difficulties since foreigners would benefit from the rise of land prices after Hungary joins the EU.

In reaction to the opposition's initiative, the government formulated its own questions on acquiring landed property and combined them with the NATO referendum (in 1997 Hungary was invited to join NATO). On August 26, the then most popular opposition party's leadership (Fidesz) changed its former stand and demanded that both the NATO and the land referendum be compulsory. In late August the cabinet - impressed by the popularity of the opposition's initiative (up to September more than 200,000 signatures had been collected) - accepted the Fidesz proposal, but stuck to its own versions of the referendum questions. Parliament debated for weeks about the questions, and the controversies were finally settled by a government decision on November 16: only the referendum on NATO membership would be held, while the one on land ownership was postponed to a later date. The government accepted that it should not propose an amendment to the land-law until the referendum was held. On November 3 radical opposition forces (with the leadership of the then vice-president of the Parliament, Ágnes Maczó) held a demonstration in Budapest demanding that parliament forbid foreigners to buy Hungarian land. The police drastically destroyed the demonstration and the Smallholders' Party excluded Maczó due to her radical stand. On November 16, some 80 per cent of the electorate voted for Hungary to become a member of NATO, and acquisition of land by foreigners was forbidden henceforth.

From the 1997 policy events we can draw the following conclusions about the structure of the Hungarian agri-policy network:

- the Hungarian agri-policy network is bi-polar in nature, it consists of a 'family farm' coalition and a 'big business' coalition;
- the coalitions include both political parties and interest representations;
- the 'big business' coalition consists of representatives of big agricultural companies (mainly former collective farms), their interest representation MOSZ (National Union of Agricultural Producers - before 1990 the National Council of Collective Farms), political parties (Socialist Party, Agrarian Union), semi-governmental organisations dominated by big farm interests (Chamber for Agriculture), and a network of experts, university professors, and academics;
- due to the events of 1997 the 'family farm' coalition is also of a bi-polar nature. The less influential radicals and the more powerful moderate organisations not only competed with each other, but in many cases co-operated against the 'big business' coalition. However, unlike the 'big business' coalition, the 'family farm' one is dominated by politicians. The moderate wing consists of political parties (MDF, Fidesz, Smallholders) and interest representation bodies (Magosz, MPSZ), while the radicals include the Party of Hungarian Justice and Life, the New Alliance for Hungary, and the radical wing of the METÉSZ. It is also worth noting that unlike the 'big business' coalition, the 'family farm' one does not have a wide expert network;
- experts and politicians of the big business coalition headed the Ministry of Agriculture (currently the Ministry of Agriculture and Regional Development) as well as its de-centralised bureaux and semi-governmental organisations (product councils, regional agricultural offices, research institutes, etc.) between 1994 and 1998. However, between 1990 and1994 and after 1998 the leading positions in these organisations were occupied by representatives of the 'family farm' organisations.

The birth of the new regional policy (NRP)

Although regional disparities were not unknown in the pre-1990 period, some unexpected outcomes of the rapid economic transformation of the early 1990s (especially the rise of unemployment) made it necessary for policy makers to formulate a new national regional policy which was thought to be an effective 'weapon' against regional differences (Csite 1997; Enyedi 1996, 1997, Nemes-Nagy 1992, 1998). The first post-communist cabinet set up a new ministry called the Ministry of the Environment and Regional Development, which was responsible for the codification of a new regional policy (NRP).

From the debate on the NRP I will highlight five of the key points:

- the critique of socialist regional and local development policy;
- the critique of the compensation solution of the late 1980s and early 1990s;
- the clarification of the new regional policy and its tasks;
- the winning over of interested parties to the development of a new regional policy;
- how regional differences were introduced as legitimate political problems.

Critique of the socialist regional and local development policy

Both the operation and the structure of the administrative system were criticised. The criticisms came in many forms, e.g.:

- inadequate aims (e.g. industrialisation in the post-industrial period);
- illegitimate policy making;
- the low efficiency of the socialist development policy;
- a shift in the development paradigm in Western Europe (the character of economic development changed from industrial to one based on complex innovation).

In regard to the development approach of 1985-95, there was less criticism of this than of the socialist solution. On one hand, this was because of feelings at the time of the change of regime (it was easier to criticise the Communists). On the other hand, the new administrative system had not yet appeared. The significant differences between the new regional policy and the transitional compensation policy could be divided into two: firstly the problems in the operation of the administrative system, and secondly the lack of development conception. The following points summarise the criticisms made about these:

- the transitional compensation development solution was inconsistent, improvised and slow;
- the local and regional development functions were managed by several ministry departments, and were therefore confused;

- there were signs that the conflicts from the communist era would reappear.

At the time of the new regional policy's foundation, the introduction of Western European conceptions of regional development took place, along with realisation of the need for a new regional policy in Hungary to enable its integration into the EU. On the basis of Western experiences the following issues were stressed:

- innovation;
- the stimulation of local resources and the injection into the innovation process;
- flexibility, many-sided methods which are sensitive to local needs and resources;
- re-industrialisation and tertiarisation as new driving forces of economic growth;
- collective action, regional development programs, small regional innovations and innovation centres.

The assignment of actors to the new regional policy also played an important part in the process of institutionalisation of the policy field. Not only was the assignment important but it also clarified the theory of the relationships between them:

- the system was under the control of the Ministry of Regional Development, but for the realisation of a new regional policy it was necessary for the Ministry of the Interior, the Ministry of Agriculture and other ministries to co-operate;
- thc other role of the NRP (which is still unclear) is that of the regions themselves;
- the third group, the counties, is where the decisions on developments are made. The county decision-makers include the representatives of small regions, the county chamber of commerce, the administrative bodies, and various other interested parties;
- the execution of development decisions happens in small regions and at regional levels, which are the subject of development projects. In the case of the small regions' organisations the formulation is based on the principle of partnership, between for example local government, the business sector and NGOs.

Part of the foundation of the new regional policy was that the European Union's NUTS system compatibility had to be installed in the regional categorisation. The small regions were categorised into four groups: rural, underdeveloped, crisis-struck industrial, and regions stricken by long-term unemployment. To separate these categories, the following statistics were used (Csatári 1996):

Underdeveloped small regions:

- average gold crown value of the cultivated lands;
- ratio of employed to the number of inhabitants in 1990;
- the unemployment rate at 31/12/1994;
- the net salary of a tax payer at 31/12/1993;
- percentage of houses connected to sewerage systems at 31/12/1994;
- amount of electricity used per household at 31/12/1994;
- turnover of shops per person at 31/12/1990;
- average qualification - number of years completed in school from the age of 10, at 1/1/1990.

Agricultural (rural) development small regions:

- rate of active agricultural employees in 1990;
- average gold crown value of the cultivated lands;
- rural index (population density where less than 12 people live in a square kilometre);
- average annual migration percentage 1990-1994.

Crisis-struck industrial regions:

- rate of active industrial employees in 1990;
- change in the number of people working in industry 1990-1993;
- proportion of long-term unemployed at 31/12/1993.

Regions of long-term unemployment regions:

- unemployment rate at 31/12/1994;
- percentage of long term unemployed at 31/12/1993;
- percentage of the long term unemployed at 31/12/1994.

In the process of forming the NRP, a heavy stress was placed on the principle of partnership. The system of subordinates and superiors was replaced by 'horizontal' relationships. A greater variety of people was involved in the preparation, making and execution of decisions.

When the period of NRP's institutionalisation is examined, the preparation and passing of the regional development law is of significant importance. Following the election of 1990, when the Ministry of Environment and

Regional Development was founded, laws concerning regional development were prepared by the Regional Development Department of this ministry. To get themselves acknowledged within the ministry caused great problems to this department of only twenty-two people. To increase the size and weight of the department, academics and university researchers were brought in. Along with the support of the scientists - who had a major influence on what regional policy-makers spoke about, and how they did so - counties and small regions which formed alliances in the beginning of the 1990s, those who could apply to the application system formed in the 1990s, and various regional development private businesses and agencies together formed a pressure group. In this process the work of PHARE is also noticeable, since it financed the regionalisation of the country (organised conferences, supported publications, research groups, etc. through whom the expert knowledge of regional development has been created).

In the early spring of 1996, the Hungarian parliament passed the first regional development law. This was a significant success for the regionalists. One part of the 'recipe for success' was the regionalists' strategy to band together social actors (scientists, policy-makers, business groups, etc.) not previously in contact with each other.

The second part of the 'recipe' was to highlight the EU's characterisation of regional development. Following the change of regime, the regional policy of the EU was interpreted positively. In this reasoning, there was another message: large parts of EU structural funds are distributed on a regional basis.

Evolution of the rural development policy network

In the evolution of rural development policy, important roles must be given to the opening of two policy windows: first the Agricultural Demonstration of 1997 and secondly the change of government following the elections of 1998.

The farmers' demonstration which started in February 1997 reinforced the efforts at preparation of a national agricultural programme. Originally the programme was to be introduced in the summer of 1997, however at the urging of government leaders and the Ministry for Agriculture it was introduced by the Agricultural Minister in March 1997. This programme included the question of rural development planning, which was new in the designs of the Ministry of Agriculture. According to interviews made at the Ministry, the then minister - László Lakos - held the expansion of production to be the most important political objective (the group which supported him was called 'The Tonnes'). At the end of 1996, Frigyes Nagy became the minister and he made the decision that one of the last sections of the agricultural programme should be about rural development. From one day to the next, rural development became one of the most important issues in political debates.

The popularity of rural development was further increased when the Independent Smallholders' Party (FKGP) made this one of its campaign issues. Following the formation of a coalition with the Hungarian Civic Party (FIDESZ) the new government united the Ministries of Agriculture and Regional Developments, and renamed it the Ministry of Agriculture and Rural Development (in Hungarian, while the official translation to English is of Regional Development). A new department was formed to deal with rural development and to prepare the national SAPARD programme.

Along with the party political struggles, however, a new process also began in the country. In the villages and the small regions, NGO organisations dealing with rural development started to form a national pressure group. In the late 1980s and early 1990s, when the Fordist rural development regime broke down, such NGOs, whose aims were to develop the settlements, were established in nearly every village. Local groups, although insulated from each other, mushroomed throughout the country. Many of the activities of local rural development were only partly financed by the state, the remaining funds coming from applications to various foundations (e.g. The Soros and Kárpátok Foundations, USAID, and PHARE).

The formation of a national network by these local organisations was implemented by experts who were acknowledged by these local groups. From the mid-1990s, an ever-increasing amount of experience was exchanged between rural developers, and new opportunities for national co-operation arose. This was the period when rural developers became familiar with the application system, project financing, and how applications should be prepared for international funds.

Rural development became a kind of social movement but it kept its independence from the interest groups surrounding political parties (e.g. the large-scale farming and family farming coalitions), and from the expert networks which wanted to increase the power of the ministries (e.g. regionalists). In the formation of the rural policy network, the period between spring and summer of 1998 was a turning point. At this time more than 200 rural NGOs formed the Rural Parliament of Hungary.

According to its founding statement:

> The organisation is formed to fulfil the aim of being accepted by the leaders of political, economic and cultural life: the countryside's resources are not only the land, its economy is not only agriculture. The national natural and cultural heritage is connected to the Hungarian countryside and the communities - all of these increase the value of resources! The small business and family business which operate in our villages and small towns, the many-sided local economy

> and market are irreplaceable, in the shadow of globalisation an ever-increasing part is the economy of the countryside, which has a large influence in the viability of the traditional agriculture (Márczis 1998: 86).

It is not possible to understand the popularity of rural development without taking into account the country's preparation for EU membership. According to Hungarian rural developers, mass production in agriculture would not receive support from the EU. However, rural and regional development as a policy idea conforming to EU conditions would receive substantial support from the EU pre-accession fund (e.g. The Sapard Program), hence it is of the utmost importance for Hungary to meet EU requirements in order to gain access to the Union, and to receive support from EU funds.

Summary

In this chapter I have approached policy changes in Hungarian rural policies from the perspective of Europeanisation. This term has been coined mainly by political scientists, and its success is closely linked with policy networks, whose emergence and transformation indicate a profound shift in European policy-making: the evolution of a multi-level system of governance. Researchers into Europeanisation focus chiefly upon the policy changes taking place in EU member countries, while non-member European states - like for instance Hungary - pass unnoticed; their experiences do not become part of wider (cross-European) discussions on the features and extent of Europeanisation.

One possible strategy to analyse non-member countries' policy changes is to compare them with those of pre-accession Portugal, Spain, Greece, Austria, etc. However, the scope of this kind of comparative study is quite limited, because these countries joined the European Union before the start of extensive structural reforms, therefore they had a say in the matters of institutional restructuring.

Hungary, the Czech Republic, Poland, etc. - the target countries of 'Eastern enlargement' - have rather weak and fundamentally uninfluential positions in accession negotiations. Having no share in EU decision-making processes has an important bearing on the way these countries are Europeanising. It is quite paradoxical that while the EU is establishing a multi-level system of governance that is, according to expectations, going to empower previously uninfluential clusters of actors, the policy actors in pre-accession countries are experiencing dis-empowerment, and the shift of decision-making competencies to the EU.

The point I make in this chapter is that since the middle of the 1990s one can understand the changes which have taken place in Hungarian rural policies as manifestations and consequences of Europeanisation dominantly via framing, and less significantly via positive integration. In the process of framing integration as one mechanism of Europeanisation, it is chiefly the transformation of perceptions and expectations of policy actors which brings forth changes in the domestic policy structures. Transformations in Hungarian agricultural policy (e.g. in farmers' registration, taxation, subsidies, etc.), or the emergence of regional and rural development policies, cannot be explained as consequences of direct instructions coming from EU headquarters, because of a lack of competencies of EU decision-making bodies to prescribe to the Hungarian government the introduction of new regulations. However, Hungarian policy actors are discursively influenced by EU bureaucracies. The increasing importance of positive integration is also not negligible: the availability of the PHARE and more recently the SAPARD programmes depends on whether the change of domestic policy structures answers EU bureaucracies' requirements.

Europeanisation of Hungarian rural policies is an outcome of actions of both domestic and EU policy actors. One segment of changes taking place in the course of Europeanisation concerns the emergence and transformation of rural policy networks. The family farm network is basically a producer network strongly influenced by right-wing political parties. Its countryside representation is of family farm-dominated agriculture rooted in the country's pre-communist peasant traditions. The big business coalition, also a producer network with many links to the post-communist Hungarian Socialist Party, is the successor of the closed agri-policy community of communist Hungary, and its main concern is competitiveness in European markets. Its representatives stress that small-scale farming will not be viable in the European Union, and that Hungarian agriculture should be based on commercial large-scale farming.

The policy network of regional developers has been one of the most successful professional networks. By taking advantage of the opportunities offered by Europeanisation, regional developers created a new policy domain, whose institutional positions are being occupied by its network members. In their understanding, the countryside is a cluster of regions hit by various social problems (unemployment, high rate of agricultural employment, low population density). The latest policy network to come into existence in the rural policy domain - the rural developers' network - is a mixture of a professional and an issue network. Its professional segment almost exclusively consists of Budapest-based academics (rural sociologists, community developers, folk artists, etc.), whose passion for folk art, peasant culture and the village community is the common issue, linking together not only professionals, but in turn local developers, roma leaders, village mayors, etc. What they stress is the economic and cultural diversity of rural settlements and areas, and the belief that the Hungarian rural future should not be built exclusively upon mass agriculture, but on local traditions.

So far, Europeanisation of the Hungarian countryside has concerned only the rural policy domain. Framing integration brought changes on the discursive level of policy-making (the emergence of new countryside representations) and on the institutional structure of policy implementation. Therefore, the beneficiaries of Europeanisation for the time being are professionals and civil servants who have been the core agents in generating changes in rural policies. Whether, and to what extent, other rural actors benefit from the process of Europeanisation is an open question for the future.

Note

Earlier versions of this chapter were presented to the COST A12 Rural Innovation Working Group I 'Theoretical Issues' Seminar (28-29 May 1999, Budapest, Hungary), to the ESRS XVIII Congress (22-25 August 1999, Lund, Sweden), and to the 'Ten Years After' Conference of the Hungarian Sociological Association (23-25 September 1999, Budapest, Hungary). I want to thank Michel Blanc, Gilles Allaire, Krzistof Gorlach, Leo Granberg, Lutz Laschewski, Tony Varley and Raul Comes Lopez for helpful comments and suggestions.

12 The role of associations in local development – the case of Garfagnana*

GIORGIO OSTI

Introduction: a three-legged development

The aims of rural development have long been identified as coupling the conservation of natural resources with the economic well-being of local populations. It has appeared that this dilemma can be resolved by so-called 'sustainable development', a concept that has informed numerous projects. A third component, termed 'culture', 'identity' or 'sociality', has recently been introduced. It is now generally agreed by scholars and institutional actors that development must include this third dimension as well. However, the notion carries a variety of meanings. When conjugated in terms of cultural identity it may give rise to paradoxical outcomes: for example, exaggerated political localism or merely tourist-oriented folklore It appears more fruitful to conceive it as 'sociality', as co-operation, as a capacity for social aggregation. Put in these terms, the emphasis is more on how a community stays together, how it takes decisions, how it manages to solidarise, rather than on its mere identification in terms of certain shared features (the Parsonian approach).

If the third 'leg' of development is conceived in this way, associations become an empirical referent of prime importance. These are entities that in formal terms are relatively simple (private, democratic and non-profit), but at the sociological level they display a wide variety of forms, goals and functions – an intriguing and often impenetrable object of analysis.

This chapter will delineate a theoretical framework in which to incorporate the role of associations in development. This framework is then used as a schema from which to derive a series of indicators. Finally, a specific case study is examined in order to determine whether the analytical categories are able to depict the variegated dynamics of local associations.

Association and development: some theoretical notions

Three approaches to the study of associations seem of particular relevance when endeavouring to grasp the role of such organisations in development.

The first derives from de Tocqueville and Putnam, the second from Gramsci and the institutionalists, and the third from scholars of the so-called 'third sector'. This framework certainly does not include all the schools and tendencies, but it provides a point of departure for the gathering and analysis of empirical data.

For Putnam (1993), the economic well-being and good administration of a community depend on the richness of the associative fabric, and on the fact that it is scarcely hierarchised. What matters, in fact, is not solely that there should be numerous associations or numerous members of them; it is also important that their internal relationships should be democratic and fair. Associations with a markedly vertical hierarchy, associations dominated by influential individuals, associations tied to churches with stable hierarchies (especially the Catholic church) do not perform their primary task of creating the climate of reciprocal trust, the spirit of co-operation, that Putnam calls the 'social capital' of a community. According to Newton (1996: 16), a linking theme runs from de Tocqueville to Stuart Mill to Putnam whereby the causal connection is mainly 'from joining to trusting'.

In the view of Putnam and others (e.g. Fukuyama 1996), there are at least two assumptions of importance to the theme of development. The first is the one just outlined – being able to associate freely and fairly generates trust in society. The second is that the trust thus created is transmitted to the institutions, to impersonal relationships, to large organisations. Mutti (1998a) argues that Putnam starts from the implicit principle of the 'overflowing glass'. When associative interactions exceed a certain threshold, some of them spontaneously overflow to the institutions and extraneous organisations. Metaphor aside, the principle is clear: a person accumulates trust within the association such that it spills over into wider society. The sum of all these experiences generates generalised trust.

Putnam has been repeatedly criticised for employing a one-way causal explanation: trust, he maintains, spreads from associations to the institutions, but he does not consider the process in reverse from the institutions to civil society (Mutti 1998b; Newton 1999; Ramella 1997, Tolbert *et al* 1998; Trigilia 1999; Warner 1999). This criticism stems from the observation that public institutions, especially local ones, have acquitted themselves very well in facilitating efficient development processes. Frequent reference is made to the administrations of Third Italy, particularly those in areas of 'red' subculture, which have proved able to aggregate political demand and to co-ordinate competing interests (Floridia, 1994). However, this analysis does not concern itself with the specific role of associations. Predominant, in fact, is the idea of an 'institutional network (political parties, Church, interest groups, welfare, cultural, recreational facilities) co-ordinated by the dominant [political] force which also controls the local government and links with the central political system' (Trigilia, 1986: 47). It is the *integration* of the local system that is the true reason for the economic-social development of those areas.

This organic model can be traced back to Gramsci, and in particular to his notion of hegemony (Jessop 1997). Gramsci conceived of a close connection between civil society and political society. The one could not be assimilated into the other, but both had to be included in a global strategy to conquer society. The capitalist society described in Gramsci's writings is able to perpetuate itself precisely because it controls and colonises civil society. A political movement intent on eliminating capitalism must give priority in its action to civil society, the arena in which the values that legitimate the social order are produced. Gramsci was aware of the autonomous capacity of associations to create their own space for action.[1] However, this was and must be closely bound up with the overall political project of changing society. In other words, the function of development is to unite the associative and political dimensions into a single organic model. This synthesis, however, does not take place at the structural level, nor even at the strictly political one. Rather, it comes about at that of *cultural direction*, or in other words, the capacity to hegemonise civil society and the values that underpin it (Bedeschi 1997).

Therefore, while the model *à la* Putnam envisages a direct causal connection from civil society to the political institutions, the Gramscian model sees a closer interdependence between the two: if development is to come about, associations and institutions must interact closely in pursuit of an overall political project.

This model seems able to account for numerous cases of local development where civil society and political institutions were deeply imbued with a single and dominant 'central image of the world'. The latter did not so much operate at the level of contents (specific development goals) as provide a communitarian cement – as the enabler of exchanges, and in the final analysis as a generator of trust (Bagnasco 1999). Now that such political homogeneity seems to have disappeared or diminished, at the more profound level of belonging (Ramella 1997; 1998) the problem arises of understanding the role that associations can play in creating development.

Onc answer, again on the basis of Gramscian philosophy, derives from the concept of *civicness*. In Putnam's case, it is 'natural' for the associative *spirit* to spill into the political one; it is this that cultivates the sense of shared well-being which overflows to society at large. Instead, according to the notion of civicness many associations are anchored in conservative, clientelistic or classist political models (Clarke 1998); some others still function as instruments of social distinction – that is, they serve to signal the privileged status of certain social groups (Ramella 1994; Trigilia 1995). No automatism here, therefore; rather, close attention should be paid to analysis of the functionality of civil society to reactionary political goals – as Gramsci pointed out in his time. The focus should be on the contribution of associations to *democratisation*, to those specific actions that positively affect the public good but without colluding with or depending on political parties or public institutions. The ideal type of this role is the environmental association, which is by definition concerned with the common good (Pellizzoni and Osti 1999) and has proved effectively able to balance public participation with independence from political power (Diamanti 1995; Ramella 1995). This function is not automatic, however, either for environmental associations or for ones similar to them. The thrust towards greater democratisation depends on numerous factors, most notably political culture, a conception of the common good, a value set on justice. Some associations therefore possess a specific cultural humus which enables them to affect the mechanisms of development, and especially to obtain further increases in participation and decision-making by citizens.

Criticisms have been brought against this conceptualisation by a number of third-sector theorists, who expressly argue that the Gramscian view of civil society should be discarded (Donati 1997), given that it envisages the primacy of the political dimension.[2] The starting-point of these criticisms is precisely civil society and third-sector organisations. This *incipit* is not coincidental: it is an assertion of the 'originative' and 'original' nature of associations (Maccarini 1996). The latter should not be interpreted on the basis of the market or of the institutional system, but in the light of their inner 'vision'. Otherwise functional interpretations will predominate: the third sector viewed as a means to reduce transaction costs with respect to the market, and as a means to reduce the costs of social services with respect to the welfare state. Instead, associations are examples of originative sociality which should be valued as such, without reference to functionality or, worse, control by the public authority. The development of the collectivity is therefore conceived as the liberating of the enormous self-managed potential of base groups free to establish, or otherwise, relationships with the public sphere. Hitherto, cultural obstacles have hampered the full deployment of this base sociality, which is expressed not only as aggregative capacity and mutuality but also as a response to concrete problems of society. It is for this reason that associations may turn themselves into the deliverers of important social services, at which point a dialectic begins with the public authority, and the associations participate on an entirely equal footing in the definition and management of common goods (cf. Scaglia and Fazzi 1999).

This position has many features in common with Putnam's. For both theories, development starts from the association. The differences derive from the cultural backgrounds of their proponents, although curiously they conduct their research in the same area (Italy). For the de Tocqueville-Putnam school of thought, associations are directly connected to institutions. The former certainly precede the latter, but they then merge in a climate connoted by trust, so that there is no contraposition between the associative and political dimensions. According to the third-sector theorists, associations must emancipate themselves from political control. Institutions are therefore, at least historically, an adversary against which associations should defend themselves or at least mark out clearly-defined boundaries. This position is therefore closer to the Gramscian one, but only insofar as they share the same historical context. In Europe, associations freed themselves from the absolute powers of the national monarchies in the 19th century. They arose as proponents of greater democracy and the protection of emerging social groups (the bourgeoisie and the working class) in opposition to the constituted powers. Freedom of association was a conquest wrested from royal absolutism. There was, therefore, a tradition of state control from which associations progressively had to free themselves. Accordingly, one may talk of a conflictual view of the relationships between civil society and the state shared by both Gramscian theory and third-sector scholars. However, this view splits when one examines historical events in detail, especially Italian ones.

Italy has two distinct features that make the relationship between civil society and the state particularly problematic The first is the 'Roman question'[3] and the Catholic Church's dominance of charitable works; the second is the control of associationism by the mass political parties, which have exploited them to build consensus – and this also applies to the parties on the left, especially the Communist Party. The criticisms now interweave in complex fashion (Zaninelli 1996), but apparently most persuasive is an interpretation *à la* Putnam which sharply distinguishes between democratic associations free of hierarchical structures and associations serving the purposes of the Catholic Church, political parties, or powerful economic lobbies.

These three approaches provide criteria with which to measure the contribution made by associations to development. Putnam posits two constraints: democracy and freedom. If associations are democratic and free from outside influences, they exert a positive effect on the collectivity as a whole. However, Putnam does not clarify the transition from the aggregative and mutualistic phase to one in the interest of the collectivity. But given his liberal approach, it is likely that he concentrates on individual members of an association, on the individuals who develop a civil conscience through associative experience. This sensibility automatically directs their interest to public affairs, and if they are administrators, to serving society more disinterestedly and efficaciously. An important aspect is multiple membership, as Simmel pointed out. The necessary condition for the creation of social capital is the pluralism of associations and relative memberships. The more an individual belongs to non-overlapping circles, the more his/her self-interest is diluted, the broader becomes his/her awareness of problems, and the less tensions and occasions for conflict arise.

According to the Gramscian institutional approach, a precondition for development is the formation of associations predicated on citizenship. Not all associations are like this, for their nature depends on the historical period, the area in which they operate and on their type. An association, too, must be aware of the existence of a political project for change to be pursued jointly with the public institutions. The primacy should therefore be assigned not to individuals but to the association integrated into a larger political project.

According to the third-sector approach, development will be the outcome of a robust ability to organise themselves by base groups: it will depend on their degree of autonomy from the state and from the market. The prime factor is therefore not political but the aggregative power of groups, their ability to develop reciprocity relationships. These relationships also provide answers to many of the problems that afflict citizens (mutual services). The free and independent association thus ensures development, or better its 'share' of development. However, in this case too it is unclear how the transition to the public sphere is achieved.

Perhaps only the Gramscian approach clarifies the matter (a sort of integrated development). However, in this case the primacy given to the political clashes with the self-image of associations. In fact, they do not envisage any political end, even an indirect one; indeed, they are suspicious of any form of politicisation. It is therefore necessary to find other criteria with which to define their role in development, drawing on the three theoretical approaches outlined thus far.

This role can be conceptualised as involving a process by which associations *come out of themselves*. The emphasis should be on a positive act of *exiting from mutuality*. This is most evident in voluntary associations, when their purpose is to provide an impartial service to third parties. In this sense, although they do not have political ends, they place themselves at the service of the collectivity. Mutual help associations – those, that is, set up to deliver a service to their own members – create development if they move away from this practice and *also* undertake initiatives for non-members. In fact, it is not sufficient to join together, albeit democratically, to have a social purpose. It is not enough to exist or to pursue a certain social end to be an agent of development. Also required is explicit action intended to aggregate and mobilise third parties. Moreover, if rural development means territorial re-equilibrium, then associations create it when they are aware of the disparities and target their actions on less advantaged areas. Otherwise, they will be confined to a mutualistic perspective which is certainly positive but not explicitly directed at development.

A further aspect of the Gramscian theory is the concept of *mediators*, or 'third actors which guarantee impersonality, communication and exchange among outsiders, the elaboration and recognition of the generalised other' (De Leonardis 1997: 184). Associations and their members should refer to 'third' democratic institutions responsible for public well-being and transparent in their functions (Donolo 1999). The self-transformation of associations is thus viewed with reference to external institutions which adopt correct procedures. Also 'autonomist' analysts of the third sector remark on this aspect. An association contributes to the public good if it agrees to be validated by a third actor. This is the significance given to the social balance-sheet and to certification procedures (Maccarini 1999). The problem thus shifts to the origin of, and the control exerted by, these third entities. According to the autonomists, the latter should spring from the sphere of the associations themselves. They should remain independent of any state control and should 'present' themselves on neutral ground accessible to all. For the Gramscians, what the nature of these entities should be is not clear, although it is evidently necessary that they should be effectively impartial and general. Ultimately, the dilemma centres on the role of the state: for the autonomists it is one of the many actors in play; for the Gramscians it is still the reference standard for any partnership procedure.

In conclusion, the criterion of self-transformation by associations seems to take material form in universal services, in procedures to publicise their work, in membership of third institutions. The crux of the matter is *thirdness*: there are those who do not contemplate it (Putnam), those who see it as the outcome of self-regulation by associations themselves, and those who see it as membership of an external organisation endorsed by the public authority.

An analytical framework

The insights furnished by these three approaches to the study of associationism may now be used to construct an analytical framework which, as far as it is possible, includes the many ambivalences seen thus far. Associations can be distinguished into four types, according to their function in rural development (Table 12.1).

The first type consists of associations which cater to the needs of individuals by organising services and practical activities (cf. Parodi 1998). If these associations are to be effectively able to generate development, they must respect the criterion of the universality of their services, or at least of extremely general access to them. Otherwise they will give rise to discrimination or social segmentation. Other forms of this deleterious tendency are use by the public authorities of non-profit organisations to reduce costs, and the commodification of voluntary work and associations in the market (Howlett and Locke 1997).

The second type of association has a recreational function. It provides relationality for its own sake, relationships with others for the pleasure that derives from them; and its action entirely lacks instrumentality. Simmel's (1997) sociability is an example. In this case the task of the association is to satisfy the natural human need to play, to be entertained, to set everyday concerns aside. All recreational associations should perform this function. The range of activities organised is extremely wide, although each area has its own specific traditions. Recreation is often interwoven with religious activities, especially in rural areas. Typical is the festival of the local patron saint, when the sense of belonging to the community is fostered. The ambivalence of this function is more subtle: it concerns Bourdieu's (1983) 'status distinctions'. Bourdieu has argued that non-instrumental activities are in fact performed in order to achieve a personal status unobtainable from the prestige of one's professional position.

Table 12.1: Relations between the development of rural areas and associationism

Functions	*Ambivalence and degeneration of functions*	*Indicators*
1) associations deliver innovative, universal services with high realisational content (*service function* → new needs)	total delegation by the public authority; services below cost, exploitation unfair competition	delivery of universal services, free or at popular prices (sports, cultural welfare services
2) associations develop sociality, play, pleasure of being together (*ludic function* → human expressiveness)	opportunity for 'status distinctions' and therefore social segmentation	organisation of festivals, fairs, fêtes, excursions, games, exhibitions, singalongs …
3) associations create generalised or systemic trust associations of (*economic function* → lower transaction costs)	particularism, clientelism, lobbying, corporativism (solidarity but within small and closed groups)	multiple memberships in associations of different kinds
4) associations facilitate participation by the citizen in the public sphere (*political function* → opinion, decision, control)	purely 'cognitive' mobilisation (without concrete experiences, without relationality, mass-media oriented)	participation in public arenas, presentation of and observations on public projects, presence in the press, presentation of projects

The third type of association is as envisaged by Putnam in that it serves to fluidise social relationships. Its purpose is to facilitate horizontal exchanges, the multiple memberships which then generate trust between private and public agents. The indicators of this function can be discerned at the individual level by measuring the overlaps among several associative memberships, and at the aggregate one by the degree of trust in economic and political-administrative systems in which the associations operate. In this case the ambivalence arises from the gaps among the networks, from the fact that, although there are multiple memberships they largely overlap, giving rise to outright power elites. At bottom, the ambivalence is again the same: associative membership is treated as an exclusive resource which gives access to privileges.

The fourth kind of association has an eminently political function, one close to the Gramscian tradition. In this case, the association performs its role in at least four ways:

- by providing political training for citizens and future administrators;
- by acting directly in the political arena as a fully-fledged political party, as a pressure group, as a consultancy and planning body for the public administration;
- by seeking to protect public goods;
- by facilitating indirect participation in the public sphere and by monitoring respect for the law. This diversified function may degenerate into institutionalisation, the formation of organisational 'iron cages', pure image-enhancement through exploitation of the media, or otiose rhetoric and intellectualism.

These distinctions among the four types of function are analytical: an actual association may perform more than one function, some perform all of them.

A survey of the associations in Garfagnana

The research described here was carried out in 1998-9 in Garfagnana, a valley located in the province of Lucca and lying between the Apennines and the Apuan Alps (Tuscany). By 'Garfagnana' is meant both an administrative unit and a cultural one. The valley, in fact, is a Mountain Community (public body) and its residents identify themselves with that geographical delimitation. Like most of the province of Lucca, the valley has a 'white' political tradition, unlike the rest of the (red) region of Tuscany. It has around 30,000 inhabitants and sixteen communes. The area was long a poor one with high rates of emigration and a declining population in the valley bottom (Andreoli and Tellarini 1989). Only in the last two decades has the population re-stabilised, with immigration slightly exceeding emigration. The valley's economic marginalisation has been reduced – as in many other peripheral areas – by a wide variety of factors, but most notably the return of emigrants to the valley, the development of green tourism, which has attracted foreign visitors, a large number of commuters to and from the lowlands, and the recent availability of good pensions for the retired. As a disadvantaged area, the valley has been targeted by policies for territorial re-equilibrium implemented by various institutions, from the regional government to the European Union. An important economic role is played by agriculture which, unlike in many other valleys, has maintained its vitality by marketing typical local products (Rossi and Rovai 1998). The most serious problem is the small upland communities, where the population is very elderly, commercial and public services have been curtailed, and there are no productive activities apart from agriculture.

The study was also exploratory in character. It was largely an occasion to test theoretical notions and to evaluate indicators. In its first phase, the valley's associations were surveyed on the basis of the official lists. This was followed by some forty interviews using a semi-structured questionnaire administered to associations in the valley's largest town and in other smaller communes, some of which were decidedly in a disadvantageous position. During its third phase, the survey concentrated on a number of associations found to typify (empirical types) ongoing development processes in the valley.

According to the official lists and direct research, there are around two hundred associations in Garfagnana (Paolicchi and Biancheri 1995), which means that there is an average of twelve associations per commune and one for every 150 inhabitants approximately. It should be pointed out that many of them are no longer active or only meet once or twice a year. Nevertheless, there is a surprising number of them, which gainsays the crude view of associationism as a typically 'urban' phenomenon (Osti 1993).

In Garfagnana, as in every other part of Italy, the majority of associations are sports clubs; even the smallest hamlet usually has a football team. There are also large numbers of choirs and brass bands, blood donor associations, and associations providing transport for the sick or handicapped (*Misericordie, Pubbliche Assistenze*). The calendars of the events scheduled in the valley's various villages yield useful information: those for the summer of 1999 listed more than 250 events between religious festivals, sports competitions, theatre performances and concerts, cultural excursions and lectures on the valley's history and nature.

Classification by type of association is always rather difficult. The majority of those in Garfagnana are recreational (19 out of 40), although this category also includes religious or more strictly cultural activities (6). It is easier to identify sports clubs (7) and 'social-welfare' associations. There are nine of the latter: four of them are blood donor associations, three are *Misericordie*, one an association of voluntary hospital workers, and one an association working with drug addicts. Two associations are parish choirs. The parishes act as the co-ordinators and organisers of numerous social initiatives, but not all of these are constituted as autonomous associations. Finally, two associations provide civil defence in emergencies, and one of them runs a facility known as *uso civico*.

The settlement structure of the valley is widely dispersed among hamlets and villages. One purpose of the survey was to verify the presence of associations in even very small communities, and it was found that these too have associative forms, known as *comitat i paesani* (village committees).

Memberships are rather high: an average of 56 persons. Nine associations have no more than 20 members, and five had more than 100. Of the latter, two have premises in the most populous commune in the valley. Of the other three, one was set up by emigrants and two are sports clubs. It is easier to identify the extremely small associations. These are of two types: the *comitati paesani*, and associations with highly specific cultural interests (one of them, for example, organises performances in the local marble quarries). The *comitati paesani* have very distinctive features: besides being small, they are informal, with no outside connections, and they usually organise one or two events a year, in particular the festival of the village's patron saint. However, they have a considerable capacity for organisation and mobilisation: in some cases, they co-ordinate even complex undertakings like the restoration of churches and historic buildings.

Apart from the *comitati* there are no other associations with such a high degree of informality and isolation. Some rare associations declare that they are not enrolled on any public register nor have any relationships with a parent-association. The others belong to national or regional umbrella organisations.

A further aspect which sheds light on the social impact of associations is their age. All those in Garfagnana have been founded very recently. Only three of them were established before 1970: two brass bands and a blood donor group. Ten were created in the 1970s, mainly on an informal basis. But the majority have been set up since 1980. What does this imply? Firstly, that associationism existed before Italy's 'industrial take-off' (Tommassini and Sardelli 1995) but it consisted of only a handful of activities. Secondly, associationism has exploded with growing affluence in general, and with the decline of emigration for Garfagnana in particular. This in some way contradicts the explanatory hypothesis of this study. It is the achievement of a certain material well-being that triggers the proliferation of associations of every kind. The latter is therefore more an effect than a cause of development. However, this interpretation should be treated with caution, and in any case it does not have absolute value, given that some (a few) associations existed before the valley's economic revival.

The activities of the valley's associations have already been described. Highlighting the many-faceted nature of these associations is the fact that only six of them do not declare a second activity, and only 22 do not declare a third one. The most frequent combination is 'sport plus village festival' and 'social activities plus civil defence'. Of considerable interest is the combination of sports activities with management of sports facilities. In this case, the association shows a certain maturity by delivering a permanent service to non-members as well. The formula is simple: the commune stipulates a contract with the sports club, which runs the facility in part for its own members and in part for other groups. Assuming the management of sports facilities serves various purposes, not least of which is ensuring a constant flow of funds to finance the activity that really interests the association. In more infrequent cases, there is a desire to create links with other small associations in different hamlets and villages.

A weakness in the survey that must be mentioned concerns financial flows. Associations are reluctant to give information about their budgets. There are several reasons for this: some associations seek financing only from time to time; others argued that the figures were trifling; more often interviewees declared that they did not know anything about the association's balance-sheet. It was not the aim of the research to investigate this particular aspect, yet the vagueness of the answers signals the precariousness of these groups and, perhaps, a misunderstanding of the economic dimension. It is believed, in fact, that being non-profit means that precise book-keeping is not required and that inflows of resources are not predictable. Consequently, there are – in terms still to be verified – 'gaps' in the treatment of the finances of associations.

The only aspect of finance on which data were collected, albeit only approximately, was the percentage distribution of revenues. Of the 36 associations that furnished information, four received more than 50 per cent of their funding in the form of public grants, and a further ten received more than 50 per cent of their funding from the public authorities but on contract to deliver services. The remaining nineteen declared that they did not receive any public contributions in the form of either grants or contracts. The ten associations 'dependent' on public funding are almost all sports clubs which also ran sports facilities. The financial relationship between an association and the local authorities tends to be 'polar': either no relationship at all or ample coverage.

The frequency of activities is another problematic variable, in this case because the situations are highly disparate. Firstly, very few associations meet once a week: only two declared 52 meetings a year; some had meetings every fortnight (5); the majority once a month (13); the others even less frequently. It is evident that a frequency of one meeting a month suggests that rather simple tasks are addressed, or that there are so few members that frequent formal meetings are not necessary. This impression is borne out by other data. When asked how many members were regularly involved in activities, the interviewees were very generous: the majority of members were classified as 'constant'. However, when they were asked how many hours a week members devoted to the association, matters were different. Almost half the associations did not reply. It is evident that the figure depends on the activity concerned. Blood donor sessions were held 3 or 4 times a year and presumably convivial meetings were also organised. In terms of type of activity, it was the sports clubs that indicated weekly hours of involvement, and even of substantial extent (around 10 hours). What matters is the type of activity and, especially, the combination of activities. For example, the blood donor group also undertook initiatives of a social nature.

In order to clarify the contribution of these associations to development, they were asked about the beneficiaries of their activities. Almost all of them replied 'the community', and almost none of them mentioned individual categories of persons like the elderly or the disabled. The reply is indicative of the associations' perception of their contribution to local society. The community was cited because these were initiatives that were largely non-selective, especially those of recreational associations. The communities in the valley do not display marked class or social divisions, so that the initiatives undertaken by associations are effectively universal. Garfagnana is not characterised by a strong polarisation of landed property, with large landowners on the one hand, and farm labourers on the other. This factor undoubtedly makes the community and its associations more cohesive.

In order to evaluate the contribution of associations to the community, the interviewees were asked to indicate the 'structures' in which initiatives open to the public were organised. Seventeen associations indicated a sports facility, four indicated a cultural facility (a music school or a theatre), three the parish hall, two the bar, one the playground, and one the clubhouse. This prompts three observations. Firstly, the majority of associations indicated a structure or place where they can hold events open to all. Secondly, this place was the sports facility, so that sport is mainly a social and democratic activity.[4] Thirdly, the most exclusive form of association – the private club – was almost non-existent, which confirms the scant class divisions in the associationism of this valley. It is more difficult to assess selection on religious grounds. In this case, however, little mention was made of the 'parish'. There are no religious minorities, and lay/Catholic conflict appears, as in the rest of Italy, less fierce than it used to be.

Next analysed are indicators of the relationships of associations with other bodies, both public and private. In synthesis the results are as following: the commune is the entity with which associations most frequently have contact, followed by the umbrella organisations and the federations to which they belong, and other associations engaged in the same activity. Of some importance is the role of the Mountain Community and of the provincial administration, although at a secondary level. Finally, there are few contacts with the health authorities and the regional administration. Almost non-existent are contacts with banks, the university and the nature park administrations. Only in four cases do associations have contacts outside the valley, while there are almost no relationships with specialised agencies. It is therefore evident that most of these associations have a very simple structure and a local network of reference.

Almost all the associations felt that they were appreciated by the community in which they operated, and they believed that their work was seen as valuable. To the question of whether there were forms of discrimination against those wishing to join an association, the unanimous reply was that there was no such discrimination. In a couple of cases, mention was made of episodes when people had been excluded because of their political ideas. Demographic differences were more evident. In the 40 associations surveyed, female members accounted for 30 per cent of the total. Women outnumbered men in only three associations, which were social-welfare associations in the valley's chief town. A curious finding is that men predominate in the *Misericordie*, which also operate in the social-welfare sector. Finally, there is a marked predominance of males in the sports clubs and in the blood donor groups. Members aged over 60 amounted to 13 per cent of the total, while in Garfagnana the population of the same age was 23 per cent (figure for 1991). Both membership by sex and membership by age do not differ greatly from the national averages (IREF 1998: 149). However, whilst in the case of sex there seems to be a certain bias against women, in the case of the elderly this is much less apparent.

The overall relationship of the associations with the political-institutional system should yield further information on their role in rural development. It has been seen that, according to the theories analysed earlier, much depends on the fine boundary between autonomy and integration. It is therefore necessary to verify whether there exists a sort of critical participation in the public sphere. Before doing so, however, one must determine the level of democracy internal to the association itself. In this regard, the replies to the questionnaire were few and disparate, and it is difficult to give a unitary interpretation of them. What is certain is that some associations have problems apparently due to two main factors: the excesses of those with greatest commitment (personalism, narrow-mindedness, opportunism, and so on) and the demands of those who participate little but nevertheless want to have a say in running the organisation. More than this is difficult to determine, not least because of the natural reluctance of interviewees to talk about problems in which they are personally involved.

We may now turn to external political features. According to the interviewees, the relations of their associations with public institutions were distributed as follows:

10 cases	'no relations'
7 cases	'we only formally attend some meetings'
19 cases	'exchanges of information and services'
1 case	'we have frequent relations but with different positions and opinions, sometimes conflictual'
3 cases	no reply

When asked to sum up their relationship with the public institutions in one word, three associations came up with *protest* (collecting signatures, sending letters, events), none mentioned *lobbying* (pressure applied to parties, agencies, officials), thirteen *education* (training courses, educational activities, conferences, lectures, etc.) and fourteen *testimony* (organising events that speak for themselves). The two classifications depict a relatively clear-cut situation: a good half of associations work closely with the public authorities but mainly on educational, non-political and symbolic terrains.

Integration with the political sphere also comes about through the presence of members with positions in the public administration to which they have been elected or appointed. At the time of the survey there were fourteen associations who had members with such positions, and six had had them in the past. Only in six cases was the present or past absence of such members explicitly mentioned (eleven associations did not reply to the question).

Eleven of the forty interviewees believed that their association was a training ground for future managers in the public administration. The range of associations that gave an affirmative answer to this question was very wide, but the replies were relatively uniform. Associationism trains people to deal with common problems, it informs them about how organisations work, and sensitises them to the needs of others.

These results clearly indicate that the majority of associations have had a negative experience of politics, which they have not yet digested. Almost all the interviewees said that the time when the parties exploited associations for electoral purposes was past. They also believed that all political goals were unacceptable, and they thereby overlooked those, like education, which an association may legitimately pursue. In other words, the vehement reaction against interference by the political parties has induced many associations to reject any form of political activity.[5]

A further indicator is the attitude towards the problem of protecting public life against corruption, misgovernment, patron-client relationship, or inefficiency. Here the most frequent reply was 'the association cannot do anything, these are problems beyond our control' (16 cases). But a substantial number of interviewees also chose 'the association can co-ordinate with other associations in joint action' and 'the association can work exclusively in education and sensitisation' (10 cases each). The item that received least preferences was 'the association can place its members in the public administration' (4 cases). The overall result in part replicates what has already been found: the majority of associations sharply separate themselves from the public sphere, some of them pursue educational goals, and very few consider a role within (or against) the public institutions. A totally new aspect is co-ordination among associations as an instrument of political action.

Conclusions

The foregoing analysis suggests a number of considerations with regard to the four functions in Table 12.1.

- There appears to be little risk that associations give rise to social discrimination. The only form found – between males and females – concerns society at large more than the area surveyed, although this does not make it any less serious or to be overlooked. Otherwise one can accept Putnam's thesis of a varied and diffuse associative fabric without explicit impediments to internal democracy. This fabric is indubitably a factor for social well-being which fosters small-scale, local trust in the valley. This bears out the hypothesis of the play and socialising function of associations. Not to be ruled out is a connection with the economic well-being achieved by the valley in recent years. However, the results do not point to any causal relation from associations to development. Rather, it seems that economic stability, political capacity to draw on external resources, and associations have all grown together.

- The universal services delivered by associations are of two types: first, non-material ones of general benefit like culture and entertainment; second, social and sporting services which require more targeted delivery to specific persons. As regards the former there is little to say: they are generally free services available to all which hinge on the identity of the village (festivals, fêtes, and so on). Some cultural associations like dance and music schools are more exclusive; sometimes they border on non-profit organisations in that they have little associative experience behind them and charge high rates. There are a number of such associations in the valley, and they are the most distant from the principle of the universal service, or one provided for marginal social groups and areas. This differentiates local society rather than contributing to its development. As regards welfare and sports associations, the universality of their services seems guaranteed.

There was no evidence of 'self-established protocols', according to the 'autonomist' approach definition, by which an association regulated its action and made its work public. In small communities, communication with the public relies on spontaneous mechanisms. It cannot be taken for granted, however. For example, there is much work to be done on the economic transparency that every association intending to have public accountability must put in place. In several cases associations had opted for contracts with the public authorities, as exemplified by the covenant established between *Misericordie*, *Pubbliche Assistenze* and the Tuscan regional government for transport of the sick, urgent or otherwise (Giovagnoli 1999). In this case, close agreement had been reached on the provision of a universal and transparent service by the voluntary organisation, and on co-ordination funding by the public authority.

Development as attention to marginal areas (or to factors which cause imbalances in economic growth) does not seem to have much bearing on associations, which grow where there is already a certain level of well-being. It was evident that in the more isolated small villages it is very difficult to set up and ensure the survival of associations.

The sports clubs are the associations that most closely matched the principle of delivery of a general service, and which are sufficiently widespread for public tasks to be assigned to them. For these associations, the process of functional and financial integration with the public authority is at an advanced stage. The reference is the commune. When this is very small or very scattered, problems of the right size for facilities arise. Attempts to co-ordinate the sports clubs have to date been largely unsuccessful.

Some associations have responded to rural fragmentation with versatility. In areas with scattered populations (and the progressive closure of public and private services), an association performing several functions is a valuable resource because it guarantees the flexibility and rapidity of responses. It should be borne in mind, however, that integration of services proceeds along separate lines: recreational-sporting on the one hand, health-civil defence-blood donation on the other.

- The results concerning public participation, according to the Gramscian perspective, are quite clear: the predominant concerns are education and example-setting. Some associations see co-ordination as a means to solve shared problems. There was general resistance against direct involvement in politics, either with the instruments typical of movements (i.e. protest) or with those of deliberative democracy (the creation of occasions to participate in schools, in the health service, in environmental protection, etc). As in other studies (Colombo 1999: 105), it again emerges that Italian rural areas have a scant propensity to create *ad hoc* spaces for participation.

A large number of associations seem determined to eschew politics. In general, this determination seems due to past colonisation by the political parties, and to scepticism that change can be achieved *of* or *through* the political system. Curiously enough, numerous local administrators are members of associations, so that the latter have many links with politics available to them. But this is not enough to bring about broad civic participation. Associations are probably still tied to the dichotomous view of either complete separation from the political system or complete identification with it. For that matter, opportunities for extra-electoral participation are limited, and there is no association in the valley – as far as one can tell – set up for this purpose.

Participation takes material form in *doing for all* rather than *deliberating on behalf of all*. The associations with a certain presence in the public arena are those which handle emergencies (health and civil defence). In this case, co-ordination with the public authorities is unavoidable, but it also requires constant appraisal of the choices most appropriate for the community.

- Ecological conditioning affects the valley. In the smallest settlements associative activity is almost non-existent apart from the football team. However, there are signs that associations are beginning to specialise (e.g. in combating drug abuse or in sports other than football) in the valley's largest inhabited centre. Demographic factors are influential, therefore, but not predominant. There are, in fact, instances in which the reverse is the case: associations with an international range of action in relatively small communes; associations in outlying communes which co-ordinate their work; unexpected cases of activism in small mountain hamlets. Nevertheless, ecological conditioning operates, and it seems that the federations have neither the ability nor the resources to 'cover' the most marginalised areas.

To conclude, the valley's associations are responding to the question of development at the fundamental level of *sociality*. A large number of persons is mobilised by small associations to create occasions to meet, entertain themselves, and help each other. Thus the path indicated by Putnam's approach seems most pertinent: the norm is many, small and unconnected associations, whose development function is implicit or linked to single members. The promotion of development as self-transformation, as the assumption of public responsibilities, is only of minor importance: there is some timid co-ordination among associations in the valley, some agreements with the public authorities, some direct management of services. The public figure of the association is suspended in limbo between the full and acknowledged assumption of responsibility and response to the purely expressive needs of individuals. In this regard, neither the Gramscian nor the 'autonomist' perspectives seem pertinent to the majority of associations: they do not seek to be part of a broad and integrated process of development, nor do they want to create their own co-ordinating organisations to deal with public institutions and firms on an equal footing. Future research must clarify whether this depends simply on the territorial scale - rural associations are necessarily more scattered - or on the fact that associations develop along lines different from those envisaged by the two theoretical models.

Research conducted on behalf of the Osservatorio per lo Sviluppo Rurale della Toscana, University of Pisa.

Notes

[1] Between 'the two great superstructural levels, that of civil society (that is, the organisms vulgarly called private) and that of political society or the State', Gramsci (1997, p. 12) introduces a middle way which rejects both the identification of the state with civil society (corporative conception) and their complete separation, which he maintains gives rise to economicism (Gramsci, 1997, pp. 277-81).

[2] The antecedents of this view are various, not strictly sociological, bodies of theory: the social doctrine of the Roman Catholic Church, associative democracy (Hirst, 1994), and sectors of ecologism (Carter, 1996).

[3] The breakdown of relations between the Papal State and the Kingdom of Savoy after the latter seized Rome in 1870. As a consequence Catholics were prohibited from participating in public life.

[4] Uslaner (1999) finds that it is sports clubs that create the most 'generalised trust' in that they educate people to compete according to shared rules as they should in public life.

[5] Substantial independence from the parties and the trade unions is also evident among young association members in the red areas of Tuscany (Ramella, 1998).

13 From agricultural to rural co-operation in France's Midi vineyard: network theory for a renewal of innovative collective action

YUNA CHIFFOLEAU

Introduction

Quality products are generally presented to French farmers as an alternative to mass agricultural production, which could help them face world economic competition and changes in European consumers' preferences (Allaire, Boyer 1995). This requires significant technological and organisational changes, which have to be tested and co-ordinated in the local production systems. Languedoc-Rousillon wine co-operatives, which represent 75 per cent of the wine production in this region and 90 per cent of vine-growers, constitute an appropriate sociological case study for this question.

The wine production sector in Languedoc-Roussillon has undergone dramatic changes which made it shift, within ten years, from:

- a dominant position in the regional economy and in the national wine sector, where vinification co-operative firms (VCFs) were engaged in mass production of table wine and were tightly interconnected in regional and national political networks, with a legal framework protective of their interests, to:
- a more chaotic situation in which former protections have disappeared and where technical and commercial requirements have dramatically evolved towards new qualities (Laporte, Touzard 1998).

Those ten years of 'quality revolution' enabled VCFs in Languedoc-Roussillon to secure an even stronger position in the commodity chain, due to new products known and traded world-wide, but simultaneously 'dysfunctions' (Crozier, Friedberg 1977) appeared in collective action that co-operative officials themselves recognise. Currently, even if many members understand the new context and its demands, many are not really involved in the 'collective project' (the rate of attendance in the Annual General Meeting is noticeably low). Above all, their relationship with the VCF is strictly economic, and that, according to the officials, impedes the organisation's capacity for innovation. In these conditions, members are either passive or ready to leave (Bajoit 1988), especially since new privately owned wineries, paying attractive prices for fresh grapes to process in wine, have appeared in the region. Co-operative managers are looking for solutions to these dysfunctions because the process of moving towards quality production is incomplete and its continuation depends on co-ordinated future investments by 'faithful' (Hirschman 1970) members.

The loosening of the co-operative tie in VCFs is, of course, a specific case of a general process that affects all kinds of association, and leads to a drop in political commitment (abstention in votes) and to a growth of consumerist individualism (Balandier 1989). Moreover, according to some economists, co-operative organisation does not fit with an 'individualist modern society' (Fulton 1995). If modernity brings new conditions for collective action, we will try to show that this lack of commitment is also linked to the evolution of VCFs themselves and not only to the 'corporate life cycle' (Mueller 1988). Of course VCFs were created at the beginning of the last century, in a context of protest movements and union demands that has disappeared (Martin 1994). This partly explains why members are now more apathetic, but the late evolution of VCFs, within the framework of the 'quality revolution', is also a key factor.

This chapter aims to show how research grounded in network theory, using an interactionist model of the co-operative organisation (presented in the next section), can reveal how such changes are dismantling the prevailing professional relations in the VCF. But it also highlights that the changes are simultaneously opening up new opportunities to develop relations between people afresh. These opportunities, analysed in terms of the model we propose, provide new theoretical and practical possibilities to imagine a reshaped VCF, adjusted to the current environment, thus introducing more general questions about renewal of 'innovative collective action' within productive organisations.

Theoretical background and method

An interactionist model to research collective action

A model of the VCF is presented here which is intended to be both heuristic and operational. It has been elaborated from the point of view of a comprehensive (Weber 1968) and interactionist sociology within which economic phenomena do not play a distinct part but are 'embedded' in social interactions (Granovetter 1985).

As may be observed in Languedoc-Roussillon, VCF seems at first sight to be an enterprise, acting at a commune level, marketing a finished product - wine - that it has processed out of raw material – grapes - supplied by the shareholders. Unlike some economists who look at the organisation as a combination of actors, rules and resources (Ménard 1990), we consider that the VCF is, above all, an 'action system' (Crozier, Friedberg 1977), or more precisely a set of interacting actors, built on a fundamental exchange: grape versus payment. In line with White (1992), society is theorised as a network woven by interactions that trigger and participate in the building of dynamic 'identities', individual or collective, which are stabilised through control.

- The first hypothesis is that members, i.e. grape producers, build 'knowledge for action', at least for grape production, within peer groups (Darré 1996). Interactions within the network of those producers, managing their individually-owned assets, tend to produce a collective identity: that of a co-operative seen as a producers' association.
- The second hypothesis (following White 1992) is that interactions between the managers of the collectively-owned assets (VCF officials and waged managerial staff) and other firms on the wine market or other institutions related to wine production (scientific and technical agencies, political agencies) tend to produce a different identity: that of a co-operative seen as an enterprise.
- Those identities are located on and evolve within two adjacent planes, connected through a specific domain, an 'interface' (Long 1989) where they meet on the occasion of the exchange 'grape versus payment'. Thus, the model contains three elements (two planes with fuzzy boundaries and an interface) belonging to different 'worlds of production' (Salais, Storper 1992).
- Such a situation, in which different 'worlds' (partly ignoring each other) meet, is construed as a specific type of collective action, 'action together' (Livet, Thévenot 1994), which requires control over interactions in order to persist. This control is the result of different types of co-ordination

(Orléan 1994), particularly through the establishment of technical and organisational rules (Coleman 1990), such as rules for grape quality appraisal or for payment computation.

- Collective action is applied to different objects (material or contractual). In the VCF case, the two main objects of 'action together' are the grape itself and the co-operative project, materialised in the different types of wine that are aimed at. The VCF rules are based upon the 'qualification' (Boltanski, Thévenot 1991) of those objects and should result from negotiation between members, according to the principle 'one member, one vote', whatever their economic or political weight (Giroux 1992). In fact, a great number of members abstain from voting and the rules are often decided by the managerial staff alone, more or less in collaboration with the Board of Directors (Nicolas 1993).
- In concrete terms, those rules affect the different types of actions (besides the 'action together') of grape producers (including non-members) that relate to the VCF: individual action, like grape production in one's own plot, or 'common action', like technical dialogue or work exchanges between peers, may change along with the dynamics of those rules.

Research question and working hypotheses: an approach through networks

At the end of 1996, some VCF officials in Languedoc-Roussillon were worrying about the apparent fall in members' commitment, most of whom no longer participated in annual general meetings. They worried also because members did not always follow technical advice aimed at promoting a new model of grape production.

Some 'free-riders' (Olson 1965) were faking acceptance of the rules, for example by mixing, before delivery, and after they had been assessed on the plot (see below), different qualities of grapes to improve the final payment. In addition, growing numbers of members were searching for another model of grape production. Yet to deal with new market requirements, managers of collectively owned assets need to summon a real 'collective action for innovation' that revolves around a quality-oriented production project. For that reason, they called for help from researchers.

Two working hypotheses framed our research, which started in Spring 1997. First, we considered that 'collective action for innovation' must be based on a relative consensus between members and managers about the objects of collective action: the two 'identities', as outlined above, must be comparatively near. That is to say, members and managers must share, at least implicitly, an 'institutional proximity' (Kirat, Lung 1999) about the objects of collective action, that is *'a common space of representations, rules and thinking models'* that makes the productive project workable. Nevertheless, within the large and heterogeneous populations that characterise wine co-operatives, we assume relations to objects may vary, thus inducing an 'institutional distance' between people which threatens the emergence of a common identity. More precisely, on the basis of the networks approach by 'equivalence' (Lorrain, White 1971), we propos to distinguish 'strategic blocks': sets of individuals who share a 'similar relation' to the objects of the 'action together'. We assume these people do not necessarily have relationships between themselves: the 'strategic block' is not seen as a 'group', in the sense of an interacting collection of individuals, but as a set of 'equivalent people' who share an 'institutional proximity' relating to the objects of action. Moreover, this does not necessarily mean the people involved all share economic and/or geographic proximity. In our view, the capacity of a group to develop a 'collective action for innovation' depends on the institutional distance between the 'strategic blocks'.

Second, following an interactionist perspective (Mead 1932), we assume there must be some interpersonal relations between members themselves on one hand, and between members and officials on the other, in order to keep a moderate 'institutional distance' about the objects of collective action between the different 'strategic blocks'. Research on collective action highlights how social networks take part in the regulation of organised action: interactions encourage the development of a 'common knowledge' base for collective action (Lewis 1969). At the same time, they promote learning of both the rules of the organisation and the work procedures (Ponssard 1994). Finally, they give rise to the development of individual practices useful for the common project, or, at least, compatible with its main orientations (Lazega 1994). Furthermore, relationships contribute to developing and keeping trust within a group (Orléan 1994) and encourage the exercise of a collective pressure that limits recourse to 'heroic sanctions' (Coleman 1990).

Finally, 'collective action for innovation' may rely on two interdependent items: relations to objects of collective action and interpersonal relations between people featuring different relations to these objects. We supposed therefore that the dysfunctions about which officials complained in 1996 could be related to recent changes in the relations both between people and between people and the objects of action. Moreover, prompted by the officials themselves, we assumed that changes were consequences not only of general social evolution but also of the 'quality revolution' in the Languedoc wine sector.

Method

Our method was the result of combining three elements:

- *a twofold-analysis of 'institutional proximity/distance' and 'professional relationships' between members*: 'professional relationships' are the exchanges related to the vine-growing activity and carried out within the framework of both an organisation and a 'system of activities' (Paul *et al* 1994);
- *a case study*: micro-sociology calls for a limited scale of investigation, but we also think one VCF can be an 'enlightening case' (Mitchell 1983). That is why we chose a comparatively big VCF (Puisserguier, Biterrois) which moved early towards quality production and was still deeply involved in the regional transformation of the local viticulture. Moreover, officials of this VCF demanded a thorough analysis of the current phenomenon;
- *a historical perspective*, comparing the situation in 1997 with that prevailing at the end of the 1980s, before implementation of the changes towards quality, in order to test the hypothesis the officials proposed: that those changes widened the ideological and physical gap between members and officials.

Our research work has been implemented in two main steps, as in an 'action-research' process (Lewin 1951). The first step was to delineate 'strategic blocks' in 1997 among the large and heterogeneous membership (roughly 750 individuals, either official shareholders or moonlighters, the latter being numerous in French vineyards). We developed this categorisation on the hypothesis that a specific relation to an object - which, from a theoretical standpoint, is called its 'qualification' - induces some specific technical and social practices relating to it. We assume these practices can be identified by observation, reviews of listings or interviews with key persons that enable the researcher to build a primary typology, which has to be improved by further work to highlight the specific 'qualification' associated with particular practices.

From a set of technical and social practices relating to the co-operative project in 1997, we defined for each practice several possible behaviours. On this basis, we distinguished nine 'strategic blocks', or sets of people exhibiting the same behaviour relating to the same practice in 1997. The data were produced by ethnographic methods: living in the village, observation, non-directive interviews (Olivier de Sardan 1995). In particular, with a snowball sample of 70 people, intentionally biased towards those hardly known by officials, we developed 'life story interviews' (Bertaux 1980) that focused on practices. During the interview, we focussed on relationships which could have encouraged or helped the changes in technical or social practices mentioned by the interviewee. Discourse analysis enabled us to specify the relationship of the interviewee with the objects of 'action together', particularly through his/her definition of 'quality grape' and 'quality wines'.

After the first round of enquiries, we produced and distributed widely at the beginning of 1998 an information leaflet which presented previously unstated points of view about 'quality', through extracts transcribed word-for-word from the current discourses of our primary interviewees. A form was enclosed with this leaflet, to be sent back to us.

In the second stage, two individuals per 'strategic block' (a total of 18 persons), chosen according to the number of their relationships - one having many, the other few - were interviewed in more detail about their practices in 1990 and the recent evolution of their professional relations, through a combination of several tools:

- use of the leaflet as a 'name generator' (Laumann 1966) through the following question: '*With whom did you speak of this leaflet and how? Did you speak with him/her before?*';
- adaptation of the 'position generator' technique used by some network analysts (Lin 1999), by asking interviewees to specify their relationships with people - with whom we knew or supposed they were or had been in contact, because of neighbourhood, for example - belonging to specific categories: 'institutions' (technical groups, leisure associations, family...) on one hand, 'strategic blocks' on the other. In that way interviewees, who generally discuss first (or even only) their 'strong ties' (their most frequent relationships, according to Granovetter's definition; Granovetter 1973) are prompted to speak about their 'weak ties' too (McCallister, Fischer 1978);
- distribution to selected individuals of a questionnaire relating to this leaflet: they were asked to hand it to the producers they would meet in a period where exchanges are usually numerous (that of sanitary treatments on grapes) and afterwards to give it back to us;
- graph representation of the 'zone' of our interviewee (depicting both his/her own relations and the relations between people with whom he/she is linked: Barnes 1972)[1] in 1990 and in 1997 and presentation of this to him/her to encourage further specification or modification of some elements. Use of such a data production technique had been made possible by the trust progressively developed with the 18 interviewees, by immersion in the village, maintenance of regular relations with them,

but also because the leaflet has been widely appreciated, showing the open-minded posture of the researchers and thus reducing reticence.

In this second stage, because of changing conditions affecting the context of our longitudinal work (orientation of the production project towards valorisation of the new 'quality products') and because we had been observing since 1998 the creation of new professional relationships and the development of new types of relations to the objects (sometimes due to the leaflet diffusion), we have been simultaneously stocking up data relating to what we consider as a third period, that is to say since 1998.

Professional relationships were coded in the following way, for each of the three periods:

- content: exchange of general information, work, equipment, technical advice,
- nature: 'institutional' (the two people involved in the relation belong to the same 'institution': family, farming association, leisure association...) or 'informal',
- frequency: low, normal, high (graded from 1 to 3 and defined by the interviewee himself),
- intimacy: distant, normal, close (graded from 1 to 3 and defined by the interviewee himself).

Relative to these criteria, our interviewees associated 'normal' and 'close' relations with *reliable* ones, defining both as *quite durable* and *those which can be quite easily mobilised.* 'Distant' relations appeared on the other hand as those they are not sure they could count on if the need arose, and with an uncertain future. We finally kept two levels of intimacy, 'reliable' and 'less reliable' relations.

We assessed the evolution of professional relations from a quantitative standpoint by looking, for each period, at both the number of ties figuring in the 18 zones of our interviewees and the ties between all the persons involved in these 18 zones. The 18 zones variously involved a total of 81 people between 1990 and 1999. Then, as networks analysts usually do (Degenne, Forsé 1994), we tried to specify the relations of each of these 81 persons with each of the other 80 (which required a few new enquiries). We recorded the data in a squared matrix of 81 columns and 81 rows, that allowed data processing by a specific software for network analysis called Ucinet V©.[2] Seventeen of the 81 people belonged to the Board of Directors or to a commission linked to the co-operative in 1997.

Finally, we distinguished three periods: before 1990, between 1991 and 1997 and since 1998, and for this set of 81 individuals we identified the technical and social practices of each in 1990, 1997 and 1999, relating to the objects of the 'action together', and the professional relationships between them.

Results

'Quality revolution' and conflicting 'strategic blocks' multiplication

Complex, radical and high-speed changes in wine co-operatives. During the 'table wine' period, the group was diverse but relatively cohesive. The VCFs had built a strong professional, cultural and ideological identity through protest movements (producers' uprising in 1907; see Mann 1998) in the beginning of the century. This identity endured, and VCF members, as well as officials, seemed to define the objects of collective action in the same way: grapes had to be produced in large quantities with high alcohol strength, while the co-operative project consisted of lobbying the officials in charge of the wine sector at the national level to maintain a legal framework to shelter profitable production of 'mass wine', although it was decreasingly suited to this (Bartoli, Boulet 1990). Most of the members subscribed to the co-operative project which was logical since it worked economically, and since the efforts for intensification on one hand, and demonstrations on the other, were consistent with society's general idea of progress and with co-operative founder culture (Martin 1994).

In the mid 1970s, political action at the local level no longer sustained economic returns, and the second and third generations of members after the 'founding fathers' were unacquainted with protest culture (Maffre-Baugé 1986). VCFs had to deal with new market requirements involving a wide range of complex and radical innovations within both farms and firms from the end of the 1980s (Laporte, Touzard 1998). Following expert advice, managers encouraged producers to plant new varieties of grape, imported from Bordeaux or Burgundy, and to develop new growing techniques (tying up vines on wires and sticks, for example). To prompt them to follow these technical 'prescriptions', most VCFs paid a better price for new varieties as a first step, and subsequently developed a payment system based on 'quality levels assessment'. In our case, for example, since 1992 grape quality has been evaluated directly on the plot, three weeks before the harvest, through the ratio between the 'leaf area exposed to sunlight' and the weight of the grapes. This ratio is measured on a 'representative' vine-plant and assessed by a control commission, consisting mainly of members of the Board. Producers are paid according to the category in which the commission puts them. This payment system has been increasingly implemented in Languedoc VCFs in recent years.

In most of the bigger organisations in Languedoc, managing teams, forced to become greater autonomous decision-makers in order to deal with high-speed market movements, had been developing two main orientations: the production of 'cultivar wines' (from only one new grape variety), and integration of the VCF in wider structures (unions of co-operatives) (Jarrige 1998).

Questionable changes and membership recombination as source of differences. Grape classification appeared immediately as a source of conflict within the co-operative population. Historically rooted in socialism, the basic idea of the co-operative movement is that each member should be treated equally (Desroche 1976). Some members do not agree to abandon this, and even if most are not against evolution from 'equality' to 'equity', rules and tools for quality classification are much questioned, especially as they determine the final payment:

- the ratio between the leaf area and the weight of grapes is the first criterion taken into account, whereas the biological condition of the fruit, and the alcohol strength, are only addressed afterwards. Most members think however that one or both of these latter criteria are more important for 'quality'. The ratio is also difficult to measure in practice and is very likely to vary from that found in the 'representative' plant;
- commissioners are members of the Board of Directors, vine-growers themselves in the village, deeply anchored in local social networks, and so in a position to be influenced in their judgement or to take advantage of their status.

More generally, this classification assumes 'grape quality' is determined by current market requirements, and emphasises new varieties, although some members think they are not suited to the Languedoc area. This affects how 'quality wines' are defined; for some, they are those which match current demand, for others they have to be developed over a long time and consist in a localised production from blending 'traditional' varieties.

Differences over what grapes and wine should be had been encouraged during the years following implementation of the new 'model' by low returns on investments, even bankruptcies, for the 'pioneers' who started production of new varieties early. A better price was not enough to balance lower yields and higher operational costs, so forcing the producers to extend their vine-growing area. Yet this extension, as well as being expensive and risky, impedes quality work, especially as new varieties already require more work and more production time. Many members believe the new farming system which they have to implement to get a return on investment is paradoxically inconsistent with quality production. On the other hand, technical prescriptions and business choices have changed, even contradictorily, year by year and/or between neighbouring VCFs, further encouraging producers to differ over grape and co-operative project qualification.

Differences are also linked to the emergence of new actors within the membership in recent years: unemployed people from cities, attracted by increasing rewards in viticulture when other activity sectors were decreasing in the region; new rural inhabitants, in search of social links and/or 'distinction' (Bourdieu 1979) through a top-quality and environment-friendly grape production (Chiffoleau 1999). Their relations with the objects of the collective action are different from those of regular or long standing members, particularly as they come from other *milieux* and are more affected by dramatic changes in global society. More generally, population movements, increased travel, higher levels of academic or professional education, contribute to the diversity of relations to objects of collective action among the whole membership. 'Institutional distance' has also increased between members themselves, thus questioning the (assumed to be cohesive) 'producers' association identity'.

From conflicting qualification to the delineation of 'strategic blocks'. The sets of practices which define the blocks and the block boundaries changed between 1990 and 1997, due to differences in the officials' prescriptions and in the behaviour of members at each period. However the blocks defined in 1997 are still relevant in 1999, even if they are evolving as we show below.

In 1990, most of the membership seemed to be involved in the same innovation trajectory. We may distinguish three blocks according to the moment they started replanting new varieties and to their level of involvement in 1990. The *avant-garde*, linked to private cellars, had started in the 1970s and were replanting moderately in 1990. The *clan members*, members of the Board or close to them, were just beginning, but on a large scale. Some *base members* were following. But for all of them, high yields and farm extension were needed in order to compensate for loss of income from the fall in the price of table wine. A fourth block is the *little ones* who generally preferred to give up vine-growing rather than invest.

Between 1990 and 1997 these four strategic blocks split into nine, because of the movement of members in several directions. Discourse analysis confirms that the nine specific combinations of behaviours in 1997 are consistent with a very similar qualification of the objects of action in the context of a 'quality wines' project. People in a given block do not necessarily share the same socio-economic status (age, academic standard, means of production ...), but can be classified into two or three 'social trajectories' (Dubar, 1998), relating to the vine-growing activity and specific to this block.

- For some, mainly members of the most important vine-grower families, 'the object' is what is (or seems to be) better paid, even if they agree it is not always relevant or sustainable: thus, they monitor their production to reach a high return and attend general meetings to get into the Board of Directors' good books. But they take little or no care of anything which does not relate to payment. In 1990, they belonged to the *clan members* block and in 1997, we identify them as the *strategists*.
- For others, grape and co-operative projects have to stay as they were in the 'table wine' period, which in practice corresponds to more comfortable and profitable ways of producing grapes. They aim at high yields with high alcohol strength. Among them, we distinguish those who ultimately produce a good 'table wine grape' (the *apparent resisters*, mostly elderly ex- or current farm workers) from those who often mix harvests from different plots to reach the maximum amount and alcohol strength permitted or paid (*real resisters*, including members of the Board in the 1980s and new vine-growers 'by obligation').
- Against these more or less dangerous 'free-riders', some want or try to be 'faithful' (average vine-grower families, ex-part-timers or part-timers moving towards full-time farming, often current members of public agricultural extension groups). But, based on their innovative behaviours, we could distinguish the *disappointed ones*, who developed radical changes as early as the new prescriptions, ran into debt, and stopping innovating, from the *cautious ones* who implemented the innovations later or more slowly.
- Others believe grape and wine production must be compatible with - even aim at - a good quality of life or environment. They subscribe to some of the managing team choices (yield limitation, for example), reject some others (ratio, VCFs merging) and develop activities that some others consider 'superfluous' (ditch cleaning), 'without any bearing' (festivities) or 'against quality production' (diversification). Among them, we distinguish the *70s pioneers*, who developed innovations during the table wine period, had and still have many relationships with private cellars, and the *little garden part-timers*, mostly new rural inhabitants, sometimes belonging to wine tasters' groups.
- Others, children of the *avant-garde*, part-timers in the wine business, young farm workers or new vine-growers (seeking, by retraining in viticulture, 'more freedom'), think the managing team does not take the best decisions, evolves too slowly or is too prescriptive. Thus, they

develop their own technical practices, used in other vineyards or Languedoc private cellars and which they learn from private cellar owners, technical firms, the media or Internet. Even if most of them are still attached to co-operative values (*the revolutionaries*), some are likely to leave (*the adventurers*).

Finally, in 1997 'blocks' were numerous and developed some practices incompatible with or in contradiction to the current 'collective project'. Although not true of all practices, such diversity questions management team decisions and weakens the economic organisation, particularly in a context of demanding markets and the emergence of attractive competitors (wineries). Since this development of numerous and conflicting 'strategic blocks' concentrates the professional relationships between people belonging to the same block, evolution towards a more consensual 'collective project' and compatible practices is more difficult to develop.

General evolution

Number/frequency: The number of relationships decreased by about 30 per cent between 1990 and 1997, and only 55 per cent of relations active in 1990 remained in 1997. The number of relations dropped particularly dramatically for individuals selected for their high number of ties. Indeed, according to producers themselves, deep differences on 'what grape and wines have to be' in the 'quality revolution' context, in addition to mistrust of a 'not convincing model', encouraged the break-up of some relationships, particularly with elected Board members presently serving as 'judges'. Grape classification also induces hierarchy among those who are 'judged', then comparisons and jealousies that lead to conflicts, withdrawal and atomisation of the membership. More simply, the new farming system, which is excessively time consuming, does not allow people to keep or to develop frequent relations with others. The basic 'common action' in viticulture, harvests, has been evolving towards an individual activity, given the mechanisation allowed by farm reorganisation (bigger plots, large space between rows, new plants on wires). Finally, new actors entering the professional group have difficulty in developing relations with others. In particular, new rural inhabitants are not considered 'peers' by long-standing producers, even if they are not absentee landlords or 'vine-growers for pleasure' as in the past: most of these 'new type' of part-time farmers work comparatively big estates with high technical knowledge (Chiffoleau 1999). Since 1997, the number of relations has been increasing again, notably by a reactivation of about 10 per cent of the broken ties.

Table 13.1: Evolution of professional relations between 1990 and 1999[a]

	1990	*1997*	*1999*
Total number of relations in the matrix	695	494 (-29 %)	527 (+7%)
Mean density of the matrix *(number of current ties / number of possible ties)*	0.107	0.076	0.081
Mean frequency of relations	1.29	1.24	1.25
Number of 'reliable' relations	489	129	297
(% of total)	(70%)	(26%)	(56%)
Number of 'institutional' relations	82	205	250
(% of total)	(12%)	(41%)	(47%)

[a]*Calculation relative to the squared matrix 81 x 81 issued from the sample of 18 people representative of strategic blocks diversity in 1997*

Reliability: Between 1990 and 1997, the number of 'reliable' relations decreased by 75 per cent; they represent only 26 per cent of active relations in 1997. Since 1998, the ratio of 'reliable' relations is increasing again.

Type: Professional relationships have become increasingly 'institutional' since 1990. Three-quarters of the relationships created between 1990 and 1997 are linked to the VCF: they involve two people each of which belongs to either the Board of Directors or to one of its commissions or technical groups. Most of the new relationships in 1999 are developing through cultural or environmental associations which have been emerging locally for two or three years. Moreover, between 1990 and 1997, most of the relations within families and between or with members of the Board showed two opposite evolutions: strengthening or breaking up (several exits from the Board, for example, followed by the break-up of relationships).

Content: From 1998, our interviewees spoke about 'exchanges of different resources', or 'collaboration on a project' (recruitment of a common worker, rehabilitation of reverted lands ...), whereas they did not mention this type of content before.

Between 1990 and 1997 two stages may be distinguished: after a first stage of development within the group, exchanges of equipment/work and technical advice decreased and were often established with people outside the group (technical experts or workers in private enterprises, private cellar owners).

Table 13.2: Evolution of professional relations by types of content between 1990 and 1999[a]

	1990 %	*1997* %	*1999* %
Exchange of equipment / work	30	20	10
Exchange of information	40	55	20
Exchange of technical advice	30	25	20
Exchange of different resources, Collaboration on a project	*Type of content which does not appear in discourses about ties relative to these periods*		50

[a]*Calculation relative to the squared matrix 81 x 81*

Furthermore, interviewees themselves stressed that, in a given context of action, not all their professional ties are equally useful in view of what they have to deal with. Specifically, for the three defined periods, vine-growers converged towards recognition of one content as more 'relevant' than others, which they linked to the main action they had (or were supposed) to do. But if people said *a posteriori* that in 1990 replanting new grape varieties was 'what they had to do', in practice most of them did not think that at this period, as we show further.

Table 13.3: Evolution of the context of action and distinction of a 'relevant content' for professional relations at each period defined, according to the producers

period	*action prescribed or supposed*	*'relevant content' of tie*
1990-1992	Planting of new grape varieties	Exchange of work and equipment
1993-1997	Implementation of new vine-growing techniques	Exchange of technical advice
1998-2000	Management of a biggest farm with new varieties requiring more technical knowledge labour time and investments	Exchange of different resources, collaboration on a project

Confrontation with 'strategic' blocks

Each period is assessed for the 81-person set and relative to the blocks we may distinguish at this period, that is the four in 1990 and the nine in 1997 and 1999. If we look at the evolution of the professional relationships between people belonging to the same strategic block ('intra-block' ties), we get the following results:

Table 13.4: Evolution of the professional relations involving people belonging to the same strategic block[a]

	1990 %	*1997* %	*1999* %
Intra-block ties/total	60	27	25
Intra-block ties/'reliable' ties	60	72	40
Intra-block ties/'relevant content' ties	80	63	34

[a]*Calculation relative to the squared matrix 81 x 81*

Discussion

Research hypotheses assessment and improvement

According to our hypotheses, the high number of blocks in 1997 and the concentration of 'relevant content' ties within the same block may explain the dysfunctions about which the officials complained. From 1992, a similar definition of 'what grape and wines have to be', highlighted through exchanges, encouraged in each block specialisation of practices strengthened by 'relevant content' exchanges concentrated in the block itself. Moreover, at this period, exchanges of technical advice with people outside the group increased, encouraging more diversification of practices between blocks.

In 1990, 80 per cent of the relevant content exchanges also involved people of the same strategic block. Most of these actually involved people belonging to either the *avant-garde* or *clan members* blocks. Retrospectively, that explains why in 1990 there were in practice few plantings of new varieties outside these two blocks. Indeed, in 1990, information about the innovation (planting new grape varieties) was not widely distributed by the *clan members*, as its implementation was supported by public aid which they considered limited in amount (*ex-official, private communication*). In practice the 'quality wines' project started to diffuse in 1992, after the recruitment of a new manager. In

1990 most of the *base members* and *little ones* were either ill-informed or not informed at all about either the innovation or the conditions of its implementation. Moreover, information exchanges hardly crossed the blocks. This also explains the crisis that followed as members realised the management team had kept privileges and information hidden at the beginning of the 1990s. Finally, in 1990 most of the membership was not evolving within the framework of a 'quality wines' project and was still cohesively in the 'table wine' system. *Base members* and *little ones* were not really conscious of the changes which had to be implemented. For the *clan members*, planting of new varieties aimed, at this period, at getting public subsidies rather than developing a 'quality wines' project (*ex-official, private communication*). Only the *avant-garde* were thinking about a new productive system, but in a period of such transition they also had to keep the previous system in most of their vineyard.

Between 1997 and 1999, beyond a small increase in the total number of ties, we observe that relationships between blocks increased slightly. In 1999 they constituted 66 per cent of ties having a 'relevant content' and 60 per cent of those classified as 'reliable'. One fourth of ties between blocks did not exist in 1997, particularly involving people we selected at the beginning for their low number of relations. Yet since mid-1999, VCF officials have noticed, more adoptions of the technical prescriptions proposed via the payment system, improvement of individual performances particularly relating to work organisation, financial management, or quality of the harvest, and the development of new practices and activities at the individual level or in groups of two or three people who were not linked at the beginning of the 1990s. Officials observe at the same time a reduction of free-riding or exit threats. These observations support our initial hypotheses about the capacity of a group to develop an 'action together', which we linked to relations between blocks.

Finally, analysis of the data allowed us to specify our hypotheses as follows: at any given time the capacity of a group to develop an 'innovation together' may be assessed by the ratio between the number of both 'relevant content' and 'inter-blocks' relations and the total number of ties. We propose this ratio as a 'collective innovation capacity indicator': it has to be measured for a specific sample, representing the diversity of relations to the objects of action defined at this time (that requires specifying the effective context of action), and considering 'relevance' of a content in relation to the given time. The validity of this ratio has, of course, to be tested (Chiffoleau 2000*a*), notably because it calls for past times appraisal (Menger 1997). Its evolution, nevertheless, is not only consistent with past events, but also with the dynamics observed in recent months, relative to the 'action together' in the VCF we studied. Following this approach we try to show how the evolution after 1997 is linked to a change in the context of action, that both induces new types of relationships and also new relations to the objects of action.

Current evolution: two worlds simultaneously on the move

The world of the producers' association. First made to plant new varieties and develop new vine-growing techniques, producers have been progressively driven to manage an increasingly complex activity system, on a bigger farm requiring not only more technical knowledge, but also more investments and labour time, thus requiring organisation and anticipation in work and financial management. Faced with new needs, even new problems (for example water management, due to vineyard consolidation), producers are looking for help and this induces re-activation of exchanges within the membership, and above all, development of ties between people in different blocks. The diversity of 'social trajectories' facilitates complementary exchanges, such as technical for financial advice, or work exchanges, between long-standing producers and new part-time farmers. Need may thus encourage relations between people who do not share an 'institutional proximity' relating to the objects of the collective action, understood in the framework of the VCF 'quality revolution' and its link to the market. Why are such exchanges occurring? These new types of need could not be fully or easily satisfied by people of the same block or people outside the group, unlike in the previous period. Beyond technical advice or harvesting machines, new part-time farmers are in search of social links in the village where they chose to live, whereas long-standing producers can not (or are reluctant to) call for help, in matters of work and financial management, from their peers who do not have time or good faculties, and outside services are expensive, for instance to clean ditches. Fairs for tourists, or wine promotions in local markets or supermarkets introduced or renewed by *little garden part-time farmers*, are rewarding and encourage some *disappointed ones*, for example, to take part in these common activities and exchange again with other blocks. More generally, Languedoc wines have been more and more profitable for a few years, the first results of the 'quality revolution' (Laporte, Touzard 1998), which makes relations less strained, particularly towards members of the Board.

More fundamentally, exchanges between blocks are encouraged by the emergence among many members of a new 'rationality' that gives another meaning to viticulture, stressing its links with the territory where it is performed, and its finality, seen as economic as well as social. Nearly all the producers we met seemed worried about the development of negative attitudes to agriculture in society. Professional debates revolve very often around the question of how not to be seen as environment polluters and this is the basis

Table 13.5: Confronting the quantitative data to the facts to validate the model we propose

	1990	*1997*	*1999*
Starting point observations (according to the VCF officials)	*Subscription of the membership to the new productive project*	*Resistance to prescribed innovations, free-riding, voice*	
Number of strategic blocks	4	9	9 (in evolution)
Number of both 'relevant content' and 'inter-block' ties/number of current ties = 'collective innovation capacity indicator'	6%	9%	33%
Retrospective Analysis	*Few innovations outside two strategic blocks*		
Facts noticed by officials from mid-1999 (and confirmed by longitudinal work)			*Individual and collective performances improvement*

[a]*Calculation relative to the squared matrix 81 x 81*

for a new consensus about their production activities and other functions that are coming into view. Moreover, territory and its development are the objects not only of informal exchanges driven by need, but also of new formal associations, in which individuals belonging to different strategic blocks take part. Some of these take great care of their plots and production, whereas others confess that they do not always do what should be done. If *strategists* stress need and the profitability of such associations, others highlight another 'sense' to their commitment. Some young men, for example, coming from vine-grower families, influenced by their wife who works outside or by the media, do not consider themselves 'farmers', 'as their father do'. Some *adventurers* even broke up professional exchanges with their father in the previous period 'for this reason'. Now numerous *cautious ones* want to commit themselves to repairing their environment or promoting local products, even if their current resources do not allow them to fulfil their wishes. Taking part in such associations constitutes a first step, and they join there people from cities, *revolutionaries* or *little garden part-timers*, who do not want to remain or to become 'farmers' but to affirm themselves as 'actors in the local development process'.

Some elderly members, without anyone in their family to take over, who are not necessarily involved in the 'quality revolution' (*apparent resisters* or even *real resisters*, more or less discredited by others) want to retrieve their reputation (particularly some who were Board members in the 1980s) by reaffirming that the initial ideal of VCFs was solidarity and the reduction of inequalities. So they wish, through the transfer of their vineyard, to help newcomers, unemployed youngsters, to settle down and start a vine-growing activity. If it remains quite anecdotal for the moment, the first cases are encouraging officials to promote this dynamic by specific actions towards elderly people (see below). This casts a different light on associations that are nowadays often seen as constituted by grassroots members only driven by self-interest, with no concern for 'ethical solidarity' (Bode 1997). On the contrary, such examples hint that demand for 'social justice' is widespread in the society, far beyond sole activists groups, the traditional promoters of this idea (Hérault, Lapeyronnie 1998). They are consistent with movements for increasing local activities in current society (Laville 1990), actions with a local range, undertaken by small groups of people who are culturally heterogeneous, driven by economic interest and by a need for solidarity.

The world of the enterprise.[3] Most VCF officials confess that, now that they have invested in the transformation of the vineyard and in the processing equipment, the limiting factor lies in the image and the marketing of the product, that still need improvement. So they try to increase value through the specific assets that are part of the production process. Those are mainly related to the idiosyncratic assets of the vineyard and of the know-how that members apply to them; that is, the elements that constitute the identity of the geographical space wherein viticulture is carried out. The grapes that are produced are no longer considered mere grapes, but grapes 'from here'. Taking account of locality as a resource induces a shift in the co-operative project which must now develop new activities to control production of a 'local identity', consistent with the requirements of specific market niches that VCF aims at. To that end, VCF officials encourage producers not only to plant new varieties and to develop new vine-growing techniques, but also to improve production of traditional varieties, take care of the environment, and take part in wine promotion activities in supermarkets or through festivities.

However, to implement such a project the VCF dramatically needs new human resources and skills. Members are on average 55 years old and 2 of every 3 have no one to take over. The prescribed production model lacks the manpower to be satisfactorily run, from a technical point of view, which explains the development of environmental problems which challenge the positive image VCFs are trying to build. Moreover, qualification of the objects of collective action in relation to the territory requires still more labour. This prompts officials to imagine relief in the form of alien actors, not raised in a vine-grower environment, past or present part-time and off-farm workers who may have built other skills and networks that may prove useful for developing new activities (like agro-tourism) towards new market niches. The economic efficiency of any VCF could depend on its ability to summon those skills and labour forces through new types of strategic take-overs. At a point when private firms are beginning to enter the grape market, when Languedoc wines are more and more rewarding, part-time farmers or moonlighters, looking for 'community' as well as for a profitability, appear more 'reliable' members than full-time farmers, who are all the more likely to leave because agricultural training emphasises privatised cellars. Officials also see in the development of solidarity towards unemployed people a way to reduce desires to leave and regenerate a dying co-operative tie, among members aware that unemployment could hit some member of their family, particularly in Languedoc-Roussillon where the unemployment rate is one of the highest in France (18 per cent in 1999, compared with 11 per cent at the national level).

Moreover, to reaffirm the VCF as a 'social economy' enterprise, both competitive enterprise and local development collective actor, a resource for employment, local services, local quality products, and a healthy and attractive environment, becomes an objective in itself. Initially probably induced by instrumental rationality (Weber 1968), this flash-back to 'co-operative principles' is widely claimed by the regional officials who have taken concrete measures to help unemployed people - for example those from urban areas - to develop vine-growing. Some Board members join together to recruit a shared worker (often previously a moonlighter in the VCF vineyard), within the framework of the 'employers' groups' currently favoured by government aids; some VCFs buy farms from retiring people, then rent plots out to young newcomers with low resource levels. At the regional level, officials of the wine co-operative sector take part in some programmes oriented towards facilitating economic and social insertion, along with informal and local collective actions which have been implemented for several years, or even new ones (specific programmes to support 'progressive establishment' in viticulture of people who cannot get access to public installation grants).

The meeting of new rationalities in the 'rural co-operative'

Objects of collective action, strictly defined in relation to the market in the 1990s, are now qualified in a different way by some producers and officials: grape and wines are seen as territorial assets on one hand, as economic and social objects on the other hand. That indicates a simultaneous evolution, both of the world of the association and of the world of the enterprise, from quite incompatible instrumental rationalities that generate conflicts or network breakdowns, to one 'axiologic rationality' based on common principles (Boudon 1998).

Producers are now encouraged to involve themselves in a new framework of action, in such varied fields as environmental management, direct selling or heritage protection. Unlike before, the normative framework for these new actions is negotiated around a new qualification of the objects: a grape carrying a local identity, compatible with a market niche, and a co-operative project embedded in a strategy of local development and newcomers' rehabilitation.

In this new framework, the institutional distance between the 'strategic blocks' decreases. The blocks, where people share the same relations with the objects of the 'action together', start evolving. Insofar as the objects are now seen for and by their integration into the territory, the delimitation of the blocks is based on the relations people have with the territory and practically, on their views on its development. Our research shows that many people, partly encouraged by exchanges with others, are willing to put the objects of the collective action into this larger framework which that encourages new exchanges between them. According to our hypotheses, this evolution towards both moderate 'institutional distance' and exchange reactivation within the membership should not impede negotiation further, especially as the two features influence each other. This should encourage implementation of an innovative and localised 'action together', even if professional relations are less numerous than in the 1980s. This evolution in the context of action is helping to decrease both the physical and the ideological distance between people of different blocks, which could encourage development and organisation of a more favourable action environment. Current rules are still linked to radical prescriptions and conflicting quality assessments which impede the evolution of the organisation towards a common stabilised 'identity'.

We thus propose the implementation of a new type of VCF, driven by (interdependent) economic and social finalities, open to its environment and developing larger functions to mobilise or create new resources. Adjustments between the rationality of some producers and officials provide ideas on how to strengthen the link between the two worlds of production (enterprise/producers' association), by changing the modalities of their 'interface'. Many members, and some officials, agree that grape and wine production have to be part of the building of a new VCF identity which is anchored to and active within the local territory, and consequently have to be treated as complementary activities. Individual efforts could then be assessed and paid according to different and multiple criteria, and no longer on new variety production alone, or even 'high ratio' fulfilment. People producing traditional varieties suited to average quality table wine and taking part in wine promotion activities are as important, for the co-operative project, as 'top-quality producers' who do not have time to be involved in 'common actions' or to take care of traditional varieties. Besides, table wine remains the most important market niche for the VCF sector (60 per cent of its production) and most European consumers will go on drinking this type of wine (Aigrain *et al* 1995). It needs nevertheless to be better managed and controlled (through effective yield limitation), localised (to face foreign wines of similar standard) and driven by a parallel high-quality production.

In concrete terms, the diversity of products may be recognised, thus allowing them to be assessed and paid for according to their specific objective, and in relation to the 'activities system' of the producer, in order to strengthen the new VCF identity by their complementary contributions to the co-operative project. Beyond new rules for appraising 'participation' - in practice, more suited to developing 'equity' - it may be useful to integrate new types of 'contributors' with the specific responsibility for developing or creating complementary activities, which could help VCF economic performance. It could mean a renewal partly by people with low resources, for example, who would carry out several activities for both the co-operative and local development: viticulture + environmental management, viticulture + activities for tourists and so on. The rural co-operative, monitoring remunerative local specific activities systems and strengthened by common principles, could also remain a key actor in Languedoc contemporary society. In such a productive organisation, everybody should be able to involve himself according to his own desires and potentialities, and would be valued according to his specific contribution to the collective project. This promotes a combination between self and common interests which encourages and stabilises collective action (Popkin 1979). In other words, the rural co-operative appears an exemplary framework of the 'project-oriented cité' proposed by conventions economists in recent writings (Boltanski, Chiapello 2000).

Conclusion

Co-operatives are at a crossroads in their development: the radical changes of recent years have resulted in a problematic situation of conflict. Our network theory-grounded research tries to point out precisely in what way. Qualitative and quantitative analysis of the relations of people to the objects of action on one hand, and of professional relationships between people on the other, has enabled us to grasp past and current dynamics relating to innovative collective action, and even to propose a model likely to help the analysis of on-going processes in rural settings (Chiffoleau 2000b). The research highlights that, beyond ideological differences and group atomisation, common principles seem to be presently emerging within co-operative populations, particularly because these populations are changing. Indeed, growing numbers of producers and official are thinking about new functions for the VCF. Thus we confirm Polanyi's interpretations of merchant dynamics which both dismantle prevailing social relations and simultaneously open up new opportunities to rebuild them (Polanyi 1944).

Regional and local officials would be well advised to catch the current opportunities and implement these principles in a new form of co-operative which could match many people's expectations. We argue that the future of VCFs hinges on whether or not they can make the transition from economic goals to multifunctionality. Of course, the principle of a vine-growing activity and a co-operative with both economic and social goals is not shared by everyone. People have varying economic and ideological dispositions, and react differently when faced with radical socio-economic changes: some adopt a new rationality while others do not move (Bourdieu 1977).

The actor emerging in the rural co-operative is neither the 'mass actor' of the table wine period nor the strategic individualist who has become the emblematic figure of modern society. He is both a member of a co-operative, which combines efficiency and solidarity, and an actor in the local development process, a 'development actor citizen' (Teisserenc 1994), 'compromiser' between the domestic, civic and merchant worlds (Eymard-Duvernay 1989). Capable of integrating his private prospects into a collective project by his active and specific commitment, and also moved by an 'individualist co-operatism' (Thuderoz 1995), he may prioritise the need 'even weak or concealed, of human beings, for life in common' (Todorov 1995). Moreover, he combines through his involvement the four types of social action that Weber distinguished: traditional, affective, axiological rational and strategic rational (Weber 1968).

Finally, the rural co-operative modifies, within the co-operative population, the 'power-dependence relations' (Emerson 1962) and 'system of roles' (Reitz,

White 1989) which structured the previous local social action system, with its powerful full-time vine-growers on one hand and powerless part-time farmers on the other. Diversity in the resources of the members promotes complementarity of roles in favour of both private and collective projects and consequently, a real re-politicisation of the local action system. But the future of wine co-operation may depend, in practice, on VCF officials' willingness and capacity to renew the egalitarian founding principles by introducing rules promoting a real equity. Indeed, in late modernity, the stake may be to evolve from the modernist (Boudon 1996) 'equality then relativism' to 'equity then commitment'.

Notes

[1] According to our research hypotheses, we aimed at measuring the proportion of relations between strategic blocks at each period rather than between concrete individuals, that is why it was more efficient to look at the 'zone'. But to ensure the reliability of the data, we stopped at the zone level, limited to the relations between two persons linked with our interviewee.

[2] Software created in 1994 by S. Borgatti, M. Everett and L. Freeeman.

[3] This part is based on interviews of officials of several co-operatives and institutions, consultation of minute books and press reviews.

14 The co-operative movement in Ireland: reconstructing civil society

HILARY TOVEY

Introduction

Discussions of the 'classical' social movements of modernity usually cite class or labour movements, less often feminist movements, but rarely if ever recognise the movements which took place within the third great 'contested world' of modernity, the rural or agrarian. This chapter is about an agrarian social movement, the movement to establish producer co-operatives among farmers which began in Ireland towards the end of the 19th century. There are still co-operative agribusiness organisations in Ireland today, although one would hesitate to say that there is a co-operative 'movement'; nevertheless the impacts which the movement has had - social, economic, and cultural-ideological - on Irish life have been considerable. In discussing it here, I want principally to address two issues: the value of a social movements approach for illuminating processes of social development, particularly given the structuralist and quantitative bias which has tended to dominate 'official' sociology in Ireland; and the contribution of agrarian actors to constructing Irish civil society, which that bias has also helped to obscure.

My focus is on the early years of the movement rather than its later institutionalisation; but a brief sketch of its history up to recent years may help to contextualise my discussion. The first co-operative creamery in Ireland was set up in the south west of Ireland in 1889, although the movement for co-operation had been underway for a couple of years previously. Farmers at the time were particularly vulnerable to exploitation by middlemen and dealers. Many farm households relied on the sale of home-made butter, along with eggs, for a regular source of cash income, but the arrival in the 1880s of mechanical methods of separating cream and processing it mechanically into butter was stimulating the rapid development of privately owned creameries, whose major shareholders were often butter dealers and merchants. Horace Plunkett, the main founder of the co-operative movement, saw the 'industrialisation' of butter production as inevitable but feared that if it fell into the control of middlemen, 'the farmers who produced milk, the most perishable of all farm products, would be even more helpless than before' (Bolger 1977: 65-6). Once started, the movement spread very rapidly: within twenty years there were close to 900 co-operative societies in the country, mainly in the south and south west. Most were dairy co-operatives - creameries set up to receive milk from farmers, separate it, and process it, primarily into butter, later into cheese, milk powder and other milk-based products and ingredients (they did not process milk for liquid consumption; that activity remained in private commercial hands until the 1970s, when Irish urban settlements reached such size as to make it commercially irresistible for the co-operatives to take over the liquid milk trade as well) - but a range of other co-operative activities developed in addition.

By the 1930s, contemporary observers were suggesting that the movement was losing much of its original dynamism. The central leadership had become autocratic, and ordinary members were not taking much part in running their organisations.

> Education in co-operation was badly lacking and organisational abuses were common. Many small co-ops had lost all semblance of co-operation - share capital and share registers were neglected, returns were not made, dividends not paid. Meetings, including the mandatory annual general meetings, if held at all, were badly attended and poorly conducted. Often there was hardly any communication with the ordinary members, committees remained unchanged for many years and frequently the authority and business responsibility devolved on the manager... (Bolger 1977: 118).

Bolger indicates that economic and technical considerations were superceding social concerns: the number of co-operatives and of members contracted during the 1930s, but both share capital and turnover greatly increased. The largest co-operatives grew most, and these increasingly saw each other as business competitors rather than co-operators. Private capital continued to be relentlessly pushed out of the agribusiness sector, at least in dairying (the pattern of agribusiness development in relation to meat and grains remained rather different), but the societies experienced great difficulties in finding enough development capital to expand in the ways their increasingly professionalised and business-school-trained managers would have liked. Most of the surplus generated from the business was returned to producers in the form of higher prices paid for their milk. Smaller producers gradually ceased to be explicitly protected by the societies, although many more of them did remain in dairying than would have been the case if they had been dealing with privately owned industrial capital.

The radical reinterpretation of the notion of 'development', and of 'development policy' which took place in the late 1950s, in which protectionism was abandoned and Ireland's economic situation was reinterpreted as that of a 'small open economy' which must engage in global economic transactions to survive, strengthened these trends (Tovey 2001, forthcoming). To manage the risks from abandoning protectionism, the state identified multinational-led industrial development as crucial for the provision of employment and of opportunities for profit for native Irish capital, while agriculture and other extractive industries such as mining, forestry and tourism were identified as areas in which Ireland enjoyed potential natural advantage for global exchanges.

By the 1980s state support for the Irish food industry had shifted from farmers to the industrial food processors (see Tovey 2000). The nature of relations between farmers and food processors underwent a striking reinterpretation in state policy documents. The concept of farmers as 'members' of 'their' co-operative processing societies was replaced by the concept of farmers as 'suppliers' to the food industry; and what had previously been addressed as problems experienced by farmers (and needing state intervention and support on behalf of farmers), such as the small size of their operations, their lack of access to capital, and the seasonal nature of Irish milk production, became redefined as problems facing the food industry which could be solved only by weeding out those farmers who could not become the sort of supplier the industry needed. The food industry gained recognition not just as a better source of stable employment than farming but also as a better location for profit accumulation. Relations between the co-operatives and finance capital changed dramatically. The largest dairy co-operatives pioneered the process of finding ways around the difficulties facing Friendly Societies under Irish law when they want to raise capital for development, and launched themselves onto the Irish stock exchange at the end of the 1980s. 'Privatisation' was achieved by creating subsidiaries which were Private Limited Companies; the co-operative retained a specified level of the shares in these companies while the rest were available for institutional or individual investors.

There is a story to be told about how a social movement which set out to de-commoditise relations within the food industry ended up creating a series of indigenous Irish TNCs which compete aggressively in the global capitalist system. However that is not the story I want to tell here. As mentioned earlier, my interest here is in the early stages of the co-operative movement in Ireland, and in particular, in the movement as a significant actor in the formation of a modern civil society. Farmers who were in their thirties in the 1880s had been born into a devastated world. The Great Famine at the end of the 1840s directly or indirectly reduced the population by half, and almost totally swept away by death or emigration the most numerous agrarian class, the cottiers or labourer-landholders, along with much of their culture and social institutions. By the 1880s, legislative measures to ensure some security of tenure for tenant farmers were beginning to structure the conditions out of which a new agrarian class, of peasant proprietors, would eventually emerge by the early 20th century. But legislation alone could not reconstruct a civil society in which this new class could act. I argue here that the construction or re-construction of civil society was a self-conscious goal of the early co-operative movement in Ireland; and as such, it offers an interesting challenge to conventional distinctions in the social movements literature between 'old' and 'new' social movements.

Rural actors and Irish modernity

Western social thought has been deeply structured by the assumption that rurality, particularity and traditionality go together to make up one side of an opposition, the obverse of which is urbanity, cosmopolitanism and the Enlightenment project of modernity. The various versions of Modernisation Theory which still inform much social thinking in Ireland today, both lay and professional, share two core beliefs: that 'rural or peripheral peoples are *outside* of modernity and therefore need to be modernised or saved' (Vandergeest and DuPuis 1996: 6); and that the modernising project has historically been the project of urban, especially state, elites, who have had to implement it against the resistance of a recalcitrant ('irrational' or 'ignorant'), and largely rural, population. Rural peoples are thus seen as the object of modernisation rather than as active subjects in the formation of modernity. 'Modernisation' becomes indistinguishable from 'industrial-isation', or the restructuring of culture and of social and economic relationships necessary to allow industrial growth to take off. Rural groups are placed outside the industrialisation process and thus effectively understood as having no social history (until they become drawn into industrial society, usually as 'problem' groups). They are thought of as belonging more to nature than to society - that is, as 'uncivilised', contributing little to the formation and development of modern civil society.

Modernisation theory is of course not the only theoretical framework which we might use to understand Irish development history. Granberg and Nikoula (1995; see also Granberg and Kovách 1998; Granberg *et al* 2001, forthcoming) identify an alternative 'peasant path' to modernity, arguing that in the early stage of development of the modern European state system, agrarian structures and agrarian values had a large impact on the politics of a number of states, particularly those outside the industrialising core. Moreover, these were not historically exceptional or deviant cases: Granberg and Nikoula see them as one route towards the development of the typical welfare state later diffused widely across Europe. The variety of forms in which European welfare statism has manifested itself in the 20th century can be explained, in other words, by examining the state project which preceded each instance, including whether or not that took the form of a 'peasant state'.

The great attraction of such approaches is that they address the history of rural Europe as that of an active participant in the formation of modernity rather than a passive sufferer of 'modernisation'. However they tend to overemphasise the role of the state in pushing forward modernising projects and to identify this as the marker of peasant activity, implying that peasants can only be active players in modernity to the extent that they can establish a peasant state. Thus they underemphasise events within civil society, specifically, the activities of rural social movements. A leaning towards state-centred social science analysis, which treats the state as the key actor in the formation of modernity, may perhaps reflect the position of semi-peripheral societies which experience economic growth as a form of 'dependent development', and in which the capacity of the state to provide the conditions most likely to attract multinational capital to locate there rather than elsewhere tends to become a dominating concern of business and commercial elites (see Wickham 1986; Breen *et al* 1990). It is probably also a residue of nationalist ideology or at least of that version of it which equated independent nationhood with possession of a separate and autonomous state. In the Irish case, sociology has tended to devote little attention to studying formative processes of civil society, with the exception of research on the Roman Catholic church, and more recently, studies of the mass media. Even research on the church has often addressed this more as a 'competitor state', focussing on its institutional arrangements and structures, and its capacity or otherwise to continue to command the adherence of its followers, rather than to be a source and creator of cultural meaning and social and moral practice.

Can social movement theorising redress the balance, and illuminate better the historical processes through which Irish civil society has been formed (as well, of course, as its contemporary shaping)? Mooney (2000) argues that theories of social movements have also to a large extent tended to neglect rurality. He notes that this neglect increases as the theoretical debates have become more oriented around analysing and explaining social movements in late modernity:

> Perhaps nowhere else in the movement literature is the assumption of an urban character more pronounced than in the literature on the 'new social movements' [r]anging from the benign neglect of the rural by some social movement analytical frameworks to the explicit urban bias of Castell's (1977) view of urban movements as the locus of resistance to the colonization of the modern life world....As feminism asked of social theory 'What about the women?', rural sociologists might ask of contemporary social movement theory: 'What about the rural?' (Mooney 2000: 37).

As a social movements analyst, Mooney himself prefers a Resource Mobilisation perspective. He suggests that if we include rurality as one factor which affects resource mobilisation capacities (where rurality is taken to mean, essentially, low population density or spatial dispersion of populations) we are led into asking a number of interesting questions about the ways in which rural and urban social movements may differ. These could range from the accessibility of populations to be mobilised via different media of communication (for example, large numbers of Danish dairy farmers already had telephones in their homes in the 1890s, when the Danish agricultural co-operative movement was at its peak (Anderson 1935)), to the potential for local social solidarity created by the absence of sharply differentiated and hierarchically ranked social groups in some rural as compared with urban settings, through to the possibility of significant differences in the availability or effectiveness of different 'repertoires of collective action' (Tilly 1978) in rural versus urban communities. However when he addresses more identity-oriented issues, Mooney's argument begins to seem less convincing. He tends to fall back, in fact, into the modernisation assumption that rural societies are outside of and the objects of the development process, suggesting that a relevant difference between rural and urban is that what Habermas has called 'the colonisation of the lifeworld' is less advanced in rural settings and therefore needs less mobilisation for its defence. When we consider the nature of contemporary agriculture and food production, this seems questionable; and it might be argued that the co-operative movement was itself in some respects a movement to expand the colonisation of the lifeworld, in the sense of encouraging the penetration of Irish farming by the 'steering mechanism' of technology, even while it resisted and tried to evict another steering mechanism, that of money and the commoditisation of exchange relationships in the food production system.

Whether or not the urban-rural divide is a relevant basis on which to construct a thesis about variation within social movements is after all not my primary concern here. My object is to draw on social movement theorising in order to construct a better understanding of the impact which certain social movements have historically had on the formation of Irish civil society. I am not trying to explain how these movements were able to come into being, and to that extent the rurality of the groups involved is perhaps not particularly relevant. Nevertheless, it seems to me that Mooney does highlight an important problem: research into and discussion of rural or agrarian social movements, because they are not generally the sort of movements which theorists have in mind when they are constructing theoretical statements, does seem quite often to throw into sharp relief the limited validity of such statements and the extent of implicit assuming which goes on in their construction. I try to develop this idea further in the next section, where I discuss how we might characterise the Irish agricultural co-operative movement in terms of some concepts offered by current theories.

Social movement theorising and the Irish co-operative movement

The claim that contemporary social movements are in some important respects 'new', or distinctive when compared with those of earlier modernity, has been associated with the work of a number of leading European theorists, notably Touraine, Melucci and perhaps to a lesser extent, Habermas (see e.g. Clark and Diani 1996). A number of attempts to identify and summarise the key distinguishing features of new social movements have been made by different writers (for example Scott 1992, 1993; Beuchler 1995; Goldblatt 1996; Alexander 1996; Katsiaficas 1997). Here I pick out three interlinked points of difference which are given particular salience in such discussions.

First, it seems to be generally agreed that new social movements are distinctive in their concern to influence or transform civil society or culture. Thus Scott, for example, sees Touraine as a pioneer of 'culturalist' interpretations of new social movements which 'stress their culture rather than their political aims, locate them in civil society rather than the state, and identify their aims with lifestyle and quality of life issues rather than political and economic issues of distributive justice and rights' (1996: 82). Older social movements are understood as primarily concerned to gain access to or share in state power, as a means of redressing material deprivation and securing citizenship rights. The labour movement, 'first-wave' feminist movements and even some 19th century nationalist movements or 20th-century movements for ethnic secession, are cited as examples. A *second* key difference is that older social movements largely adhered to an instrumental form of rationality (in which the end effectively justifies the choice of any one of a wide range of means) or a focus on strategic success, but that this is either less evident within or is explicitly rejected by new social movements. Contemporary social movements, like Weber's charismatic form of organisation, embody action which is 'substantively' rational; choice of action is tightly constrained by the ends or objects sought by the movement and only those means which 'prefigure' (Mooney 2000) the ends are acceptable. For example, a movement oriented to expanding participatory democracy in society in general cannot choose to pursue strategies which require it to establish internal movement relations and structures which exclude participation in decision-making by most of its members. This is linked in turn to a *third* characteristic also generally regarded as central: new social movements are involved in ongoing efforts to construct a collective identity and shared interests among their members (this also helps to explain why much movement activity appears non-rational or non-strategic). By inference, the social movements of early modernity were more able to take shared and collective interests on the part of their members as a given; the changes which have occurred in the process towards late modernity have had the effect of weakening or destroying single-stranded structural cleavages (particularly class cleavages) which previously divided societies into clearly opposed groups. New social movements have to try to mobilise support from a plurality of different social groups and networks, usually crossing class boundaries, and to construct from this plurality a convincingly coherent, even if only temporary, collectivity.

Attempts along lines like this to construct ideal-typical conceptualisations of 'new' versus 'old' social movements have been widely criticised. However, as a framework for analysing the Irish co-operative movement they can be rather illuminating - if not quite in the ways expected. Given its historical and geographical location, the co-operative movement should fit much better into the category of a 'movement of early modernity' rather than a 'new' social movement, and indeed discussions of it usually take for granted that it began as a movement for the economic advancement of a marginalised and vulnerable social group in late 19th century Ireland (small-scale newly-constituted 'peasant proprietors') and that its aims and values were thoroughly 'materialist' (to reduce the possibilities for exploitation in agricultural trade so as to retain more of the profit from farming in the hands of the farmers themselves). However my claim is that this is much too narrow an account of the aims and orientation of the movement.

The co-operative movement began as a mobilisation outside the state and to some extent in opposition to the state. As West (1982) has put it, it 'established Home Rule in agriculture'. Commentators often emphasise the hostility which was expressed towards the co-operative movement during its first two to three decades of life by members and supporters of the Nationalist Party in Ireland (see Kennedy 1996b), but the co-operative movement is nevertheless best understood as part of the broad general movement for national independence which developed in Ireland at the end of the last century. This is clearly how its main founder himself understood it. In his Preface to his book *Ireland in the New Century* (1904), for example, Plunkett explicitly linked 'the movement for the organisation of agriculture and rural credit on co-operative lines' with 'efforts of various kinds to revive old or initiate new industries' and 'the literary and artistic movements' (p. viii) and described his aim in the book as to clarify 'the essential unity of the various progressive movements in Ireland' (p. ix). At the close of the 19th century, the failure of demands for even limited political independence created a situation in which other forms of independent development and autonomy became the more important - linguistic and cultural, literary, organisational and administrative, and economic. The co-operative movement, like the Gaelic League and the Sinn Féin movement of Arthur Griffiths, saw the political vacuum of the period as an opportunity to create social, cultural and economic structures which would enable Ireland to take some control of its own future development, whether formally under British colonial rule or not. In this sort of context, theoretical distinctions between movements which are oriented to 'capturing the state' or seizing political power, and movements which try to foster cultural development, lose most of their meaning.

Movement leaders, particularly Plunkett, did expend a lot of effort in trying to induce the British state in Ireland to support co-operation, both financially and through its agricultural policies, but the co-operative movement was nevertheless a movement which wanted to by-pass state power, in favour of empowering agrarian social groups within their own local relationships. In that respect it is comparable with contemporary LETS or some organic production movements. That it was, at least initially, clearly a top-down movement is very interesting in this context. Most discussions of new social movements appear to assume that 'newness' is incompatible with 'top-downness' but Buttel (1993) argues that most, if not all, agrarian social movements have been the creations of elites of one sort or another. We might indeed suspect that any movement which sets out to transform civil society will necessarily start as a top-down movement. Plunkett - a member of the Protestant landed gentry who spent much of his youth working on cattle ranches in America - seems perhaps a very unlikely 'intellectual' yet as I will argue further below, the role he played appears to have been very close to that of the 'movement intellectual' discussed by Eyerman (1990), or to Gramsci's idea of radical intellectual work as the articulation of a cross-class ideology which would allow a new social order to be 'qualitatively affirmed within the life of the proletariat [in this instance, read 'peasantry'] and of society in general' (Waters 1994: 185). Indeed, many of the founding members of the movement - large landowners, clergy or professionals rather than farmers - seem to have understood what they were embarked on as the creation of a new rural rural civilisation, which would encompass the re-valuing and re-empowerment of rural society, its skills and culture, in the face of a world which was becoming increasingly urbanised and industrialised. Irish society at this time being still predominantly rural, recreating rural civilisation was also a way of recreating a distinctively Irish civilisation and a way of distinguishing Irish cultural identity from that of (urban industrial) England.

The second criterion I identified above was the choice of strategic versus 'prefigurative' or substantively rational means of action. Here again the co-operative movement, in its early period at least, fits the standard account of 'new' social movements better. Co-operative societies were organised in ways which were intended to maintain a degree of equality between members and to protect in particular the members who were smaller producers. The organisational characteristics of the early societies - for example the very low level at which share prices were set, and the insistence on 'one man one vote' in decision-making rather than 'one share one vote' - pointed forward to the egalitarianism, communalism, and replacing of relations of competition and exploitation between different economic groups by relations of justice and fairness, which movement leaders hoped to see institutionalised within rural Irish society in general. In the long run, the movement did not manage to maintain these value positions; we could even argue, paradoxically, that it was the very attempt to do so which generated the conditions which led eventually to their transformation into the more orthodox capitalist corporations which most have become today - particularly the effect they had in limiting the societies' capacity to build up a reservoir of capital to use for future expansion. Nevertheless, the instrumental values of efficiency and economic success were not allowed to alter the structure of organisational relations in any decisive way for nearly the next one hundred years. (Perhaps this suggests that instead of dividing movements into 'old' and 'new' depending on the stage of development of modernity in which they are found, we might do so on the basis of the stage of development of the movement itself. Perhaps all movements start off as 'new' movements and grow into 'old' ones, at least in terms of their attitude to strategic success).

Some further instances of the prefigurative logic of the movement can be found in the dislike of centralisation, and the emphasis on participatory structures, which characterised its early leaders. The Irish Agricultural Organisation Society (now ICOS) was not established until about 5 years after the first co-operative societies had been set up, by which time some 400 very diverse societies existed. Plunkett had been notably reluctant to establish any sort of central co-ordinating body for the movement, and insisted that if there must be one, it should be representative not only of those (mainly 'gentry') who had subscribed money to the movement to get it started, but also of the farmers running the local societies themselves (Plunkett 1904: 208). In a similar way, when asked to organise Boards to advise the new Department of Agriculture and Technical Instruction which the English government established in Ireland in 1895, he insisted that these must be as far as possible democratically constituted. Plunkett believed that one significant moral and social benefit of the co-operative movement was that it had made it possible for 'the people themselves' to be involved in selecting their own representatives to help advise the department (1904: 244).

The third feature used to differentiate old and new social movements which I noted above is the place within movement activities dedicated to the construction of meaning: the creation of collective identity and a sense of shared interests. Any discussion of this aspect of the early co-operative movement is hindered by the fact that we have to rely for contemporary accounts primarily on the writings of movement leaders, who generally had fairly clear instrumental goals in mind in recording the issues and dimensions of movement activity which they did. (Perhaps all 'early modern' movements suffer from the fact that sociological or anthropological observers were not generally at hand to record what went on in them, which may explain our impression that these movements were dominated by instrumentally rational types of activity.) For example, Anderson's account of the early years of the movement (*With Horace Plunkett in Ireland*) is largely taken up with recording his own activities in carrying the gospel of co-operation to often very remote and sometimes quite hostile parts of rural Ireland and in notching up numbers of new society formations; he has little to say about the way in which local societies operated once they were established, or about the mundane activities of their farmer members.

Nevertheless, occasionally he does record events which suggest that local societies were well aware of the need to construct and to express collective identity and interests. For example, 'One really robust little group of Agricultural Societies in the neighbourhood of Thurles, whose only finances consisted of a £1 share per member, on which half a crown had been paid up, accomplished a great feat...' (1935: 88). Realising they were being exploited in purchasing superphosphate from local dealers, who generally varied the price to the purchaser depending on how astute they considered him to be and how well able to pay, they held a meeting among themselves to organise a bulk order for all the societies together, and advertised for tenders for its supply. The result was a saving of about 40 per cent on the price of fertilisers. Anderson goes on the say that:

> The arrival of the consignment, in a special train to Thurles, was treated as a great occasion, as indeed it was ...The concourse of co-operating farmers, with the horses drawing their carts gaily decorated with ribbons and headed by a local band playing National Airs, loaded up their lots and dispersed, cheering vociferously, through the town of Thurles ...The ripples that followed this first incursion into the domain of wholesale purchasing spread all over the country, like the circles on a pool into which one has thrown a stone (1935: 90-91).

Plunkett also noted that the co-operative societies were involved in promoting a wide range of different sorts of organisation at local level, not just milk processing - from the establishment of 'Agricultural Banks' to village libraries for self-education, both technical (dairying) and aesthetic (learning Irish) - and that 'social gatherings of every kind, dances, lectures, concerts, and such-like entertainments, which have the two-fold effect of brightening rural life and increasing the attachment of members to their society, are becoming a common feature in the movement ...' (1904: 199).

The argument that most 'older' social movements, when we look at them closely, turn out to exhibit features that have been claimed to distinguish 'new' social movements has of course been made by many others before, and more convincingly than I can achieve here. Cohen (1996) argues that it is more useful to organise the social movements literature instead in terms of two 'competing theoretical paradigms' - the implication being that differences which researchers have claimed to find between the social movements of early and later modernity have more to do with changing fashions in theorising than with any generalisable empirical differences. The two theoretical stances she identifies are the 'resource mobilisation ' and 'identity-oriented' paradigms. In her view, they have in fact a lot in common: even though each in a sense fails to address explicitly the issue which is most central to the analysis of the other, both share a view of collective social action as normal, involving actors who are generally rational and well-integrated into their society and who use forms of association and strategies of opposition which are 'specific to the context of a modern pluralist civil society' (1996: 175). Modern pluralist societies provide, for example, available public spaces in which to make one's protest visible; social institutions such as the mass media; civil rights (of association, of free speech, etc.); a representative political system; and a legal system which is separate from the political system. In other words, in Cohen's analysis, both of the two contemporary perspectives on social movements agree that 'it is civil society itself which forms the terrain on which such movements appear and operate' (1996: 175). Where they differ is that the Resource Mobilisation paradigm analyses collective action in terms of a model of strategic instrumental rationality, based on 'an observer's overview of the political environment that could be useful to organisers' (1996: 176). Social movements are understood to be ultimately always engaged in a struggle for access to power and to the material rewards that control over power can bring. This 'instrumentalising and decontextualising approach' (Alexander 1996: 210) which can sometimes go so far as to treat social movement ideologies as little more than 'strategies for mobilising masses' (209), presupposes, Cohen argues, what in late modern society in fact needs to be explained: that in a de-traditionalised society, the shared interests which could underpin strategic assaults on power have to be achieved through collective social mobilisation itself. The 'identity-oriented' paradigm is superior in its capacity to make sense of processes of collective identity formation (particularly in the case of groups who are not mobilised in order to increase their political or social inclusion or whose identities are not constructed by an experience of exclusion). It also allows us to consider that social movements engage in 'the politics of influence' - they target their activity not on the state but on civil society, 'to protect, modernise, democratise its institutions and social relations' (Cohen 1996: 181) and open up and defend spaces in which new identities can be created. Thus, while both theoretical paradigms treat civil society as the *terrain* of collective activity, the identity-oriented paradigm recognises that civil society is often the *target* of collective activity as well.

In place of a 'new/old' distinction, then, Cohen offers us a 'different theoretical paradigms' account. But her position is a little paradoxical. It seems clear that she finds one of these paradigms more suited to understanding contemporary social movements than the other, and that she does so on what are largely *empirical* grounds. Moreover, in arguing that, she reproduces the historical periodisation of new social movements theorising in a covert way. She need not necessarily be criticised for this - the fact that movements occur later or earlier in modernity (or at different stages of the development of civil society) might seem to be quite a good reason for expecting to find differences between them which are both theoretically and empirically relevant. What I do criticise, however, is the way in which that idea is allowed to re-enter the discussion implicitly, so that its normative or evolutionary assumptions are left unexamined. Early modern social movements are presented as movements *in* modernity, contemporary social movements as movements *in and for* modernity. To understand contemporary social movements, we must not only examine the movement conditions which are shaped and presented to activists by the stage of development reached by the civil society in which they are acting, but also recognise that late modern activists increasingly seek to alter and elaborate those conditions in a reflexive way. But why suppose that a reflexive concern to deepen and strengthen civil society will only be found where civil society itself is already quite strong, that is, in late modernity?

Arguably it makes as much sense to see that sort of concern as characteristic of movements which develop in the early stages of modernity, when civil society is itself in need of formation (see also Bagguley 1999).

Pakulski has complained that:

> The civil society theorists exaggerate the differences between the 'old' and 'new' movements. In their eagerness to present social movements as positive, democratising and emancipatory forces, they ignore those movements (e.g. Fascist) and forms of contemporary social activism (e.g. nationalist, fundamentalist) which do not fit the liberal-democratic sentiments underlying the civil society schemes. This ideological preference appears as a selective blindness which limits the analytical and theoretical sharpness of the civil society studies ... (1995: 73/4).

His target is the selectivity of what gets included from the contemporary range of possibilities as a 'new' social movement, but we might equally complain of the 'selective blindness' which prevents older, agrarian, movements from being recognised as movements to develop the norms, culture and institutions of civil society.

Social movements and the construction of meaning: the co-operative movement and Irish civil society

The creation of a peasant proprietorship in late nineteenth century Ireland occurred in two stages. The first was the Land War and the first Land Acts which provided the legal basis for the creation of peasant holdings out of the breakup of landed estates. These contributed to the elaboration of a symbolic system of national identity, in which native ownership of the land and national independence became closely intertwined (Kane 1996), but did not go far in supplying a sociocultural identity for the new 'class' of peasant farmers which was being created. That came, I argue here, with the second wave of 'peasantisation', in which the co-operative movement played a crucial part in converting a legal category into a social status group.

The movement contributed two particularly significant elements to the cultural construction of Irish civil society. First, it constructed peasant proprietors less as a class internal to Irish agrarian society, more as citizens of the new nation which was (assumed to be) moving fairly quickly towards some form of autonomous management or self-government. Second, it constructed them as the bearers of a particular version of, or project for, modernity. Again I have not the space here to develop this at any length (but see Tovey 2001 forthcoming) and can only sketch a brief outline of the sorts of evidence I have in mind. I rely heavily on the published writings of the movement's main leader, Horace Plunkett. This seems justifiable to the extent that contemporary commentators such as Anderson or Digby do not identify any other activist in the movement who was also a movement theorist in the same way as Plunkett was, and he seems to have dominated the early development of movement ideas to a striking extent. But it is not, of course, the same as being able to catch ideas as they were expressed in interaction nor to see how such ideas were understood and/or re-formulated by movement members on the ground.

Plunkett emphasised frequently that the co-operative movement was a movement to ensure the *economic* survival of the small landholders which British land policy and Irish nationist demands were bringing into existence. He was very aware that these peasant proprietors were being created in Ireland at a time when farmers in many parts of Europe, including in particular in Britain, were finding their survival severely threatened by the opening up of world markets for agricultural products due to greatly increased production in America and other parts of the New World and to the development of refrigerated forms of transport which could carry beef, for example, across large parts of the globe and keep it edible to its destination. Nevertheless he constantly re-iterated that economic improvements on their own were not enough: what was needed was a movement to cultivate the *moral* and *practical* qualities of the Irish people themselves, to reform the Irish *character*, to inculcate a spirit of self-help instead of that 'extraordinary' reliance on 'political remedies for economic ills' to which the Irish seemed so prone (Plunkett 1904: 33), and to provide 'spiritual aids' - to develop an Ireland which is 'distinctively Irish' and 'accords with the national sentiment' (41). 'Our central problem' he says at one point, is 'the rehabilitation of rural life in Ireland' (159), and the solution has to lie in 'the conscious attempt of the Irish people to develop a civilisation of their own' (166).

Plunkett's diagnosis of the core problem of Ireland at the time as a problem of 'the Irish character' was perceived by many Irish nationalists as provocative in the extreme (see e.g. West 1986) - particularly when he went on to suggest, in a sort of garbled version of Weber, that the problems of character were linked to the predominance of the Roman Catholic religion. Yet from a contemporary standpoint, we might argue that it was his vocabulary more than his theoretical analysis which was the problem. His remarks about changing the Irish character are easily translated into a discourse on the need to empower oppressed and marginalised rural groups who have lost belief in their capacity to effect change. Indeed, he himself explains 'the building of character' as a matter of seeking 'to instil into the mind of each individual the too little recognised importance of his [sic] own contribution to the sum of national achievement' (Plunket 1904: 176). More broadly, it is not difficult to rephrase his arguments in terms closer to those of contemporary civil society theory and to see him as attempting to 'reconstitute a civil society that is no longer dependent upon ever more regulation, control and intervention' (Offe 1985: 820), or to push forward the democratic conditions, the rights of citizenship and in particular the conditions to allow such rights to be realised, such as the development of free association, which Cohen (1996) and Cohen and Arato (1992) see as necessary stages in the development of bourgeois civil society.

Plunkett saw the co-operative movement as about *both* empowering the individual *and* facilitating collective association. Effectively he makes no distinction between empowering through individualisation and giving 'the new Irish peasant proprietory' some form of 'corporate existence or collective organisation and identity' (1904: 44). The associations and societies set up by the co-operative movement 'were to be primarily for purely business purposes' but 'were bound to have many collateral effects. They would open up outside of politics and religion, but not in conflict with either, a sphere of action where an independence new to the country would have to be exercised' (168). Although he clearly valued the effects the movement was having on farmers (he speaks for example of a new capacity among the new peasant proprietors to face and resolve problems which no changes in land tenure conditions, or no rent reductions, could have given them), more important to him ultimately was the way in which 'the movement has exercised a profound influence in those departments of our national life where ... new forces must not only be recognised but accepted as essential to national wellbeing ...' (208). Again the example he gives comes in a reference to the opening up of *a space outside politics* for debate and deliberation - perhaps the closest he could come to arguing, in a Habermasian way, for the necessity in the Ireland of the time of an expanded 'public sphere'.

To rely solely on the ideas expressed by Plunkett, particularly when he was reflecting back on the work of the movement some 15 years after it had started, as a source for understanding the movement's ideological impact on Irish society is certainly risky. It is clear that other actors were also involved in influencing its development over time and, if not initiators of movement thinking, they were able to play a considerable role in modifying and altering its main thrust. Research by Kennedy (1996a), for example, shows that parish priests - after an initial period of reserve and suspicion towards the movement - often played an important role in helping to found, and to maintain, local co-operative societies. However Kennedy argues that what the clergy got involved in, in most cases, was 'a conditional and selective promotion' (133) of co-operative enterprises. For example they would support societies which were engaged only in milk processing, but not multi-purpose ones which also operated co-operative shops. The reason for this, of course, was the great hostility of traders and shopkeepers to such ventures. Kennedy notes that local clergy often shared, if not kinship ties, then at least social status and political and cultural interests with the stronger traders in the community. Even more importantly, however, they held a model of an ideal society which inclined them towards the management of rural social relationships in ways which would avoid or suppress conflict. 'Social deviants such as co-operative pioneers had to be reintegrated into community structures if the old clerical power and authority patterns were to be maintained intact' (1996a: 132); this was achieved by legitimising one direction of co-operative organisation (dairy processing) but blocking other directions (which would affect the interests of traders). The priest, Kennedy notes, was 'in a powerful position to determine what activities were the legitimate concern of what particular social groups' (130). The outcome was a selective facilitation of the ideology of co-operationism which tended to stress its socially accommodative rather than socially innovative elements, and which may help to explain the dramatic decline in co-operative education and participation which observers were noting by the 1930s (see above) and which could be seen as laying the conditions for the eventual transformation of co-operatives into capitalist corporations in the 1990s.

The other element in co-operative ideology which I stress briefly here was its emphasis on accomplishing technical improvements in Irish farming, particularly dairy farming. Right from the start co-operative activists emphasised the need to reform Irish farming not just legally but also technically, and to set new quality standards for Irish food. Within two years of the establishment of the first co-operative society, movement organisers were recognising that they had stimulated a greater demand for advice on technical and business matters than they could meet from their existing resources (Digby 1949), including advice on creamery engineering, on how to test milk for its purity or its butter-fat content, on how to secure constant supplies of fuel, and salt, and boxwood for packaging butter in, and how to market butter so that one co-operative creamery would not end up competing with another or 'helpless in the hands of the wholesalers' (1949: 63). In looking for new areas into which to expand the principles of co-operation, Plunkett always looked for agricultural processes which were already industrialised or which were suited to rapid industrialisation; like the *Irish Farmers' Journal* today, he refers consistently to Irish agriculture and food production as 'our industry'. The movement saw industrialisation of farming and food as something which could be accomplished by the farmers themselves, with appropriate scientific and technical assistance, and understood itself as enabling farmers to resist the processes of 'expropriation' and 'substitution' through which industrial capital seeks to dominate agrarian labour processes (Goodman *et al* 1987). Such resistance however involved embracing technical development rather than attempts to avoid or suppress it.

Thus the co-operative movement provided farmers with a collective identity and set of cultural meanings which did not seek to defend gains they had made within traditional agrarian structures, but rather emphasised the place which they could hold within the modern society which was coming into being in Ireland. It articulated a distinctively agrarian 'project for modernity', based on the vision of a relatively uniform small-scale producer society, organised around independent, productive and efficient family farms, which remained the core ideology of most Irish farmer organisations, inside or outside the co-operative movement, up to the 1960s at least (Smyth and Healy 1996). In this vision, the small scale of farm holdings could be compensated by the high levels of technical proficiency of the farmers; family motivations and family skills, fed into the farm labour process, could obviate the need for expanded land area and result in a form of highly technologically developed, high value-added farming which, in conjunction with a co-operatively organised food processing and agribusiness sector, could guarantee the conditions for reproduction for a large number of farm households. It is a project for modernity which has often been described as 'populist' (Curtin and Varley 1991), and which has been found in widely different parts of the world during the late 19th and early 20th centuries, from the writings of the Russian agronomist Chayanov to the programmes of farm organisations in Canada and the United States to the syndicalist movements of France, and the co-operative movement in Denmark which of course had itself been an explicit model for the Irish movement.

Conclusion

Bringing a social movements approach to bear on the interpretation of Irish socio-historical development can be fruitful for both sides. What we learn when we start to think of Irish agrarian history in terms of social movements theories is that the contemporary condition of rural Ireland is not the inevitable outcome of an evolutionary process of modernisation of a previously 'uncivil' or traditional world, but the outcome of a struggle between two contending projects for modernity in which the agrarian version of Irish modernity was defeated and replaced by an essentially capitalist and urban version. Today in Irish society, agriculture has lost most of its significance other than economic: it is now simply one, decreasingly profitable or attractive, 'sector' within the economy. The idea of an 'agrarian' social world is meaningless; ecological space replaces social relations as a way of defining what is distinctive about the rural. But this is not the inevitable adaptation of the 'historically transcended' to a new situation, rather it indicates the defeat of an authentically alternative culture and world view. Thinking of rural change in this way forces us to reconsider the meaning of 'urbanisation'. It can also prompt questions about whether new 'agrarian projects for modernity' could develop in contemporary conditions, what form they might take and from where they might emanate. What role can rural groups today play within, what contribution can they make towards, the expansion of civil society in late modernity?

Turning things the other way round, we can ask what we learn about contemporary social movements theorising when we try to apply it to something like the Irish co-operative movement. The emphasis, among some movement theorists at least, on civil society as the target and not merely the terrain of movement activity turns out, I would argue, to be very illuminating even when applied to a case which seems on the face of things far removed from those usually considered among the litany of 'new' social movements.

Ultimately what that suggests is that we need to reconsider our understanding of late modernity. Not only do different societies experience 'early' or 'late modernity' at different historical moments, perhaps, but in fact many of the things we assume to be marks of maturity in modernity (such as increased reflexivity, or heightened processes of individualisation alongside a renewed preoccupation with the construction of collective identity) have been part of the human experience for longer than is often recognised.

Bibliography

Adshead, M. (1996), 'Beyond Clientelism: agricultural networks in Ireland and the EU', in *West European Politics* 19 (3), pp. 583-608

Agenda 2000, 'Pour une union plus fort et plus large', juillet 1997, *Bulletin de l'Union Européenne*, Supplement 5/97

Aigrain, P., Boulet, D., Lalanne, J.B. and Laporte, J.P. (1995), *Les comportements individuels de consommation de vin en France 1980-1995*, ONIVINS/INRA-ESR, Paris

Akademie für Raumforschung und Landesplanung (1994), 'Perspektiven der Entwicklung ländlicher Räume in Deutschland', Konzepte und Empfehlungen für die Landes-, Regional- und Kommunalpolitik, Hannover

Alexander, J. (1996), 'Collective action, culture and civil society', in J. Clark and M. Diani (eds), *Alain Touraine*, Falmer Press, London/Washington

Allaire, G. and Boyer, R. (1995) (eds), *La grande transformation de l'agriculture*, INRA Economica: collection Economie agricole et agro-alimentaire, Paris

Allen, P. (1993) (ed.), Food for the Future: conditions and contradictions of sustainability, John Wiley, New York

Anderson, R.A. (1935), *With Horace Plunkett in Ireland*, Macmillan and Co., London

Andreoli, M. and Tellarini, V. (1989), *La ricerca di nuovi equilibri nell'Alta Garfagnana*, Cnr, Progetto Finalizzato Ipra, Università di Pisa

Arce, A. and Marsden, T. (1993), 'The social construction of international food: a new research agenda', *Economic Geography* 69 (3), pp. 293-301

Armstrong, J. and Worden, N. (1990), 'The slaves, 1652-1834', in R. Elphic and H. Giliomee (eds), *The shaping of the South African society: 1652-1840*, Maskew Miller Longman, Cape Town

Austin, E.J., Deary, I.J., Gibson, G.J., McGregor, M.J. and Dent, J.B. (1996), 'Attitudes and Values of Scottish Farmers: yeoman and entrepreneur as factors, not distinct types', *Rural Sociology* 61 (3), pp. 464-474

Avery, D. (1995), *Saving the Planet with Pesticides and Plastic: the environmental triumph of high-yielding farming*, Hudson Institute, Indianapolis

Bagguley, P. (1999), 'Beyond emancipation? The reflexivity of social movements' in M. O'Brien, E. Penna and C. Hay (eds), *Theorising Modernity*, Addison Wesley Longman, Harlow, Essex

Bagnasco, A. (1999), *Tracce di comunità*, Il Mulino, Bologna

Bajoit, G. (1991), 'Exit, voice, loyalty ... and apathy: les réactions individuelles au mécontentement', *Revue Française de Sociologie*, XXIX, pp. 325-345

Balandier, G. (1989), 'Le lien social en question', *Cahiers Internationaux de Sociologie,* LXXXVI, pp. 5-13

Balogh, S. and Pölöskei, F. (1979), *Agrárpolitika és agrárátalakulás Magyarországon,* Akadémiai, Budapest (Farming policy and farming transformation in Hungary)

Bánlaky, P. (1996), 'Ugrómókus, zöldbáró - most micsoda? Agrárértelmiség a változásokban: utak és sorsok', in *A falu* XI (4), pp. 25-34 (Farming intellectuals in the transformation: paths and fates)

Barbic, A. (1994), 'Agricultural Change, Rural Society and the State in Central and Eastern Europe', in D. Symes. and A.J. Jansen (eds), *Agricultural Restructuring and Rural Change in Europe*, Wageningen

Barnes, J.E. (1972), *Social networks*, Addison Wesley, Reading, Massachusetts

Barr, N. and Cary, J. (1992), *Greening A Brown Land. The Australian search for sustainable land use*, Macmillan, Melbourne

Barrientos, S. (1997), 'The hidden ingredient: female labour in Chilean fruit exports', *Bulletin of Latin American Research* 16 (1), pp. 71-81

Barrientos, S., Bee, A., Matear, A. and Vogel, I. (1999a), *Women and Agribusiness. Working miracles in the Chilean fruit export sector*, Macmillan, Basingstoke

Barrientos, S., McClenaghan, S. and Orton, L. (1999b), *Gender and codes of conduct: a case study from horticulture in South Africa*; Draft Report funded by UK Department for International Development (ESCOR)

Barthe, L. (1995), L'Élaboration du Systéme Agro-Industriel de Roquefort, in J.L. Bonniol and A. Saussol (eds), *Grands Causses: nouveaux enjeus, nouveaux regards*, Féderation pour la Vie et la Sauvegarde du Pays des Grands Causses, Millau

Barthez, A. (1993), 'Une agriculture en mutation', *Données sociales*, pp. 466-475

Baudrillard, J. (1983), *Simulations*, Semiotext[e], New York

Baudrillard, J. (1998), *The Consumer Society: myths and structures*, Sage, London

Bauman, Z. (1991), *Modernity and Ambivalence*, Polity Press, Cambridge

Bazin, G. (1994), 'Aides Directes et Disparites des Revenus Agricoles', *Économie Rural*, no. 220-221, pp. 192-196

Beck, U. (1991), 'Der Konflikt der zwei Modernen', in U. Beck, *Politik in der Risikogesellschaft*, Frankfurt a.M.

Beck, U. (1992), *Risk Society: towards a new modernity*, Sage, London

Becker, H. (1997), 'Dörfer heute - Ländliche Lebensverhältnisse im Wandel 1952, 1972 und 1993/95', in *FAA*, Bd.307, Bonn

Bedeschi, G. (1997), 'Società civile. Storia del concetto', *Enciclopedia delle Scienze Sociali*, pp. 98-104, Treccani, Roma

Bee, A. and Vogel, I. (1997), 'Temporeras and Household relations: seasonal employment in Chile's agri-export sector', *Bulletin of Latin American Research* 16 (1), pp. 83-95

Belasco, W. (1993), *Appetite for Change*, Cornell University Press, Ithaca

Belk, R.W. (1995), 'Studies in the new consumer behaviour', in D. Miller (ed.), *Acknowledging Consumption: a review of new studies*, Routledge, London

Benoît, G. (Directeur du Parc National des Cévennes) and Valéro, G. (Qualité France) (1997), 'Agriculture et Environnement dans le Parc National des Cévennes: le lancement de l'opération "agneau de parcours"', *Bulletin de la Société Languedocienne de Géographie*, no. 3-4, pp. 101-121

Berlyne, D.E. (1960), *Conflict, Arousal and Curiosity*, McGraw-Hill, New York

Berov, L. (1974), 'The Economic Development of Bulgaria down the Ages', *Profizdat*, Sofia

Bertaux, D. (1980), 'L'approche biographique, sa validité, sa méthodologie', *Cahiers Internationaux de Sociologie*, LXIX (2), pp. 198-225

Bertram, H. (1996), 'Familienentwicklung und Haushaltstrukturen' in H. Bertram, H. Nickel, O. Niedermayer and G. Trommsdorff (eds), *Städte und Regionen - Räumliche Folgen des Transformationsprozesses*, Opladen

Bessière, J. (1998), 'Local Development and Heritage: traditional food and cuisine as tourist attractions in rural areas', *Sociologia Ruralis* 38 (1), pp. 21-34

Beuchler, S. (1995), 'New social movement theories', *Sociological Quarterly* 36 (3), pp. 441-64

Beus, C.E. and Dunlap, R.E. (1991), 'Measuring Adherence to Alternative versus Conventional Agricultural Paradigms: a proposed scale', *Rural Sociology* 56 pp. 432-460

Bill, R. (1993), 'Der bernische Forstdienst einst und jetzt', *Schweizerische Zeitschrift für Forstwesen* 144 (7), pp. 495-502

Birch, S.M. (1993), *A Guide to Deciduous Fruit in the Cape*, Unifruco, Bellville, Cape

Bjørkum, E., Lien, M. and Kjærnes, U. (1997), *Kjøtt, forbruker-oppfatninger og offentlig debatt. En litteraturstudie*, Rapport nr. 3, Statens Institutt for Forbruksforskning, Lysaker

Blöchliger, H.-J. (1994), 'Explorative Framework' in Organisation for Economic Co-operation and Development (OECD), *The contribution of amenities to rural development*, OECD, Paris

Bocock, R. (1993), *Consumption: key ideas*, Routledge, London

Bode, I. (1997), 'Le difficile altruisme des groupes d'intérêt: le cas du syndicalisme CFDT et de la mutualité ouvrière', *Revue Française de Sociologie*, XXXVIII, pp. 269-300

Bolger, P. (1977), *The Irish Co-operative Movement - its history and development*, Institute of Public Administration, Dublin

Boltanski, L. (2000), 'The project-oriented Cité', communication to Columbia University, 05/08/2000

Boltanski, L. and Chiappelo E. (2000), *Le nouvel esprit du capitalisme*, Gallimard, Paris

Boltanski, L. and Thévenot, L. (1991), *De la justification: les économies de la grandeur*, Gallimard, Paris

Borgula, K. (1996), 'Schweizerische Waldwirtschaft unter Beschuss durch den WWF', *Schweizerische Zeitschrift für Forstwesen* 147 (4), pp. 303-307

Börzel, T.A. (1997), 'What's So Special About Policy Networks? - An exploration of the concept and its usefulness in studying European governance', in *European Integration online Papers* (EIoP) 1 (16)

Boudon, R. (1996), 'Relativisme et modernité', *Revue Européenne des Sciences Sociales*, XXXVIV (106), pp. 169-192

Boudon, R. (1998), 'Au delà du modèle du "choix rationnel"?', in B. Saint-Servin, E. Picavet, R. Fillieule and P. Demeulenaere (eds), *Les modèles de l'action*, P.U.F., Paris

Bourdieu, P. (1977), *Algérie 60: structures économiques et structures temporelles*, Éditions de Minuit: collection Grands Documents, Paris

Bourdieu, P. (1979), *La distinction, critique sociale du jugement*, Éditions de Minuit, Paris (Italian edition, 1983; English edition, 1984)

Boyd, M. (1991), *Organization, Performance and System Choice*, Westview, Boulder

Brauer, K., Willisch, A. and Ernst, F. (1996), 'Intergenerationelle Beziehungen, Lebenslaufperspektiven und Familie im Spannungsfeld von Kollektivierung und Transformation', in L. Clausen (ed.), *Gesellschaften im Umbruch* -Verhandlungen des 27. Kongresses der Deutschen Gesellschaft für Soziologie in Halle an der Saale 1995, Frankfurt a.M./New York

Breen, R., Hannan, D.F., Rottman, D.B. and Whelan, C. (1990), *Understanding Contemporary Ireland*, Gill and Macmillan, Dublin

Bruckmeier, K. (2000), 'LEADER in Germany and the discourse of autonomous regional development', *Sociologia Ruralis* 49 (2)

Bruckmeier, K. and Grund, H. (1994), 'Perspectives for Environmentally Sound Agriculture in East Germany', in D. Symes and A. J. Jansen (eds), *Agricultural Restructuring and Rural Change in Europe*, Wageningen

Bruijn, C.A. de and Dagevos, J.C. (1999), *Massa-individualisering en de agrofood-sector* [Mass individualisation and the agro-food complex], NRLO, Den Haag

Bryant, C.R. and Johnston, T.R.R. (1992), *Agriculture in the City's Countryside*, Belhaven Press, London

Bryman, A. (1988), *Quantity and Quality in Social Research*, Unwin Hyman, London

Buck, D., Getz, C. and Guthman, J. (1997), 'From farm to table: the organic vegetable commodity chain of northern California', *Sociologia Ruralis* 37 (1), pp. 3-19

Bulgarian Business (weekly newspaper for economic information), 3-9 June 1991, 10-16 June 1991, 18-24 November 1994

Buller, H. (2001), 'Is this the European model?', in H. Buller and K. Hoggart (eds), *Agricultural Transformation, Food and Environment - Perspectives on European Rural Policy and Planning* Vol 1, Ashgate, Aldershot

Buller, H. and Hoggart, K. (2001) (eds), *Agricultural Transformation, Food and Environment - Perspectives on European Rural Policy and Planning* Vol 1. Ashgate, Aldershot

Bundesamt für Bauwesen und Raumordnung (BBR) (1998), *Aktuelle Daten zur Entwicklung der Städte, Kreise und Gemeinden*, Bd.1, Bonn

Bundesamt für Statistik (BFS) and Bundesamt für Umwelt, Wald und Landschaft (BUWAL) (1997), *Umwelt in der Schweiz 1997*, Bundesamt für Statistik, und Bundesamt für Umwelt, Wald und Landschaft, Bern

Bundesamt für Umwelt, Wald und Landschaft (BUWAL) (1996a), 'Neue Wege der forstlichen Planung', *Umweltmaterialien* no. 45, *Wald*, Bundesamt für Umwelt, Wald und Landschaft, Bern

Bundesamt für Umwelt, Wald und Landschaft (BUWAL) (1996b) *Handbuch Forstliche Planung. Reihe Vollzug Umwelt*, Bundesamt für Umwelt, Wald und Landschaft, Bern

Bundesamt für Umwelt, Wald und Landschaft (BUWAL) (1996c), Praxishilfe: Fallbeispiele der überbetrieblichen forstlichen Planung, *Reihe Vollzug Umwelt*, Bundesamt für Umwelt, Wald und Landschaft, Bern

Bundesanstalt für Arbeit (1999), *Werkstattberichte*, Nürnberg

Bundesforschungsanstalt für Landeskunde und Raumordnung (BfLR) (1995), *Laufende Raumbeobachtung. Aktuelle Daten zur Entwicklung der Städte, Kreise und Gemeinden 1992-1993*, Bonn

Burch, D., Lyons, K. and Lawrence, G. (1999), 'What do we mean by "green"? consumers, agriculture and the food industry', *Consuming Foods, Sustaining Environments*, Proceedings from the 6th Annual Agri-food Research Network, Central Queensland University, Rockhampton

Burguière, A. (1975), *Bretons de Plozévet*, Flammarion, Paris

Bußmann, F.-E. (1998), 'Dorfbewohner und Kommunalpolitik - eine vergleichende Untersuchung in 14 Dörfern der Bundesrepublik Deutschland unter besonderer Berücksichtigung der länderspezifischen Gemeindeordnungen und der Verwaltungsstrukturen', FAA, Bd. 309, Bonn

Buttel, F. (1993), 'Environmentalisation and Greening: Origins, Processes and Implications', in S. Harper (ed.) *The Greening of Rural Policy: international perspectives*, Belhaven, London

Buttel, F. (1993), 'The production of agricultural sustainability - observations from the sociology of science and technology' in P. Allen (ed.), *Food for the Future - Conditions and Contradictions of Sustainability*, John Wiley, New York

Calvet, Y. (1995), *Synthèse des Études sur la Frequentation Touristique de Parc National des Cévennes 1962-1994*, Parc National des Cévennes, Florac

Campbell, C. (1995), 'The sociology of consumption', in D. Miller (ed.), *Acknowledging consumption: a review of new studies*, Routledge, London

Campbell, H. (1996), 'Recent Developments in Organic Food Production in New Zealand: organic food exporting in Canterbury', Department of Anthropology, Dunedin

Campbell, H. and Coombes, B. (1999a), 'Green Protectionism and Organic Food Exporting from New Zealand: crisis experiments in the breakdown of Fordist trade and agricultural policies', *Rural Sociology* 64 (2), pp. 302-319

Campbell, H. and Coombes, B. (1999b), 'New Zealand's organic food exports: current interpretations and new directions in research' in D. Burch, J. Goss and G. Lawrence (eds), *Restructuring Global and Regional Agricultures: trans-formations in Australasian agri-food economies and spaces*, Ashgate, Aldershot

Campbell, H., Fairweather, J. and Steven, D. (1997), 'Recent Developments in Organic Food Production in New Zealand: Part 2, Kiwifruit in the Bay of Plenty', Department of Anthropology, Dunedin

Capefruit (1998), internet site www.capefruit.com, Unifruco/Capespan, Cape Town

Carter, N. (1996), 'Worker Co-operatives and Green Political Theory', in B. Doherty and M. de Geus (eds), *Democracy & Green Political Thought*, Routledge, London

Castells, M. (1998), *The rise of the network society* (Vol. 1 - The information age: economy, society and culture), Blackwell, Oxford

Chambers, R. (1983), *Rural development: putting the last first*, Longman, Essex

Chamboredon, J.C. (1980), 'Les usages urbains de l'espace rural: du moyen de production au lieu de récréation', *Revue française de sociologie* 21 (1), pp. 97-119

Chiffoleau, Y. (1999), 'La pluriactivité en coopération viticole: une chance pour l'emploi en Languedoc-Roussillon?', *Economie Rurale* no. 253

Chiffoleau, Y. (2000a), 'Historical approach of professional networks in wine co-operative firms in Languedoc (Southern France): a contribution to the renewal of innovation analysis in rural settings', Sunbelt XX, Vancouver, International Conference on Social Networks / INSNA, April 13-16 2000

Chiffoleau, Y. (2000b), *Réseaux et pratiques de l'innovation en milieu coopératif*, Thèse de Doctorat de Sociologie, Université Paris V - René Descartes, Paris

Clark, J. and Diani, M. (eds) (1996), *Alain Touraine*, Falmer Press, London/Washington

Clarke, G. (1998), 'Non-Governmental Organizations (NGOs) and Politics in the Developing World', *Political Studies* XLVI, pp. 36-52

Clarke, R. (1991), 'Organic foods: for better or worse', *Food Australia* 43 (1), pp. 12-14

Cloke, P. (1997), 'Country Backwater to Virtual Village? Rural Studies and the Cultural Turn', *Journal of Rural Studies* 13 (4), pp 367-375

Cohen, J. L. (1996), 'Mobilisation, politics and civil society: Alain Touraine and social movements' in J. Clark and M. Diani (eds) (1996), *Alain Touraine*, Falmer Press, London/Washington

Cohen, J.L. and Arato, A. (1992), *Civil Society and Political Theory*, MIT Press, Cambridge, Massachusetts

Cole, A. and John, P. (1995), 'Local Policy Networks in France and Britain: Policy Co-ordination in Fragmented Political Sub-Systems', in *West European Politics* 18 (4), pp. 89-109

Coleman, J.S. (1990), *Foundations of social theory*, The Belknap Press of Harvard University Press, Cambridge, Massachusetts

Coleman, W.D. (1993), 'Reforming Corporatism: the French banking policy community, 1941-1990', in *West European Politics* 16 (2), pp. 122-143

Colombo, M. (1999), 'La mobilitazione ambientale. Comitatismo e associazionismo nell'area genovese', *Studi di Sociologia* XXXVII (1), pp. 95-122

Comité Departmental du Tourisme (1996), Observatoir Departemental du Tour-isme de la Lozére, Mende

Conzelmann, T. (1998) '"Europeanisation" of Regional Development Policies? Linking the Multi-level Governance Approach with Theories of Policy Learning and Policy Change', in *European Integration Online Papers* 2 (4)

Coombes, B. and Campbell, H. (1998), 'Recent Developments in Organic Food Production in New Zealand. Part 4: The Expansion of Organic Food Production in Nelson and Golden Bay', Department of Anthropology, Dunedin

Corrigan, P. (1997), *The sociology of consumption: an introduction*, Sage, London

Courson, J.P. (1990), 'L'hypermarché se rapproche, l'épicier quitte le village', *Données sociales*, pp. 202-205

Creed, G. (1997), *Domesticating Revolution: from socialist reform to ambivalent transition in a Bulgarian village*, Pennsylvania State University Press, Pennsylvania

Crosnier, C. (1996), 'Administrer la Nature: enjeux biologiques et sociaux dans le Parc National des Cévennes', in P. Baudot et al., *Impact de l'Homme sur les Milieux Naturels: perceptions et mesures*, Éditions de Bergier, Grasse

Crozier, M. and Friedberg, E. (1977), *L'acteur et le système: les contraintes de l'action collective*, Seuil, Paris

Csatári, B. (1996), *A magyarországi kistérségek néhány jellegzetessége*, MTA RKK-Phare, Kecskemét (Some features of the Hungarian microregions)

Csite, A. (1997), 'Vidékfejlesztési rendszerek Magyarországon 1970-1996: Az Új Regionális Politika', in *Szociológiai Szemle* no. 1, pp. 79-96 (Rural development systems in Hungary 1970-96)

Csite, A. (1998), 'Constructing the Miserable Countryside in Hungary in the 1990s', in L. Granberg and I. Kovách (eds), *Actors of the Changing European Countryside*, Institute for Political Science, Budapest

Csite, A. (1999), 'Politikahálózatok és mez_gazdaság', in A. Hamar (ed.), *Agrá-rátalakulás Magyarországon a '90-es években I-II*, PHARE Térségfejlesztési ismeretek fels_fokon III-IV. kötet. KGF Szolnok - MTA RKK: Szolnok. II. kötet pp. 163-205 (Policy networks and agriculture)

Csite, A. and Granberg, L. (2000), 'From the Village Action Movement to the LEADER: the emergence of the Finnish rural policy network'; Working

Paper Series of the Research Institute of the Swedish School of Social Sciences (forthcoming)
Csizmadia, E. (1962), *Két út - két világ*, Akadémiai, Budapest (Two paths - two worlds)
Curtin, C. and Varley, T. (1991), 'Populism and petit capitalism in Ireland', in S. Whatmore, P. Lowe, T. Marsden, C. Curtin and T. Varley (eds), *Rural Enterprise*, David Fulton, London
Dagevos, J.C. and de Meere, F. (1997), *De spruitjescultuur voorbij: consumenten en veranderingen in de vraag naar voedingsmiddelen* [Beyond musty smells: consumers and changes in food demand], LEI-DLO, Den Haag
Dagevos, J.C. and Hansman, H.J.M. (1999), 'Oriëntatie op consument gevoelige aangelegen-heid' [Orientation on consumers is a sensitive affair], *VMT*, no. 7, pp. 11-13
Dalen, E. (1997), *Telefonundersøkelse om kunnskap om og holdninger til mattrygghet*, Markeds- og Mediainstituttet, Oslo
Dalen, E. (1998) *Kjennskap og holdninger til norsk landbruk*, Markeds- og Mediainstituttet, Oslo
Darré, J.P. (1996), *L'invention des pratiques*, Karthala, Paris
David, M.G., Foulon, A. and Gissot, C. (1992), 'Les relations des personnes âgées avec leur entourage', *Sociétés contemporaines* 10, pp. 111-126
De Klerk, M. (1992), 'Prospects for commercial agriculture in the Western Cape (Cape Town)'; Working Paper 11, Economic Trends Research Group, Development Policy Research Unit, University of Cape Town, Cape Town
De Leonardis, O. (1997), 'Declino della sfera pubblica e privatismo', *Rassegna italiana di sociologia*, XXXVIII (2), pp. 169-193
Degenne, A. and Forsé, M. (1994), *Les réseaux sociaux*, Armand Colin, Paris
Delanty, G. (1999), *Social Theory in a Changing World - Conceptions of Modernity*, Polity Press, Cambridge
Desroche, H. (1976), *Le projet coopératif: son utopie et sa pratique, ses appareils et ses réseaux, ses espérances et ses déconvenues*, Éditions Économie et Humanisme, Paris
Diamanti, I. (1995), 'Le associazioni e la politica', in C. Trigilia (ed.), *Cultura e sviluppo. L'associazionismo nel mezzogiorno*, Donzelli, Roma
Diederen, P., de Vlieger, K., Beers, G., de Man, A.P. and Engelbart, F. (1996), *FLAK 2010: flexibele agrarische ketens in de 21ste eeuw* [Flexible agricultural chains in the 21st century], NRLO, Den Haag
Digby, M. (1949), *Horace Plunkett - an Anglo-American Irishman*, Blackwell, Oxford
Donáth, F. (1977), *Reform és forradalom*, Akadémiai, Budapest (Reform and revolution)
Donati, P. (1997), 'Alla ricerca di una società civile. Che cosa dobbiamo fare per aumentare la capacità di civilizzazione del paese?', in Aa.Vv., *La società civile in italia*, Mondadori, Milano

Donolo, C. (1999), Il legame sociale e le buone istituzioni, "Quaderni di Animazione e Formazione", Edizioni Gruppo Abele, pp. 71-77

D'Silva, E. (1997), 'Why institutional reforms in forestry? Lessons from international experience', *Natural Resources Forum* 21 (1), pp. 51-60

Du Toit, A. (1992), 'The farm as family: paternalism, management and modernisation on Western Cape wine and fruit farms', Centre for Rural Legal Studies, Stellenbosch

Du Toit, A. (1993), 'The micro-politics of paternalism: discourse on management and resistance on South African fruit and wine farms', *Journal of Southern African Studies* 19 (2), pp. 314-336

Dubar, C. (1998), 'Trajectoires sociales et formes identitaires: clarifications conceptuelles et méthodologiques', *Sociétés contemporaines*, no. 29, pp. 73-85

Eder, K. (1996), *The Social Construction of Nature: a sociology of ecological enlightenment*, Sage, London

Egli, C., Lietha, A., and Geiser, U. (1997), 'Every participation process is unique: new experiences in Swiss forest planning' in B. Solberg and S. Miina (eds), *Conflict management and public participation in land management*, EFI Proceedings no. 14, European Forest Institute, Joensuu, Finland

Eisen, A. and Kaase, M. (1996), 'Transformation und Transition: zur politischen Analyse des Prozesses der deutschen Vereinigung', in M. Kaase, A. Eisen, O.W. Gabriel, O. Niedermayer and H. Wollmann (eds), *Politisches System*, Opladen

Eizner, N. (1996), 'Postface', in M. Jollivet and N. Eizner (eds), *L'Europe et ses Campagnes*, Presse de Sciences Po, Paris

Elster, J. (1998), 'Emotions and economic theory', *Journal of Economic Literature*, XXXVI, pp. 47-74

Emerson, R.M. (1962), 'Power-dependance relations', *American Sociological Review*, no. 27, pp. 31-40

Enyedi, G. (1996), *Regionális folyamatok Magyarországon az átmenet id_szakában*, Hilscher Rezs_ Szociálpolitikai Egyesület, Budapest (Regional processes in Hungary in the period of transition)

Enyedi, G. (1997), 'A területfejlesztési politika dilemmái' in *Kritika* no. 7, pp. 13-15 (Political dilemmas of regional development)

Erdei, F. (1977), *Mez_gazdaság és szövetkezet*, Akadémiai, Budapest (Agriculture and co-operative)

Erdei, F. and Fazekas, B. (1970), *A magyar mez_gazdaság huszonöt éve*, Akadémiai, Budapest (25 years of Hungarian agriculture)

Ewert, J. and Hamman, J. (1999), 'Why paternalism survives: globalization, democratization and labour on South African wine farms', *Sociologia Ruralis* 39 (2), pp. 202-221

Eyerman, R. (1990), 'Intellectuals and progress: the origins, decline and revival of a critical group', in J.C. Alexander and P. Sztompka (eds), *Rethinking Progress*, Unwin Hyman, London

Eymard-Duvernay, F. (1989), 'Conventions de qualité et formes de coordination', *Revue Économique*, no. 2, pp. 329-359

Fairweather, J. (1999), 'Understanding how farmers choose between organic and conventional production: results from New Zealand and policy implications', *Agriculture and Human Values*, 16, pp. 51-63

Featherstone, M. (1991), *Consumer culture and postmodernism*, Sage, London

Fehér, L. (1970), *Agrár- és szövetkezeti politikánk 1965-1969*, Kossuth, Budapest (Our Farming and Co-operative Policy 1965-69)

Fine, B. and Leopold, E. (1993), *The world of consumption*, Routledge, London

Fink, M., Grajewski, R., Siebert, R. and Zierold, K. (1994), 'Rural Women in East Germany', in D. Symes and A. J. Jansen (eds), *Agricultural Restructuring and Rural Change in Europe*, Wageningen Agricultural University, Wageningen

Finkelstein, J. (1999), 'Rich food: McDonald's and modern life', in B. Smart (ed.), *Resisting McDonaldization*, Sage, London, pp. 70-82

Fischler, C. (1993), *L'Homnivore*, Odile Jacob, Paris

Floridia, A. (1994), 'Modelli interpretativi e ipotesi di ricerca', in A. Floridia, L. Parri and F. Quaglia, *Regolazione sociale ed economie locali*, Angeli, Milano

Frows, J. (1998), 'The Contested Redefinition of the Countryside. An analysis of rural discourses in The Netherlands', in *Sociologia Ruralis* 38 (1), pp. 54-68

Fuat Firat, A. and Dholakia, N. (1998), *Consuming people: from political economy to theatres of consumption*, London, Routledge

Fukuyama, F. (1995), *Trust: the social virtues and the creation of prosperity*, Hamish Hamilton, London

Fukuyama, F. (1999), *The great disruption: human nature and the reconstitution of social order*, The Free Press, New York

Fuller, A.M. (1990), 'From Part-time Farming to Pluriactivity: a decade of change in rural Europe', *Journal of Rural Studies* 6 (4), pp. 361-371

Fulton, M. (1995), 'The future of Canadian agricultural co-operatives: a property rights approach', *American Journal of Economy*, no. 77, pp. 1144-1152

Gabriel, Y. and Lang, T. (1995), *The unmanageable consumer: contemporary consumption and its fragmentation*, Sage, London

Geiser, U. (1996), 'There are no encroachments here; land use dynamics and land use discourses along the Western Ghats in South Kerala', Project Paper, Workshop on Land use dynamics in Kerala: an actor-oriented approach, Centre for Development Studies, Thiruvananthapuram, Kerala, India

Geiser, U. (1999), 'Towards sustainable land use and livelihoods in North-West Pakistan; learning from the Kalam Integrated Development Project' (KIDP 1981-1998) draft, Zurich

Geiser, U. (forthcoming), 'To "participate" with whom, for what (and against whom): forest fringe management along the Western Ghats in Southern Kerala', in R. Jeffery and B. Vira (eds), *Participatory approaches in natural resources management*, Macmillan, London

Giddens, A. (1984), *The Constitution of Society. Outline of the Theory of Structuration*, Polity Press, Cambridge

Giddens, A. (1990), *The Consequences of Modernity*, Polity Press, Cambridge (French edition, 1994; Norwegian edition, 1997)

Giddens, A. (1994), *Beyond Left and Right - the Future of Radical Politics*, Polity, Cambridge

Gifi, A. (1985), *Princals*, Department of Data Theory, Universiteit Leiden, Leiden

Giovagnoli, F. (1999), 'Il caso Misericordie', in L. Fazzi and E. Messora (eds), *Modelli di welfare mix*, Angeli, Milano

Giroux, N. (1992), 'Participation and strategic decision-making in a co-operative', *Annals of Public and Co-operative Economics* 63 (1), pp. 5-24

Gojard, S. (1994), Le jardinage dans l'enquête Modes de vie de L'INSEE, mémoire de DEA, ENS/EHESS

Goldblatt, D. (1996), *Social Theory and the Environment*, Polity Press, Cambridge

Goldman, S., Nagel, R. and Preiss, K. (1995), *Agile competitors and virtual organizations: strategies for enriching the customer*, Van Nostrand Reinhold, New York

GoNWFP, IUCN Pakistan (1996), Sarhad Provincial Conservation Strategy, Sarhad Programme Office, Peshawar

Goodman, D. (1999), 'Agro-food studies in the "Age of Ecology": nature, corporeality, bio-politics', *Sociologia Ruralis* 39 (1), pp. 17-38

Goodman, D. and Watts, M. (1994), 'Reconfiguring the rural or fording the divide?: capitalist restructuring and the global agro-food system', *The Journal of Peasant Studies* 22 (1), pp. 1-49

Goodman, D., Sorj, B. and Wilkinson, J. (1987), *From Farming to Biotechnology - a theory of Agro-Industrial Development*, Blackwell, Oxford

Government of India (1988), 'National Forest Policy, 1988', in L.K Jha, *India's forest policies; analysis and appraisal*, Ashish Publishing House, New Delhi

Government of Pakistan/Journalists Resource Centre (GOP/JRC), IUCN Pakistan (1991), *The Pakistan National Conservation Strategy*, Government of Pakistan/Journalists Resource Centre; and IUCN Pakistan, Islamabad, Pakistan

Grabher, G. and Stark, D. (1997) (eds), *Restructuring Networks in Post-Socialism. Legacies, Linkages, and Localities*, Oxford University Press, Oxford

Gramsci, A. (1997), Antologia dai "Quaderni dal carcere" (ed. A. Montanari), Einaudi, Torino

Granberg, L. and Kovách, I. (1998), 'Beyond socialism, beyond the peasant state?', in L. Granberg and I. Kovách (eds), *Actors on the Changing European Countryside*, Institute for Political Science, Budapest

Granberg, L., Kovách, I. and Tovey, H. (2001, forthcoming), 'Introduction', in L. Granberg, I. Kovách and H. Tovey (eds), *Europe's Green Ring: Peasantisation and Depeasantisation in the European Periphery*, Ashgate, Aldershot

Granberg, L. and Nikoula, J. (1995) (eds), *The Peasant State: the state and rural questions in 20th century Finland*, University of Lapland Social Sciences Publications, no. B.20, Finland

Granovetter, M.S. (1973), 'The strength of weak ties: a network theory revisited', *American Journal of Sociology*, no. 78, pp. 1360-1380

Granovetter, M.S. (1985), 'Economic action and social structure: the problem of embeddedness', *American Journal of Sociology* 91(3), pp. 481-510

Gray, J. (2000), 'The Common Agricultural Policy and the re-invention of the rural in the European Community', *Sociologia Ruralis* 40 (1), pp. 30-52

Grignon, C. (1980), 'Styles d'alimentation et goûts populaires', *Revue Française de Sociologie* 21 (4), pp. 531-569

Groenewald, C.J. (1993), 'Social development of the South African agricultural labour force', paper presented at the conference 'Comparative Research on Welfare States in Transition', 9-12 September, Oxford, England

Grundmann, R. (1997), 'Politiknetzwerke und globale ökologische Probleme: Der Fall der Ozonschicht', in *Politische Vierteljahrschrift* 38 (2), pp. 247-273

Grundmann, S. (1997), 'Territorialplanung in der DDR: Indikatoren zur Analyse regionaler Disparitäten - Die sozial-räumliche Struktur der DDR in den 80er Jahren', in A. Becker (ed.), *Regionale Strukturen im Wandel*, Opladen

Grundmann, S. and Schmidt, I. (1988), *Wohnortwechsel, Volkswirtschaftliche und soziale Aspekte der Migration*, Berlin

Grunert, K., Baadsgaard, A., Larsen, H. and Madsen, T. (1996), *Market Orientation in Food and Agriculture*, Kluwer Academic Publishers, Boston

Gunter, B. and Furnham, A. (1998), *Children as consumers: a psychological analysis of the young people's market*, Routledge, London

Gupta, H.S. (1997), 'Legal provisions for joint forest management', *Indian Forester* (June), pp. 572-576

Guthman, J. (1998), 'Regulating meaning, appropriating nature: the codification of California organic agriculture', *Antipode* 30 (2), pp. 135-154

Habermas, J. (1981), *Theorie des kommunikativen Handelns* (2 volumes), Edition Suhrkamp Neue Folge Band 502, Frankfurt a.M.

Hagedorn, K., Lütteken, A. and Laschewski, L. (1999), 'Towards institutions of sustainable agriculture', paper presented at the Rural Development Conference in Poprad (Slovakia), December 10-14

Hainz, M. (1999), 'Dörfliches Sozialleben im Spannungsfeld der Individualisierung', *FAA*, Bd. 311, Bonn

Hajer, M.A. (1995), *The Politics of Environmental Discourse: ecological modernisation and the policy process*, Clarendon Press, Oxford

Ham, A. van den (1998), 'Meer natuur op landbouwbedrijven' (More nature on farms), Agricultural Economics Research Institute (LEI-DLO), Report 3, 168, The Hague (in Dutch)

Hansman, H. (1999), 'De consument gevangen in cijfers: zoektocht naar het bestaan van consumentenbeelden' [The consumer caught by figures: searching for consumer images], LEI, Den Haag

Hansman, H.J.M. and Dagevos, J.C. (2000), '"Catching" consumers: applying a consumer images approach', Proceedings IAMA World food and agribusiness congress 1999, http://www.ifama.org/1999_forum_papers_htm

Harcsa, I., Kovách, I. and Szelényi, I. (1995), 'The Price of Privatisation', in *Review of Sociology of the Hungarian Sociological Association*, pp. 47-64

Harper, S. (1993) (ed.), *The Greening of Rural Policy: international perspectives*, Belhaven, London

Hassall and Associates (1990), *The Market for Australian Produced Organic Food*, RIRDC, Canberra

Hérault, B. and Lapeyronnie, D. (1998), 'Le statut et l'identité: les conflits sociaux et la protestation collective', in O. Galland and Y. Lemel (eds), *La nouvelle société française: trente années de mutation*, Armand Colin, Paris

Hervieu, B. (1994), *Les Champs du Futur*, Éditions Juilliard, Paris

Hill-Lanz, S. (1994), 'Women on Farms. A report on women farm workers in the Western Cape', Lawyers for Human Rights, Pretoria

Hill-Lanz, S. and Mitchell, M. (1997), 'Women on farms growing stronger', National Women on Farms Programme: 1996 Conference Proceedings, Women on Farms Projects, Stellenbosch

Hind, A.M. (1997), 'The changing values of the co-operative and its business focus', *American Journal of Agricultural Economics* 79 (4), pp. 1077-1082

Hirschman, A. (1970), *Exit, voice and loyalty*, Harvard University Press, Cambridge, Massachusetts

Hirst, P. (1997), *From Statism to Pluralism. Democracy, Civil Society and Global Politics*, UCL Press, London (Italian edition, 1999)

Hobsbawm, E. and Ranger, T. (1983) (eds), *The Invention of Tradition*, Cambridge University Press, Cambridge

Hofmeyr, I.-M. (1995), 'Women on farms: living and working in South Africa', NWFP Occasional Paper - December 1995, Lawyers for Human Rights, Stellenbosch

Holbrook, M.B. and Hirschman, E.C. (1982), 'The experiential aspects of consumption: consumer fantasies, feelings, and fun', *Journal of Consumer Research*, 9, pp. 132-140

Howlett, S. and Locke, M. (1997), 'Governance in the voluntary sector and the reciprocal relationship in the governance of localities', NCVO, South Bank University, 4-5 September

Humm, A.M. (1999), 'Auf dem Weg zum sozialistischen Dorf? Zum Wandel der dörflichen Lebenswelt in der DDR und der Bundesrepublik Deutschland 1952-1969', *Kritische Studien zur Geschichtswissenschaft*, Bd. 131, Göttingen

Inglehart, R. (1990), *Culture shift in advanced industrial society*, Princeton University Press, Princeton

Ingold, T. (1992), 'Culture and the Perception of the Environment', in E. Croll and D. Parkin (eds), *Bush Base: Forest Farm*, Routledge, London

Institut National de la Statistique et des Études Économiques (INSEE) (1998), *Les campagnes et leurs villes*, Collection Contours et Caractères, Paris

Institut National de la Statistique et des Études Économiques (INSEE) (1992), 'Recensement de la Population de 1990, Population-Activités-Ménages, la France et ses Régions', INSEE, Paris

International Union for the Conservation of Nature (IUCN) (1997), *Beyond fences: seeking social sustainability in conservation*, IUCN Gland, Switzerland

Ivanics, A. (1982), *Hogyan gondolkoznak, dolgoznak, élnek a termel_szövetkezeti vezet_k?,* Mez_gazdasági Kiadó, Budapest (How do our collective farm leaders think, work and live?)

James, A. (1993), 'Eating green(s) discourses of organic food', in K. Milton (ed.), *Environmentalism. A View From Anthropology*, Routledge, London

Jarrige, F. (1998), 'Ancienne institution, nouveaux enjeux, la coopération agricole aujourd'hui: réflexions sur le cas des coopératives vinicoles du Midi', in G. Miclet, L. Sirieix and S. Thoyer (eds), *Agriculture et alimentation en France: en quête de nouvelles légitimités*, Economica, Paris

Jeffery, R. (1998), 'Introduction' in R. Jeffery (ed.), *The social construction of Indian forests*, Manohar, New Delhi

Jensen, R. (1999), *The dream society: how the coming shift from information to imagination will transform your business*, McGraw-Hill, New York

Jessop, B. (1997), 'A neo-Gramscian approach to the regulation of urban regimes: accumulation strategies, hegemonic projects and governance', in M. Lauria (ed.), *Reconstructing regime theory: regulating urban politics in a global economy*, Sage, London

Johnston, R.J., Gregory, D. and Smith D.M. (eds), *The Dictionary of Human Geography* (third edition), Blackwell Publishers Ltd., Oxford

Jollivet, M. (1997), *Vers un rural postindustriel*, L'Harmattan, Paris

Jollivet, M. and Mendras, H. (1971) (eds), *Les Collectivités Rurales Françaises*, Colin, Paris

Juhász, P. (1983), 'Az agrárértelmiség szerepe és a mez_gazdasági szövetkezetek', in *Medvetánc* 1. (The role of the agrarian intellectuals and the collective farms)

Juhász, P. (1988), 'A településfejlesztési koncepció és a faluosztályok elmélete', in *Tér és társadalom* 2 (2), pp. 3-18 (The local development white paper and the theory of village classes)

Kalam Integrated Development Project (KIDP) (1990), Proceedings and recommendations of the workshop on forest issues. Kalam Integrated Development Project, Kalam, Swat, Pakistan

Kane, A.E. (1996), 'The centrality of culture in social theory: fundamental clues from Weber and Durkheim', in S.P. Turner (ed.), *Social Theory and Sociology (The Classics and Beyond)*, Blackwell, Oxford

Kassim, H. (1994), 'Policy Networks, Networks and European Union Policy Making: a sceptical view', in *West European Politics* 17 (4), pp. 15-27

Katsiaficas, G. (1997), *The Subversion of Politics: European Autonomous Social Movements and the Decolonisation of Everyday Life*, Humanities Press, New Jersey

Kaule, G. *et al* (1997), Regional Guidelines to Support Sustainable Landuse by EU Agri-Environmental Programmes (AEP), Stuttgart

Keliyan, M. (1994), 'The Role of the State in the Transition of Bulgarian Agriculture to a Market Economy', in D. Symes and A.J. Jansen (eds), *Agricultural Restructuring and Rural Change in Europe*, Wageningen

Kelly, K. (1998), *New rules for the new economy: 10 ways the network is changing everything*, Fourth Estate, London

Kemilev, A. (1946), 'The National Income of Bulgaria', *National Economy*, Year 1, No 1, November 2-16, Monthly Journal of High Economic Chamber, Sofia

Kennedy, L. (1996a), 'The early response of the Irish Catholic clergy to the Co-operative Movement', in *Colonialism, Religion and Nationalism in Ireland*, Institute of Irish Studies, QUB, Belfast

Kennedy, L. (1996b), 'Farmers, traders and agricultural politics in pre-independence Ireland' in *Colonialism, Religion and Nationalism in Ireland*, Institute of Irish Studies, QUB, Belfast

Kind, G. (1997), 'Territorialentwicklung und Territorialplanung in der DDR: Ergebnisse und Auswirkungen auf die Raumstruktur Deutschland', in A. Becker (ed.), *Regionale Strukturen im Wandel*, Opladen

Kinnear, S. (1999), 'Overview of the Organic Industry in Australia', presented at Farming for the Future. Organic Produce for the 21st Century. Mackay, Queensland, 22-23 September

Kirat, T. and Lung, Y. (1999), 'Innovation and proximities: territories as loci of collective learning process', *European Urban and Regional Studies* 6 (1), pp. 27-38

Kjærnes, U. (1996), 'Kjøttimport og forbrukerinformasjon', Arbeidsnotat nr. 3-1997, Statens Institutt for Forbruksforskning, Lysaker

Knill, C. and Lehmkuhl, D. (1999), 'How Europe Matters. Different Mechanisms of Europeanisation', in *European Integration online Papers* 3 (7)

Knudsen, A.J. (1999), 'Deforestation and entrepreneurship in the North-West Frontier Province, Pakistan', in S.T. Madsen (ed.), *State, society and the environment in South Asia*, NIAS Man and Nature in Asia Series no.3, Curzon Press, Surrey

Kooten, G.C. van (1993), *Land Resource Economics and Sustainable Development: Economic Policies and the Common Good*, University of British Columbia Press, Vancouver

Korsnes, O., Andersen, H. and Brante, T. (1997) (eds), *Sosiologisk leksikon*, Universitetsforlaget, Oslo

Kovách, I. (1988), *Termel_k és vállalkozók*, MSZMP Társadalomtudományi Intézet, Budapest (Producers and entrepreneurs)

Kovács, K. (1996), 'The Transition on Hungarian Agriculture 1990-1993: general tendencies, background factors and the case of the "Golden Age"' in R. Abrahams (1996) (ed.), *After Socialism. Land Reform and Social Change in Eastern Europe*, New Directions in Anthropology Series vol. 6, Berghahn Books, Oxford, pp. 51-84

Kovács, K. (1998), 'Bomlás és sarjadás az agrárgazdaságban', in *Replika* no. 33-34, pp. 177-192

Kozhucharova, V. (1999), 'The Bulgarian Village and its Paradoxes: the advent of late modernity' in V. Kozhucharova (ed.), *The Bulgarian Village and Late Modernity*, Bulgarian Society for Rural Sociology, Institute of Sociology, Bulgaria Rusticana, Sofia

Kritzinger, A. and Vorster, J. (1995), 'The labour situation in the South African Deciduous Fruit Industry', University of Stellenbosch, Stellenbosch

Kritzinger, A. and Vorster, J. (1996), 'Women farm workers on South African deciduous fruit farms: gender relations and the structuring of work', *Journal of Rural Studies* 12 (4), pp. 339-351

Kritzinger, A. and Vorster, J. (1997), 'The conceptualisation of farm family business: deciduous fruit farming in South Africa', *Sociologia Ruralis* 37 (1), pp. 114-133

Kritzinger, A. and Vorster, J. (1998), 'Women on South African farms: empowerment across or along race and class divisions?', *Sociologia Ruralis* 38 (3), pp. 331-350

Kumar, K. (1993), 'Modernity' in W. Outhwaite and T. Bottomore (eds), *The Blackwell Dictionary of Twentieth-Century Social Thought*, Blackwell, Oxford

Kumar, K. (1995), *From post-industrial to post-modern society; new theories of the contemporary world*, Blackwell, Oxford

Lagarenne, C. and Tabard, N. (1998), 'Inégalités territoriales de niveau de vie', INSEE Première, no. 614, INSEE, Paris

Lambert, Y. and Brémaud, M.A., 'Religion et société dans une commune rurale de Bretagne: Limerzel', *Archives de l'observatoire du changement social*, vol. 3, pp. 255-289

Lange, P.A.M. van, Otten, W., De Bruin, E.M.N. and Joireman, J.A. (1997), 'Development of prosocial, individualistic, and competitive orientations:

theory and preliminary evidence', *Journal of Personality and Social Psychology* 73 (4), pp. 733-746
Larmet, G. (1993), 'Alimentation et relations sociales dans une commune de Bretagne rurale', mémoire de DEA, ENS/EHESS
Larmet, G. (1999), L'organisation des pratiques d'approvisionnement alimentaires', Document de travail CORELA, INRA, no. 99-04
Larmet, G. (2000), 'The social differentiation of food provisioning according to social category and income (1971-1991)', Document de travail CORELA, INRA, no. 00-01
Larrain, J. (1993), 'Modernization' in W. Outhwaite and T. Bottomore (eds), *The Blackwell Dictionary of Twentieth-Century Social Thought*, Blackwell, Oxford
Laschewski, L. (1998a), 'Continuity and Change - Agricultural Restructuring in East Germany', *Eastern European Countryside* 4, Torún, pp. 37-48
Laschewski, L. (1998b), *Von der LPG zur Agrargenossenschaft - Untersuchungen zur Transformation genossenschaftlich organisierter Agrarunternehmen in Ostdeutschland*, Berlin
Laschewski, L. (in collaboration with A. Schulze) (1998c), 'Konsum im Wandel - Wege und Perspektiven ostdeutscher Konsumgenossenschaften', *Berliner Beiträge zum Genossenschaftswesen*, Institut für Genossenschaftswesen, Berlin
Laschewski, L. (2000), 'Vertagtes Ende oder erfolgreicher Wandel?', in K. Hagedorn, and A. Eisen (eds), *Lernstücke - Genossenschaften in Ostdeutsch-land*, Berlin
Laumann, E.O. (1966), *Prestige and association in a urban community*, Bobbs-Merill, Indianapolis
Laville, J.L. (1990), 'Économie et solidarité: pour une problématique du changement institutionnel', *Cahiers Internationaux de Sociologie*, LXXXIX, pp. 289-312
Lawrence, G. (1987), Capitalism and the Countryside: the rural crisis in Australia, Pluto Press, Sydney
Lawrence, G. (1996), 'Contemporary Agri-food Restructuring: Australia and New Zealand', in D. Burch, R. Rickson and G. Lawrence (eds), *Globalization and Agri-food Restructuring: perspectives from the Australasia region*, Avebury, Aldershot
Lawrence, G. (1999), 'Agri-food Restructuring: a synthesis of recent Australian research', *Rural Sociology* 64 (2), pp. 186-203
Lawrence, G., Lockie, S. and Lyons, K. (1999), '"Healthy for you, healthy for the environment": corporate capital, farming practice and the construction of "green" foods', *Rural Society* 9 (3), pp. 543-554
Lazega, E. (1994), 'Analyse de réseaux et sociologie des organisations', *Revue Française de Sociologie*, XXXV(2), pp. 293-315
Le Jeannic, T. (1997), Trente ans de périurbanisation: extension et dilution des villes', *Économie et Statistique*, no. 307, pp. 21-41

Léger, D. and Hervieu, B. (1979), *Le Retour à la Nature, au fond du forêt ... l'état*, Seuil, Paris

Lewin, K. (1951), *Field theory in social science*, Harper and Row Editions, New York

Lewis, D.K. (1969), *Convention: a philosophical study*, Harvard University Press Cambridge, Massachusetts

Leynaud, E. (1985), *L'État de la Nature*, Édition du Parc National des Cévennes, Florac

Lhuillier, C. (1993), *Observer pour Agir: Pastoralism, Environnement et Developpement sur le Causse Méjan*, Observatoire du Causse Méjan, Xe Plan d'État, Région Languedoc Roussillon, INRA-ESR, Montpellier

Lien, M., Bjørkum, E. and Bye, E.K. (1998), 'Kjøtt, holdninger og endringer', Rapport nr. 4-1998, Statens Institutt for Forbruksforskning, Lysaker

Lilliston, B. and Cummins, R. (1998), 'Organic vs "organic": the corruption of a label', *The Ecologist* 28 (4), pp. 195-200

Lin, N. (1999), 'Building a network theory of social capital', *Connections* 22 (1), pp. 28-51

Linder, S. (1971), *Arme élite!: toekomst en tijdgebrek* [Poor élite!: future and lack of time], Wolters-Noordhoff, Groningen

Linder, W. (1996), 'Forstorganisation – aus der Sicht der bestehenden Organisationsstrukturen', *Schweizerische Zeitschrift für Forstwesen* 147 (1), pp. 41-48

Listhaug, O. (1995), 'The Dynamics of Trust in Politicians', in H.-D. Klingemann and D. Fuchs (eds), *Citizens and the State*, Oxford University Press, Oxford

Listhaug, O. (1998), 'Confidence in political institutions: Norway 1982-1996', paper presented at 'The World Values Surveys Stockholm Workshop', NTNU: Trondheim

Listhaug, O. and Wiberg, M. (1995) 'Confidence in Political and Private Institutions', in H.-D. Klingemann and D. Fuchs (eds), *Citizens and the State*, Oxford University Press, Oxford

Livet, P. and Thévenot, L. (1994), 'Les catégories de l'action collective', in A. Orléan (ed.), *Analyse économique des conventions*, P.U.F., Paris

Lockie, S. (1999), 'The state, rural environments, and globalisation: "action at a distance" via the Australian Landcare program', *Environment and Planning* A, 31, pp. 597-611

Lockie, S., Lyons, K. and Lawrence, G. (forthcoming), 'Constructing "Green" Foods: corporate capital, risk and organic farming in Australia and New Zealand', in *Agriculture and Human Values*

Long, N. (1989), *Encounters at the interface*, Agricultural University, Wageningen

Long, N. and Long, A. (eds), *Battlefields of knowledge; the interlocking of theory and practice in social research and development*, Routledge, London

Lopiano-Misdom, J. and De Luca, J. (1998), *Streettrends: how today's alternative youth cultures are creating tomorrow's mainstream markets*, Harper Business, New York

Lorrain, F. and White, H.C. (1971), 'Structural equivalence of individuals in social networks', *Journal of Mathematical Sociology*, no. 1, pp. 49-80

Lovisolo, R. (1994), 'Harmonisation of Trade Aspects of Organic Produce', paper presented at The Tenth Scientific Conference of the International Federation of Organic Agriculture Movements, Lincoln University, New Zealand

Lovisolo, R. (1997), 'The Codex Alimentarius Commission - what it means to Australia and the labelling of organic food', presented at Australian Institute of Agricultural Scientists and Technologists National Conference, Ulverstone, Tasmania

Luginbuhl, Y. (1991), 'Le Paysage Rural: la coleur de l'agricole, la saveur de l'agricole, mais que reste-t-il de l'agricole? De l'agricole au paysage' *Études Rurales*, 121-124, pp. 35-41

Luhmann, N. (1979), *Trust and Power*, Wiley, Chichester

Luhmann, N. (1988), 'Familiarity, Confidence, Trust: problems and alternatives', in D. Gambetta (ed.), *Trust: making and breaking co-operative relations*, Blackwell, Oxford

Lyon, N. (1994), 'Organic farming to attract new markets', *Australian Farm Journal* 4 (10), pp. 62-63

Lyons, K. (1998), 'Understanding organic farm practice: contributions from ecofeminism', in D. Burch, G. Lawrence and J. Goss (eds), *Australasian Food and Farming in a Globalised Economy: recent developments and future prospects*, Monash Publications, Melbourne

Lyons, K. (1999), 'Corporate environmentalism and organic agriculture in Australia: case study Uncle Tobys', in *Rural Sociology* 64 (2), pp. 251-265

Lyons, K. (forthcoming a), 'From Sandals to Suits: green consumers and the institutionalisation of organic agriculture', *Consuming Foods, Sustaining Environments*, Proceedings from the 6th Annual Meeting of the Agri-Food Research Network, CQU Press, Rockhampton

Lyons, K. (forthcoming b), *Situated Knowledges, Science and Gender: a sociology of organic agriculture in Australia and New Zealand*, PhD thesis, Central Queensland University, Rockhampton

Lyons, K. and Lawrence, G. (1999), 'Alternative knowledges, organic agriculture and the biotechnology debate', *Culture and Agriculture* 21 (2), pp. 1-12

Lyotard, J.-F. (1984), 'The postmodern condition: a report on knowledge', in *Theory and History of Literature*, vol. 10, Manchester University Press, Manchester

MacCannell, D. (1973), 'Staged Authenticity: arrangements of social space in tourist settings', *American Sociological Review*, no. 79, pp. 589-603

MacCannell, D. (1976), *The Tourist: a new theory of the leisure class*, Macmillan, London

Maccarini, A. (1996), 'Le associazioni sociali', in P. Donati (ed), *Sociologia del terzo settore*, NIS, Roma

Maccarini, A. (1999), 'L'universale in rete: autonomia sociale e costituzione della società globalizzante', *Sociologia e Politiche Sociali* II (2), pp. 9-48

MacKenzie, D.R. and Berrier, R.J. (1987), 'Risk and the public's acceptance of genetic engineering', in L.R. Batra and W. Klassen (eds), *Public Perceptions of Biotechnology*, Agricultural Research Institute, Maryland

Maffre-Baugé E. (1986), *Vendanges amères*, Éditions J.P. Ramsay, Paris

Maget, M. (1955), 'Remarques sur le village comme cadre de recherches anthropologiques', *Bulletin de psychologie*, 8 (7-8), pp. 375-382

Malcolm, B., Sale, P. and Egan, A. (1996), *Agriculture in Australia: an introduction*, Oxford University Press, Melbourne

Mann, P. (1998), 'Crises de surproduction et mobilisations cognitives dans le Midi viticole', Pôle Sud, CNRS-CEPEL, n°9, novembre 1998, pp. 48-56

Márczis, M. (1998), 'Az élet él és élni akar' in *A Falu* 13 (3), pp. 79-86 (The life does live and wants to live)

Margairaz, D. (1988), 'Foires et marchés dans la France préindustrielle', EHESS, Paris

Margolis, H. (1996), *Dealing with Risk*, The University of Chicago Press, Chicago

Marks, G. and McAdam, D. (1996), 'Social Movements and the Changing Structure of Political Opportunity in the European Union' in *West European Politics* 19 (2), pp. 249-278

Marsden, T. (1992), 'Exploring a rural sociology for the Fordist transition', *Sociologia Ruralis* 32 (2/3), pp. 207-229

Marsden, T. (2000), 'Food matters and the Matter of Food', *Sociologia Ruralis* 40 (1), pp. 20-29

Marsden, T.K. *et al* (1993), *Constructing the Countryside*, UCL Press, London

Marti, W. (1993), 'Waldbauliche Möglichkeiten und Grenzen im Emmentaler Privatwald', *Schweizerische Zeitschrift für Forstwesen* 144 (7), pp. 525-528

Martin, J.P. (1994), *Les syndicats de viticulteurs en Languedoc (Aude et Hérault) de 1945 à la fin des années 1980*, Thèse de Doctorat Histoire des civilisations/histoire contemporaine, Université Paul Valéry, Montpellier III

Matear, A. (1997), 'Gender and the state in rural Chile', *Bulletin of Latin American Research* 16 (1), pp. 97-106

Mathieu, N. (1998), 'La Notion de Rural et les Rapports Ville/Campagne en France: les années quatre-vingt-dix', *Économie Rurale* no. 247, pp. 11-20

Mayson, D. (1990), *The Rural Foundation. Management and change on fruit farms: a case study of selected farms in the Elgin area*, University of Cape Town MA Thesis, Cape Town

McCallister, L. and Fischer, C.S. (1978), 'A procedure for surveying personal networks', *Sociological Methods and Research*, 7, pp. 131-148

McKenna, M., Roche, M. and Le Heron, R. (1998), 'Sustaining the fruits of labour: a comparative localities analysis of the integrated fruit production programme in New Zealand's apple industry', *Journal of Rural Studies* 14 (4), pp. 393-409

McMichael, P. (1997), 'Rethinking globalization: the agrarian question revisited', *Review of International Political Economy* 4 (4), pp. 630-62

Mead, G.H. (1932), *Mind, self and society from the standpoint of a social behaviorist*, University of Chicago Press, Chicago

Ménard, C. (1990), *L'économie des organisations*, Paris

Mendras, H. (1984), *La fin des paysans suivi de vingt ans après*, Actes Sud, Arles

Menger, P. (1997), 'Temporalité et différences interindividuelles: l'analyse de l'action en sociologie et en économie', *Revue Française de Sociologie*, XXXVIII, pp. 587-633

Meurs, M., Kozhucharova, V. and Stoyanova, R. (1999), 'Bulgaria: from Co-operative Village to Agro-Industrial Complex, The Rise and Fall of Collective Agriculture', in M. Meurs (ed.), *Many Shades of Red. State Policy and Collective Agriculture*, Rowman & Littlefield, Boulder, Colorado

Meyer-Engelke, E., Schubert, H. and Heuwinkel, D. (1998), *Beispiele nachhaltiger Regionalentwicklung*, Stuttgart

Michelsen, J. (1996), 'Organic farmers and conventional systems: the recent expansion of the organic food market in Denmark', *American Journal of Alternative Agriculture* 11 (1), pp. 18-23

Miegel, M. (1994), 'Wie realistisch und wünschenswert ist die Angleichung der Lebensverhältnisse in West und Ost?', Biss public 14, 5-9

Miles, S. (1998), *Consumerism - as a way of life*, Sage, London

Miller, D. (1995) (ed.), *Acknowledging consumption: a review of new studies*, Routledge, London

Mitchell, J.C. (1983), 'Case and situation analysis', *Sociological Review* 31 (2), pp. 187-211

Monjauze, A. (1971), 'Un Parc National Adapté aux Vieilles Civilisations', Options Méditerranéennes, 9 Octobre, pp. 87-93

Monk, A. (1998), 'The Australian organic basket and the global supermarket', in D. Burch, G. Lawrence, R. Rickson and J. Goss (eds), *Australasian Food and Farming in Globalised Economy: recent developments and future prospects*, Monash, Melbourne

Monk, A. (1999), 'The Organic Manifesto: organic agriculture in the world food system', in D. Burch, J. Goss and G. Lawrence (eds), *Restructuring Global and Regional Agricultures: transformations in Australasian agri-food economies and spaces*, Ashgate, Aldershot

Mooney, P. (2000), 'Specifying the "rural" in social movement theory', *Polish Sociological Review* 1 (129), pp. 35-55

Mormont, M. (1996), 'Le Rural Comme Catégorie de Lecture du Social', in M. Jollivet and N. Eizner (eds), *L'Europe et ses Campagnes*, Presses de Sciences Po, Paris

Mousset, S. (1992), 'Protection de la Nature et Agriculture dans le Parc National des Cévennes: Bilan d'une Expérience Insolite', *Annales du Parc National des Cévennes*, Tome 5, Florac

Mueller, D.C. (1988), 'The corporate life cycle', in R.S. Thompson and M. Wright (eds), *Internal organisation, efficiency and profit*, Philipp Allan Publishers, Oxford

Mulkay M. (1993), 'Sociology of science' in W. Outhwaite and T. Bottomore (eds), *The Blackwell Dictionary of Twentieth-Century Social Thought*, Blackwell, Oxford

Murdoch, J. (1996), 'A Regionalist Perspective on Rurality: economy, environment and the evolution of complex regional forms', Proceedings of the IBG, Rural Geography Conference, January, Strathclyde, Scotland

Murphy, M. (1995), 'South African farm workers: is trade union organisation possible?', *South African Labour Bulletin* 19 (3), pp. 20-26

Mutti, A. (1998a), 'I diffusori della fiducia', *Rassegna italiana di sociologia* XXXIX (4), pp. 533-549

Mutti, A. (1998b), *Capitale sociale e sviluppo. La fiducia come risorsa*, Il Mulino, Bologna

Navracsics, T. (1997), 'Területi érdekképviselet a Kádár-korszakban' in *Társad-almi Szemle* 52 (4), pp. 79-94 (Spatial interest representation in the Kádár-era)

Nemes-Nagy, J. (1992), 'Margójegyzetek a regionális politikához', in Tér és Társadalom no. 1-2, pp. 69-76 (Notes to the regional policy)

Nemes-Nagy, J. (1998), 'A tér a társadalomkutatásban', Hilscher Rezs_ Szociálpolitikai Egyesület, Budapest (Space in social research)

Newton, K. (1999), 'Social Capital and Democracy in Modern Europe', in J. van Deth, M. Maraffi, K. Newton and P. Whiteley (eds), *Social Capital and European Democracy*, Routledge, London

Nicolas, P. (1993), 'Règles et principes dans les sociétés coopératives agricoles françaises: évolution du droit et des pratiques de 1960 à 1992', in A. Zevi and J.L. Monzon-Campos (eds), *Coopératives, marchés, principes coopératifs*, DeBoeck Université

Norwegian Food Control Authority (2000), *Analyser og resultater 1998-2000. Holder maten mål?*, Statens næringsmiddeltilsyn, Oslo

Nygård, B. and Storstad, O. (1998), 'De-globalization of Food Markets? Consumer Perceptions of Safe Food: the case of Norway', in *Sociologia Ruralis* 38 (1)

O'Rourke, E. (1999), 'The Causse Méjan: changing relationships between agriculture, environment and society within a French National Park', *Landscape Research* 24 (2), pp. 141-165

Oberschall, A. (1972), *Social conflict and social movements*, Prentice Hall, Englewood Cliffs

Offe, C. (1985), 'New social movements: challenging the boundaries of institutional politics', *Social Research* 52 (4), pp. 817-67

Olivier de Sardan, J.P. (1995), *Anthroplogie et Développement*, Karthala, Paris

Olson, M. (1965), *The logic of collective action*, Harvard University Press, Cambridge, Massachusetts

Orbán, S. (1972), *Két agrárforradalom Magyarországon*, Akadémiai, Budapest (Two rural revolutions in Hungary)

Orléan, A. (1994), 'Vers un modèle général de la coordination économique par les conventions', in A. Orléan (ed.), *Analyse économique des conventions*, P.U.F., Paris

Orwell, G. (1949), *Nineteen Eighty Four*, Harmondsworth, London

Osti, G. (1993), 'Dimensione dell'insediamento e associazionismo in Italia-Siedlungsgrosse und Vereinswesen in Italien', *Annali di Sociologia-Soziologisches Jahrbuch*, 9.I, pp. 169-190, 191-215

Pakulski, J. (1995), 'Social movements and class: the decline of the Marxist paradigm', in L. Maheu (ed.), *Social Movements and Social Classes (The Future of Collective Action)*, Sage, London

Paolicchi, P. and Biancheri, R. (1995), *Aspetti sociali dell'associazionismo*, Regione Toscana, Firenze

Parade, L. (1991), 'Das Leben in den Dörfern: vor, während und nach der Wende', unpublished presentation to the 2nd Europäischen Dorf-erneuerungskongress der Europäischen ARGE in Reichenbach (Sachsen) 28 November

Parc National des Cévennes (PNC) (1994), *Programme d'Aménagement 1994-1999*, Florac

Parc National des Cévennes (PNC) (1996), *Réhabilitation et Restauration des Pelouses Séches du Causse Méjan - Programme National 'Recrée la Nature: Réhabilitation, Restauration et Création d'Écosystèmes'*, PNC, Florac

Parodi, M. (1998), 'Science sociale et "spécificité méritoires" des associations', *La Revue de MAUSS* 11, pp. 136-154

Paul, J.L., Bory, A., Bellande, A., Garganta, E. and Fabri, A. (1994), 'Quel système de référence pour la prise en compte de la rationalité de l'agriculteur: du système de production agricole au système d'activités', in M. Sebillote (ed.), *Recherches-système en agriculture et développement rural: Communications*, Symposium international, Montpellier, 21-25 novembre, pp. 46-52

Peet, R. and Watts, M. (1996), 'Liberation ecology; development, sustainability and environment in an age of market triumphalism', in R. Peet and M. Watts (eds), *Liberation ecologies; environment, development and social movements*, Routledge, London

Pellizzoni, L. and Osti, G. (1999), 'Democrazia e cooperazione nella tutela dell'ambiente', *Quaderni di Sociologia* XLIII (21), pp. 113-143

Penfold, C. and Miyan, M. (1998), 'Organic agriculture - now and the future', in D. Michalk and J. Pratley (eds), *Agronomy - growing a greener future*, in Proceedings of the 9th Australian Agronomy Conference, Australian Society of Agronomy, Wagga Wagga, pp. 669-672

Perrot, M. and de la Soudière, M. (1998), 'La Résidence Secondaire: un nouveau mode d'habiter la campagne', *Ruralia* no. 2, pp. 137-149

Peterson, J. (1997), 'States, Societies and the European Union', in *West European Politics* 20 (4), pp. 1-23

Peterson, J. and Bomberg, E. (1999), *Decision-making in the European Union*, Macmillan, London

Philo, C. and Kearns, G. (1993), 'Culture, history, capital - a critical introduction to the selling of places', in G. Kearns and C. Philo (eds), *Selling Places - the city as cultural capital, past and present*, Pergamon, Oxford

Pine, B.J. and Gilmore, J.H. (1998), 'Welcome to the experience economy', *Harvard Business Review*, July-August, pp. 97-105

Pine, B.J. and Gilmore, J.H. (1999), *The experience economy: work is theatre and every business a stage*, Harvard Business School Press, Boston

Pinkney, T. (1993), 'Modernism and postmodernism', in W. Outhwaite and T. Bottomore (eds), *The Blackwell Dictionary of Twentieth-Century Social Thought*, Blackwell, Oxford

Ploeg, B. van der (1997), 'Mentality groups in agriculture: results of a survey in Germany (Baden-Württemberg), Great Britain (Kent/Sussex) and the Netherlands (North-Holland)', paper for the XVII congress of The European Society for Rural Sociology, August

Ploeg, B. van der (1999), 'Sustainable Land Use from a Farmer's Perspective', report on the contribution by the Sociological Team (NL2) to an international research project 'Regional Guidelines to Support Sustainable Land Use (regulation 2078/92)' focussing on Waterland, a peat-meadow district North of Amsterdam, as an exemplary case for 'Wider Farm Development', Winand Staring Centre (SC-DLO), Wageningen

Ploeg, J.D. van der, and Long A. (1994), 'Endogenous development: practices and perspectives', in J.D. van der Ploeg and A. Long (eds), *Born from within; practices and perspectives of endogenous rural development. European perspectives on rural development*, vol. 1. Van Gorcum, Assen, The Netherlands

Ploeg, J.D. van der, Renting, H., and Minderhout-Jones, M. (2000) (eds), *The Socio-Economic Impact of Rural Development - Realities and Potentials*, complete issue of *Sociologia Ruralis* 40 (4)

Plunkett, H. (1904), *Ireland in the New Century*, Irish Academic Press, Dublin

Poffenberger, M. and McGean, B. (1996) (eds), *Village voices, forest choices; joint forest management in India*, Oxford University Press, New Delhi

Polanyi, K. (1944), *The Great Transformation*, Beacon Press, Boston

Ponssard, J.P. (1994), 'Formalisation des connaissances, apprentissage organisationnel et rationalité interactive', in A. Orléan (ed.), *Analyse économique des conventions*, P.U.F., Paris

Popcorn, F. (1991), *The Popcorn Report: Faith Popcorn on the future of your company, your world, your life*, Doubleday, New York

Popkin, S.L. (1979), *The rational peasant: the political economy of rural society in Vietnam*, University of California Press, Berkeley

Preiss, K., Goldman, S.L. and Nagel, R.N. (1996), *Cooperate to compete: building agile business relationships*, Van Nostrand Reinhold, New York

Pünkösti, Á. (1986a), *Az elithez tartozni: a téeszelnökök kapcsolatrendszere*, Tömegkommunikációs Kutatóközpont, Budapest (Belonging to the élite: social networks of collective farm leaders)

Pünkösti, Á. (1986b), *Az önkormányzat lépcs_i*, Tömegkommunikációs Kutatóközpont, Budapest (Stairs of self-governance)

Pünkösti, Á. (1988), *Kiválasztottak*, Árkádia, Budapest (The chosen people)

Putnam, R.D. (1993), *La tradizione civica nelle regioni italiane*, Mondadori Milano (*Making Democracy Work: civic traditions in modern Italy)*, Princeton University Press, Princeton

Ramella, F. (1994), 'Gruppi sociali e cittadinanza democratica. L'associazionismo nella letteratura sociologica', *Meridiana*, no. 20, pp. 93-133

Ramella, F. (1995), 'I caratteri della partecipazione: dirigenti, soci, utenti', in C. Trigilia (ed.), *Cultura e sviluppo. L'associazionismo nel mezzogiorno*, Donzelli, Roma

Ramella, F. (1997), 'Cittadini e produttori: civicness e sviluppo locale', *Sviluppo Locale* IV (6), pp. 5-42

Ramella, F. (1998), 'La subcultura "rossa": tra apatia e nuovo civismo', *Meridiana*, no. 32, pp. 121-151

Rangelova, R. (1998), 'Agricultural Restructuring and Rural Employment in Bulgaria: findings from a study', in *Bulgarian Village: challenges of the new century*, Bulgaria Rusticana, Trojan

Ray, C. (1998), 'Culture, intellectual property and territorial development', *Sociologia Ruralis* 38 (1), pp. 3-20

Ray, C. (1999), 'Endogenous Development in the Era of Reflexive Modernity', *Journal of Rural Studies* 15 (3), pp. 257-267

Redclift, M. and Woodgate, G. (1997), *The International Handbook of Environmental Sociology*, Edward Elgar, Cheltenham

Reitz, K.P. and White, D.R. (1989), 'Rethinking the role concept: homomorphisms on social networks', in L.C. Freeman, D.R. White and A. Kimball Romney (eds), *Research methods on social networks analysis*, Georges Mason University Press, Fairfax, Virginia

Reynaud, J.D. (1989), *Les règles du jeu, l'action collective et la régulation sociale*, Armand Colin, Paris

Rhodes, R.A.W. (1990), 'Policy Networks: a British perspective', in *Journal of Theoretical Politics* 2 (3), pp. 293-317

Rhodes, R.A.W. and Marsh, D. (1992) (eds), *Policy Networks in British Government*, Clarendon Press, Oxford

Rickson, S.T. (1997), 'Outstanding in their field: women in agriculture', *Current Sociology* 45 (2), pp. 91-133

Ritchie, M. and Campbell, H. (1996), 'Historical development of the organic agriculture movement in New Zealand', in H. Campbell, J. Fairweather and D. Steven, *Recent Developments in Organic Food Production in New Zealand*, Department of Anthropology, Dunedin

Ritzer, G. (1996), *The McDonaldisation of Society: an Investigation into the Changing Character of Contemporary Social Life* (revised edition), Pine Forge Press, Thousand Oaks, California

Ritzer, G. (1998), *The McDonaldization thesis: explorations and extensions*, Sage, London

Rodewald, B. (1994), 'Glasow - Ein Dorf im "Schatten der Grenze"', *FAA*, Ländliche Lebensverhältnisse im Wandel 1952, 1972 und 1992, Bonn, 443-468

Rodewald, B. and Siebert, R. (1995), *Arbeitsmarkt und Mobilität. Eine empirische Analyse von Dimensionen räumlicher Immobilität und Mobilität im Kontext regionaler Arbeitsmarktentwicklungen im ländlichen Raum Ostdeutschlands*, Endbericht, Müncheberg

Rogers, E. (1962), *Diffusion of innovations*, Free Press, New York

Rogers, S.C. (1991), *Shaping Modern Times in Rural France*, Princeton University Press, Princeton

Romány, P. (1983), *A szocialista átalakulás jellemz_i falun*, Kossuth, Budapest (The features of socialist transformation in villages)

Romány, P. (1996), 'A nagyüzemi átszervezés és a mez_gazdaság integrációja', in I. Orosz, L. Für and P. Romány (eds), *Magyarország agrártörténete*, Mez_gazda, Budapest (Large-scale reorganisation and agricultural integration)

Rossi, A. and Rovai, M. (1998), 'The evolution process towards the setting up of strategies of valorisation of typical local products. The case of Garfagnana spelt', in F. Arfini and C. Mora (eds), *Typical and traditional products: rural effect and agro-industrial problems*, University of Parma Press, Parma

Rouquet, C. (2000), 'La percée du téléphone portable et d'internet', INSEE Première, no. 700

Rural Industries Research and Development Corporation (1996), 'The Domestic Market for Australian Organic Produce', RIRDC Research Paper No 96/1, Canberra

Rycroft, A. and Jordaan, B. (1992), *A guide to South African labour law*, Juta, Cape Town

Sachs, W., Loske, R. and Linz, M. (1998), *Greening the North: a post-industrial blueprint for ecology and equity*, Zed Books, London

Salais, R. and Storper, M. (1992), 'The four "worlds" of contemporary industry', *The Cambridge Journal of Economics*, no. 2, pp. 169-194

Sarhad Provincial Conservation Strategy (SPCS) (1993), 'Draft strategy for the sustainable development of forestry in NWFP', Sarhad Provincial Conservation Strategy / IUCN Pakistan, Peshawar

Saunders, C., Manhire, J., Campbell, H. and Fairweather, J. (1997), 'Organic Farming in New Zealand: an evaluation of the current and future prospects including an assessment of research needs', MAF Policy Technical Paper, Department of Economics and Marketing, Lincoln University, Canterbury

Scaglia, A. and Fazzi, L. (1999), 'Il ruolo dell'associazionismo civico e delle forme organizzate di tutela e rappresentanza del cittadino', in L. Fazzi and E. Messora (eds), *Modelli di welfare mix*, Angeli, Milano

Schäfer, R., Stricker, H.-J. and Soest, D. von (1992), 'Kleinstädte und Dörfer in den neuen Bundesländern: Aufgabenfeld für die städtebauliche Erneuerung', *Schriftenreihe des Deutschen Städte- und Gemeindebundes*, H. 48, Göttingen

Schifferstein, H. and Ophuis, P.O. (1998), 'Health-related determinants of organic food consumption in The Netherlands, in *Measurements of Consumer Attitudes* 4 (1), MATFORSK, Ås

Schmidt, F. (1996), *Sozioökonomische Tendenzen im Vergleich bäuerlich geprägter Haupt- und Nebenerwerbsbetriebe aus den neuen und alten Bundesländern*, Halle/S., Diss

Schulze, G. (1997), *Die Erlebnisgesellschaft: Kultursoziologie der Gegenwart* [The experience society: cultural sociology of these modern-days], Campus Verlag, Frankfurt

Schwartz, O. (1990), *Le monde privé des ouvriers*, P.U.F., Paris

Scott, A. (1990), *Ideology and the New Social Movements*, Unwin Hyman, London

Scott, A. (1992), 'Political culture and social movements' in J. Allen, P. Braham and P. Lewis (eds), *Political and Economic Forms of Modernity*, Polity Press, Cambridge

Scott, A. (1996), 'Movements of modernity: some questions of theory, method and interpretation' in J. Clark and M. Diani (eds), *Alain Touraine*, Falmer Press, London/Washington

Scott, J.W. (1988), 'Deconstructing equality versus difference: or, the uses of poststructuralist theory for feminism', in *Feminist Studies* 14

Scully, P. (1987), *The bouquet of freedom: social and economic relations in Stellenbosch district 1870-1900*, University of Cape Town MA Thesis, Cape Town

Segalen, M. (1985), *Quinze générations de Bas-Bretons: parenté et société dans le pays Bigouden Sud, 1720-1980*, P.U.F., Paris

Seibel, W. (1997), 'Erfolgreich gescheiterter Institutionentransfer: Weshalb der Dritte Sektor in Ostdeutschland institutionelle Elastizitäten schafft, obwohl es ihn kaum zu geben scheint', in H. Wollmann, H.-U. Derlien, K. König, W. Rentzsch, and W. Seibel (eds), *Transformation der politisch-administrativen Strukturen in Ostdeutschland*, Opladen

Shell, R.C.H. (1994), *Children of bondage - a social history of the slave society at the Cape of Good Hope, 1652-1838*, Witwatersrand University Press, Johannesburg

Siebert, R. and Zierold, K. (1996), 'Gewinnen und Verlieren. Die Dörfer in den neuen Bundesländern', *FAA*, Ländliche Lebensverhältnisse im Wandel: 1952, 1972 und 1993/94, Bd. 305, Bonn, 33-39

Siebert, R. and Zierold, K. (1997), 'Veränderungen der Erwerbslagen von Dorfbewohnern im erwerbsfähigen Alter in den neuen Bundesländern', in K.-S. Rehberg (ed.), *Differenz und Integration. Die Zukunft moderner Gesellschaften*, Opladen/Wiesbaden

Siegenthaler, P. (1998), 'Neue Chancen für Forstleute als kompetente Betreuer des ländlichen Raumes', *Schweizerische Zeitschrift für Forstwesen* 149 (3), pp. 183-188

Simmel, G. (1997), *La socievolezza*, Armando, Roma

Skjerve *et al* (1996), *Matvarekvalitet, dyrehelse og internasjonal handel, Risiko forbundet med økt import av levende dyr, kjøtt og andre dyreprodukter fra EØS-området og andre land*, Den Norske Vetreinærforening, Oslo

Slater, D. (1997), *Consumer culture and modernity*, Polity Press, Cambridge

Slovic, P. (1994), 'Perceptions of Risk: paradox and challenge' in B. Brehmer and N.E. Sahlin (eds), *Future Risks and Risk Management*, Kluwer Academic Publishers, Dordrecht

Smith, M. (1992), 'The Agricultural Policy Community: maintaining a closed relationship', in R.A.W. Rhodes and D. Marsh (eds), *Policy Networks in British Government*, Clarendon Press, Oxford

Smyth, L.P.F. and Healy, S. (1996), *Farm Organisations in Ireland (A Century of Progress)*, Four Courts Press, Dublin

Social Forestry Project Malakand-Dir. (SFPMD) (1992), Social Forestry Project Malakand-Dir. Saidu Sharif, Swat, Pakistan

Stander, S. (1983), *Tree of Life: The story of Cape Fruits*, Weber, Cape Town

Statt, D.A. (1997), *Understanding the consumer: a psychological approach*, Macmillan, Houndsmills

Stephani, A. (1972), *La Culture Vivant dans le Mont Lozère: approche anthropologique d'une culture oral*, Institut National d'Education Populaire, Marly-Le-Roi

Storstad, O. (1998), 'Med hjelm and redningsvest ved matbordet', in R.Å. Nilsen and P. Veiden (eds), *Sosiologisk Fantasi*, Ad Notam, Oslo

Storstad, O. (1999), 'Medias konstruksjon av risiko. Mediapåvirkning på matom-rådet', paper no. X/99, Senter for bygdeforskning, Trondheim

Swain, N. (1993), 'The Smallholders versus the Green Barons: class relations in the restructuring of Hungarian agriculture', Working Papers Rural Transition Series No. 8, CCEES, Liverpool

Swain, N. (1998), 'A Framework for Comparing Social Change in the Post-Socialist Countryside', in *Eastern European Countryside* 4, Torun

Swain, N.J. (1997), 'Path Dependence and the Pattern of Agricultural Transformation in Eastern Europe', paper presented at the International Workshop 'Rural Restructuring and Employment in the Balkans', Greece, 18-20 April

Szabó, M. (1998), 'Agrártiltakozások - 1997', in S. Kurtán, P. Sándor and L. Vass (eds), *Magyarország politikai évkönyve 1998*, Demokrácia Kutatások Magyar Központja Alapítvány, Budapest (Rural protests)
Szelényi, I. (1990), *Városi társadalmi egyenl_tlenségek*, Akadémiai, Budapest (Urban social inequalities)
Teisserenc, P. (1994), 'Politique de développement local: la mobilisation des acteurs', *Sociétés Contemporaines* 18-19, pp. 187-213
Tennant, D. (1978), *Soil Amending Technology, Grassland Farming and New Zealand Economic Development* (PhD thesis), Department of Geography, Massey University, Palmerston
Thomashow, M. (1995), *Ecological Identity: becoming a reflective environmentalist*, MIT Press, Massachusetts
Thuderoz, C. (1995), 'Du lien social dans l'entreprise: travail et individualisme coopératif', *Revue Française de Sociologie* XXXVI, pp. 325-354
Tilly, C. (1978), *From Mobilisation to Revolution*, Addison-Wesley, Reading, Massachusetts
Todorov, T. (1995), *La vie commune: essai d'anthropologie générale*, Seuil, Paris
Tolbert, C.M., Lyson, T.A. and Irwin, M.D. (1998), 'Local capitalism, civic engagement and socioeconomic wellbeing', *Social Forces* 77 (2), pp. 401-428
Tomlinson, J. (1996), 'Markets are institutions that enable exchange to take place', in A. Kuper and J. Kuper, *The Social Science Encyclopaedia*, Routledge, London
Tommassini, L. and Sardelli, A. (1995), *Associazionismo popolare in Toscana fra ottocento e novecento*, Regione Toscana, Firenze
Touzard, J.M. (1996), 'La révolution tranquille de la viticulture languedocienne', *Revue de l'Économie Méridionale* 44 (176), pp. 5-10
Touzard, J.M. and Laporte, J.P. (1998), 'Deux décennies de transition viticole en Languedoc-Roussillon: de la production de masse à une viticulture plurielle', *Pôle Sud*, no. 9, pp. 26-47
Tovey, H. (1999), '"Messers, visionaries, and organobureaucrats": dilemmas of institutionalisation in the Irish organic farming movement', *Irish Journal of Sociology* vol. 9
Tovey, H. (2000), 'Milk and modernity - dairying in the Republic of Ireland', in H. Schwarzweller and A. P. Davidson (eds), *Dairy Industry Restructuring* (Research in Rural Sociology and Development Vol 8), JAI Press, London/NewYork
Tovey, H. (forthcoming), 'Creating and re-creating modernity: peasantisation and de-peasantisation in Ireland' in L. Granberg, I. Kovách and H. Tovey (eds), *Europe's Green Ring*, Ashgate, Aldershot
Treibel, A. (1994), *Einführung in soziologische Theorien der Gegenwart*, 2. Auflage, UTB, Leske + Budrich
Triandis, H.C. (1995), *Individualism and Collectivism*, Westview Press, Boulder

Trigilia, C. (1986), *Grandi partiti e piccole imprese*, Il Mulino, Bologna

Trigilia, C. (1995) (ed.), *Cultura e sviluppo. L'associazionismo nel mezzogiorno*, Donzelli, Roma

Trigilia, C. (1999), 'Capitale sociale e sviluppo locale', *Stato e Mercato*, no. 57, pp. 419-440

Trijp, J.C.M. van (1995), *Variety seeking in product choice behaviour*, Landbouw Universiteit Wageningen, Wageningen

Twigg, J. (1983), 'Vegetarianism and the Meanings of Meat', in A. Murcott (ed.), *The Sociology of Food and Eating: essays on the sociological significance of food*, Gower, Aldershot

United Nations (UN), DESA, CSD (1999), 'Underlying causes of deforestation and forest degradation', draft text for IFF 3rd meeting (May) Geneva

Urry, J. (1990), *The Tourist Gaze: leisure and travel in contemporary societies*, Sage, London

Uslaner, E.M. (1999), 'Democracy and social capital', in M.E. Warren (ed.), *Democracy and Trust*, Cambridge University Press, Cambridge

Vági, G. (1982), *Versengés a fejlesztési forrásokért*, KJK: Budapest (Competition for development resources)

Vági, G. (1991), *Magunk, uraim!*, Gondolat Kiadó, Budapest (Ourselves, gentlemen!)

Van den Bulck, J. (1992), 'Pillars and Politics: neo-corporatism and policy networks in Belgium', in *West European Politics* 15 (2), pp. 35-55

Van Onselen, C. (1992), 'The social and economic underpinning of paternalism and violence on the maize farms of South Western Transvaal, 1900-1950', *Journal of Historical Sociology* 5 (2), pp. 127-160

Vandergeest, P. and DuPuis, E.M. (1996), 'Introduction' in E.M. DuPuis and P. Vandergeest (eds), *Creating the Countryside*, Temple University Press, Philadelphia

Varga-Suplic, S. (1988), *A területi szövetségek és a TOT megformálása*, Kossuth, Budapest (Construction of the regional associations and the National Council of collective farms)

Vernet, J.L. (1995), 'Végétation et Paléoécologie des Grands Causses' in J.L. Bonniol and A. Saussol (eds), *Grands Causses: nouveaux enjeus, nouveaux regards*, Féderation pour la Vie et la Sauvegarde du Pays des Grands Causses: Millau

Vira, B. (1997), 'Deconstructing participatory forest management: towards a tenable typology', *OCEES Research Paper* no. 11, Oxford Centre for the Environment, Ethics and Society, Mansfield College, Oxford

Vries, W. de (1995), 'Pluri-activiteit in de Nederlandse landbouw' (Pluri-activity in Dutch agriculture), Circle for Rural European Studies, Wageningen Agriculture University, Wageningen

Wachter, S. (1987), *État, Décentralisation et Territoires*, l'Harmattan, Paris

Wandel, M. and Bugge, A. (1993), 'Til bords med forbrukerne. Forbrukernes ønsker og prioriteringer på matområdet i 90-årene', Rapport nr. 2-1997, Statens Institutt for forbruksforskning, Lysaker

Ward, N. and Almås, R. (1997), 'Explaining change in the international agro-food system', *Review of International Political Economy* 4 (4), pp. 611-29

Ward, N. and Woodward, R. (1998), 'The Europeanisation of rural development policy in the UK: the case of East Anglia', CRE Working Paper 31

Warner, M. (1999), 'Social Capital Construction and the Role of the Local State', *Rural Sociology* 64 (3), pp. 373-393

Waters, M. (1994), *Modern Sociological Theory*, Sage, London

Watts, M. and Peet, R. (1996), 'Conclusion: towards a theory of liberation ecology', in R. Peet and M. Watts (eds), *Liberation ecologies; environment, development and social movements*, Routledge, London

Weber, F. (1998), *L'honneur des jardiniers. Les potagers dans la France du XXe siècle*, Belin, Paris

Weber, M. (1968), *Economy and society*, Bedminster Press, New York (Berkeley edition, 1978; French edition, 1995)

Weber, W. (1996), 'Forstorganisation - aus der Sicht grösserer Waldeigentümer', *Schweizerische Zeitschrift für Forstwesen* 147 (1), pp. 33-39

West, T. (1982), 'Foreword' to 1982 reprint of H. Plunkett (1904), *Ireland in the New Century*, Irish Academic Press, Dublin

West, T. (1986), *Horace Plunkett, Co-operation and Politics: an Irish biography,* Colin Smythe, Gerrards Cross, Buckinghamshire

White, H. (1992), *Identity and Control*, University Press, Princeton, New Jersey

Wickham, J. (1986), 'Industrialisation, work and unemployment' in P. Clancy, S. Drudy, K. Lynch and L. O'Dowd (eds), *Ireland: a sociological profile*, Institute of Public Administration, Dublin

Wierenga, B. and van Raaij, W.F. (1987), *Consumentengedrag* [Consumer behaviour], Stenfert Kroese, Leiden

Wilson, F. (1985), 'Women and agricultural change in Latin America: some concepts guiding research', *World Development* 13 (9), pp. 1017-1035

Winkel, R. (1996), 'Regionalentwicklung in den neuen Bundesländern', in H.W. Jenkis (ed.), *Raumordnung und Raumordnungspolitik*, München

Winter, M. (1997), 'New Policies and New Skills: agricultural change and technology transfer', *Sociologia Ruralis* 37 (3), pp. 362-381

Woderich, R. (1996), 'Peripherienbildung und kulturelle Identität', in R. Kollmorgen, R. Reißig and J. Weiß (eds), *Sozialer Wandel und Akteure in Ostdeutschland*, Opladen

Wollmann, H. (1996), 'Institutionenbildung in Ostdeutschland: Neubau, Umbau und "schöpferische Zerstörung"', in M. Kaase, A. Eisen, O.W. Gabriel, O. Niedermayer, and H. Wollmann, *Politisches System*, Opladen

World Bank (1997), 'India to protect Kerala's forest resources with help from local communities', press release no. 98/1692/SAS, The World Bank, Washington D.C.

World Bank (1998), 'Kerala Forestry Project', project appraisal document; Rural Development Sector Unit, South Asia Region, The World Bank, Washington D.C.

World Wildlife Foundation (Schweiz) (1995), 'Waldwende', *Panda Magazin* 3/95

World Wildlife Foundation (Schweiz) (1998), *WWF-Faktenblatt Waldschutzgebiete*

Young, L. (1997), *World Hunger*, Routledge, London

Youngberg, G., Schaller, N. and Merrigan, K. (1993), 'The sustainable agricultural policy agenda in the United States: politics and prospects', in P. Allen (ed.), *Food for the Future: conditions and contradictions of sustainability*, John Wiley, New York

Zaninelli, S. (1996), 'Gli sviluppi storici', in G.P. Barbetta (ed.), *Senza scopo di lucro. Dimensioni economiche, legislazioni e politiche del settore nonprofit in Italia*, Il Mulino, Bologna

Zierold, K. (1995), 'Soziostrukturelle Veränderungen und soziale Differenzierungen in 4 ostdeutschen Dörfern', Endbericht, Halle (unpublished)

Zierold, K. (1997), 'Veränderungen von Lebenslagen in ländlichen Räumen der neuen Bundesländer', in A. Becker (ed.), *Regionale Strukturen im Wandel*, Opladen

Zierold, K., Laschewski, L. and Dippmann, L. (1996), 'Zwei vergleichende explorative Fallstudien in Ostdeutschland', in Landwirtschaftliche Rentenbank (ed.), *Landwirtschaft im ländlichen Raum - Funktionen, Formen, Konflikte -* , Schriftenreihe Bd. 10, Frankfurt a.M., Landwirtschaftliche Rentenbank

Zola, E. (1980), *La Terre*, Gallimard, Paris

* 9 7 8 1 1 3 8 7 2 9 0 4 9 *